THE ENGINEER IN THE GARDEN

by the same author

THE FAMINE BUSINESS
FUTURE COOK
THE FOOD CONNECTION
FOOD PLANTS FOR THE FUTURE
GLOBAL ECOLOGY
LAST ANIMALS AT THE ZOO

THE ENGINEER IN THE GARDEN

Genes and Genetics: From the Idea of Heredity to the Creation of Life

COLIN TUDGE

JONATHAN CAPE
LONDON

First published 1993

1 3 5 7 9 10 8 6 4 2

First published in the United Kingdom
in 1993 by Jonathan Cape
Random House, 20 Vauxhall Bridge Road, London SW1V 2SA

Random House Australia (Pty) Limited
20 Alfred Street, Milsons Point, Sydney,
New South Wales 2061, Australia

Random House New Zealand Limited
18 Poland Road, Glenfield,
Auckland 10, New Zealand

Random House South Africa (Pty) Limited
PO Box 337, Bergvlei, South Africa

Random House UK Limited Reg. No. 954009

A CIP catalogue record for this book
is available from the British Library

ISBN 0-224-03826-5

Printed in Great Britain by
Clays Ltd, St Ives PLC

CONTENTS

Prologue

A woman of no great age – just 120 or so – reaches from her balcony a quarter mile above the ground, brushes aside the snow, and picks from the ivy that clothes the building to the eaves, a ripe passion fruit. It is the scarlet variety – a touch of colour – almost seedless, and big as a pomegranate. Hers is a pleasant and typical urban existence; the high-rise apartments all around like cliffs and spires of green, dotted here and there with fruit that once grew only in the tropics; deer in the gardens; eagles and kites overhead; the occasional wolf, yelping in the forest on the city edge.

You may or may not find such a vision pleasing: too fanciful, perhaps; too artificial; or too smug. It is a matter of taste. You will surely prefer it, though, to another which perhaps is more likely: a desert that stretches through almost all of Africa; a rising sea that has already obliterated lowlands everywhere, from the arable fields of eastern England to the entirety of Florida; fragments of humanity under siege, desperately clinging to what they have left, spawning committees, throwing up despots, and inventing religions to fend off and to explain to themselves the horror that has overtaken them.

Or indeed – for futurism is a game that anyone can play – you might care to envisage a thousand and one scenarios of your own: of life much the same as now; of life that contrives to be the same as now, but in which there are two, or three, or five times as many people; of life like now but with half the houses empty; like now but with no animals, or wild plants; or so radically different from the present as to beggar belief – with people living effectively for ever, and dinosaurs in zoos, and the world cleaned up (or nibbled away) by life-forms reinvented, and without precedent in the history of the Universe. All that is certain is that almost any future you care to envisage could in theory come about. Provided you do not choose to reinvent the laws of physics, which seem to be beyond transgression, *anything* you can bring to mind could probably be achieved.

Or rather – any future you may care to bring to mind could overtake us: for we, the British, the Americans, the Australasians, the Africans and the Asians – we the human species – are not in control. We aspire, or many of us do, to live in democracies. Even despots love the word 'democracy'. In the name of democracy, we hold protracted and immensely expensive elections. Of course this is worthwhile, for some governments are clearly less disastrous than others.

But even the world's most committed democracies have a quality that is merely cosmetic; because, in the end, the lives of individual people, and the destiny of humanity as a whole and of our fellow species, is only in part – only in superficial part – determined by governments, elected or otherwise. For at least two million years, since human beings first began seriously to develop technologies, our individual lives and our overall fate, even our evolution, has largely been shaped by those technologies; and in the main those technologies have effectively followed their own logic. One idea has led to another, and each new idea has shaped society afresh. Governments, kings, emperors, like the rest of us, have for the most part merely adjusted to what had become available. Nobody has truly *determined* the course of events. The most successful have been those who have adjusted fastest to whatever new technique is most powerful.

So long as the technologies were feeble, and so long as human beings were a rare and scattered species, this *laissez faire* attitude to our own ingenuity mattered very little. But the technologies soon became very powerful; and because of that, we soon ceased to be rare and scattered. Our ancestors had fire, a million years ago. It is truly amazing what can be achieved with a stone axe and a bone-tipped spear. Within the past few tens of thousands of years, it seems, our ancestors obliterated vast suites of other animals, including the big herbivores of Europe and North America and the sabre-tooths that preyed upon them. A few thousand years ago, before the Romans came, ancient Britons de-forested much of Britain, and created the modern moors and heaths. The beautiful but stark islands of the Mediterranean were forested until a few thousand years ago. There was never a *policy* to bring about such vast and permanent ecological change. It was just the way things turned out, as our ancestors followed their noses, exploited the tools they had to hand, made new ones, and solved their day-to-day problems. The technology employed was of the kind that now looks so quaint in local museums. With stone axes and an aptitude for fire our ancestors altered the entire world, long before any of today's societies – or any of those in written history – had come into existence.

We have come a very long way since the stone axe. We have not simply created vastly better technologies. We have devised quite new *ways* of creating technologies. We no longer rely simply upon ingenuity and common sense. For the past several thousand years, at least, many kinds of philosopher have practised what may loosely be called 'science'; and in the past 300 years, in the western world, the sciences have been refined into a series of disciplines, and of methods, which produce deep and more or less certain insights into the mechanisms of the world and which, increasingly, far transcend common sense. Out of this new science has come a new species of technology which can properly be called 'high-tech'; the kind of technology that *cannot* be created, or even conceived, without the extra insights of formal science. 'Ordinary' technologies, which relied simply upon human invention, were powerful enough, and still predominate in everyday life. They embrace even the steam engine, which in its earliest forms involved no *bona fide* science at all. But when science and technology work in harmony – each feeding upon, and developing the other – the combination is very powerful indeed. The resulting machines are stunning; to date, the microchip is perhaps the apotheosis. The effect of high tech upon the world at large is commensurately huge. The rate of change outstrips the transformations that our ancestors brought about, a hundred or a thousand fold.

Now we are busily developing a science, and a resultant group of high technologies, that could change the world and our attitude to it more radically than any we have seen before: the science and technologies of genetics, and in particular of what has been called (I believe misguidedly) 'genetic engineering'. As I hinted above, the science and technologies promise – or threaten – nothing less than the creation of life, or the indefinite prolongation of the life we already have.

If this science and these technologies are deployed adroitly, they could do more than any other sciences or techniques to solve our present problems, and avert the pending disasters of the future. More than any other science, they could help us to provide agricultures that can feed ten billion people, but do so humanely and without destroying the rest of the environment. They could provide methods of manufacture that preserve fossil fuels and reduce, or even reverse, the pending greenhouse effect; and transform the economies of present-day 'Third World' countries, and truly help to create a new 'world order'. They could provide us with medical techniques that could defeat most infections within a century from now, and probably begin a serious attack upon the cancers while at the same time providing benign but certain methods of contraception that at last will enable people to regulate their own families as so many obviously want to do, even if they have no money or access to high-flown clinics. For the conservation of animals and plants, the science and high technologies of genetics are already *vital.* Without their adroit application, we cannot hope to save more than a token proportion of our fellow creatures from extinction. On the other hand, if we fail to remain in control, then the science and the new technologies could equally be employed for ends that we may well consider evil, or at least deeply sinister; and which, if things go wrong, could trigger a chain of biological destruction that could outstrip any we have seen before.

As things are, there are two prime reasons why we cannot hope to deploy science and technology astutely; to seize what is there to be seized, and avoid the potentially horrendous side-effects. First, we simply have not defined what it is we want to achieve. We still have the attitudes of the new stone age; that is, we are still content merely to follow our noses. Later in this book I will discuss the notion of natural selection, and explore its deep flaw, which is that natural selection does not look ahead, and in general is bound to favour short-term advantage over long-term. Our stone-age ancestors did not plan, they *evolved*, according to natural selection. If we simply do as they did, we are bound to accumulate long-term problems, and if we do not address them, they will become overwhelming. In fact, the ecological ills of the world which have now become such a cliché represent, in large part, the accumulated side-effects of 10,000 years of agriculture, along with another 500,000 or so of over-efficient hunter-gathering. But, except for the occasional world conference, where politicians compete for a few days to appear far-sighted, world politics as practised is not distinguished by any particular sense of direction. In general, we do what is there to be done, as our Cro-Magnon ancestors did, and then adapt as best we can to whatever new circumstances we happen to have created.

Second – which is closer to the subject of this book – we cannot hope to deploy science and technology for the public good in the foreseeable future because, quite simply, science and high tech are not in the public domain. People talk about all kinds of things, from football to opera, from sex to politics, but only professional scientists commonly talk about science. To be sure, this is far more true in Britain than in many other countries. In Britain, indeed, educated people still take a pride in *not* knowing any science. The term 'intellectual' is still reserved for those whose learned discourse is confined entirely to literature. The mood is changing slightly, but until very recently (and certainly when I was at University) scientists were considered an odd and slightly dangerous bunch, and professors of physics who were foolish enough to reveal what they did for a living at cocktail parties were likely to spend the rest of the evening talking to the potted plants. We live in a society dominated by science and technology, but we do not live in a science 'culture'. Science and technology are treated not as a flowering of human creativeness, subject to human frailty, but like the weather: a fact of existence, beyond our control.

Indeed – in Britain at least – science and what are vaguely termed 'the humanities' are loosely divided into what C.P. Snow in the late 1950s called 'the two cultures', and the schism between the two is deep and pernicious. In the century in which science and high tech have made their greatest impact, and brought the world to a point of departure that may be terminally destructive but could still be idyllic, most people have remained unaware of most of the changes and the new ideas, and the people who are acknowledged as intellectual leaders take a positive pride in *not* understanding the source of the most significant change.

Yet without understanding, there can be no control; or at least none of the fine and subtle kind that is now required. Furthermore, the necessary understanding requires more than the passing of exams, for science and the high technologies that spring from it can be deployed adroitly only when science has become a natural and accepted part of our culture. It is not simply a 'method' for exploring the material facts of the Universe: its ideas change the way we look at all of life. Science must indeed be placed within broader contexts: integrated, in so far as this is possible, into all aspects of philosophy, including economics, politics, ethics, aesthetics, and indeed into religion. The point is not to put science on a pedestal, so that all other disciplines may pay homage, but to bring science and the high technologies that it generates back under human control; or rather, to begin to exercise control for the first time in our very long history.

The purpose of this book is to contribute to the scientific culture, and to do this in particular by looking at the science that I personally find the most interesting (this is my book, after all) and at the technologies which, in the decades and centuries to come, have the power to change the world for good or for ill more profoundly and more quickly than ever before.

Specifically, I want to describe the new science of genetics, and the technologies it produces, largely just to show how interesting it all is. Unless you find science interesting, you will not allow it to permeate your thinking. But there is no problem with this. Science is not a penance. It contains some of the greatest excitements that any pursuit has to offer: aesthetic as well as intellectual.

I also want to show what the new and astonishingly powerful technologies might achieve, both for good and not so good, and to suggest, however presumptuous it may seem, what the criteria should be for judging what is good.

Finally, in the last few chapters, I want to suggest ways in which we might begin to exercise sensible control, and at last begin to deploy technologies for proper ends. To be sure, I am not advocating that we should set up any particular bureaucracies, or make any particular changes of law. I want rather to address the kinds of attitudes of mind that are needed, and the kinds of problems that truly have to be addressed.

One way to write a book like this is to plunge straight in, with a list of agricultural or medical triumphs and disasters. But understanding is the important thing, and there can be no understanding without a feel for the underlying science. Besides, the science is the most interesting. So I will begin at the beginning. It is appropriate to start at the dawn of modern biology: in the middle of the nineteenth century, with Charles Darwin.

CHAPTER I

The Puzzle of Heredity and the Idea of the Gene

Heredity matters. It matters to each of us that our children, our parents, our families, are *ours*. It has mattered too much and too often in history that some of us belong to this family, or this race, and other people belong to another; so much, indeed, that human history (and probably much of prehistory) has been punctuated by genocide, which literally means the elimination of a people, but in practice tends to imply the elimination of people perceived to be different; at least, when the people attacked are *not* perceptibly different, as in civil war, the sin is commonly felt to be the greater. Heredity matters to animals as well; many have evolved ways of avoiding incest, at least between siblings (though father/daughter relationships are harder to avoid); and many spend their time or risk their lives in helping their offspring or their siblings, while treating non-relatives of their own species as rivals. Many plants avoid incest by chemical means; they can mate successfully only with others that are of the same species, and yet are subtly different from themselves.

Genetics is the study of heredity – or, more specifically, of the *mechanisms* of heredity. Because heredity matters, genetics matters. Indeed, of all the life sciences, genetics probably matters the most of all.

Genetics, then, lies at the heart of myths, of culture, and of the most basic instincts of people and other animals. Human concern with heredity is as old as humanity itself, older, indeed, for it also occupied all our animal ancestors. This book is an overview of the modern science of genetics. I will begin just before that science properly began: with Charles Darwin, and in particular with his pivotal publication of 1859, *On the Origin of Species by Means of Natural Selection.*

HEREDITY AND CHARLES DARWIN

John Maynard Smith once opined that if Charles Darwin had not written *Origin of Species* he would still be remembered 'as the greatest of all field naturalists'.

On several counts, then, Darwin can reasonably be said to have been the greatest biologist of all time. He was certainly not the first to conceive the notion that living things evolved, one from another, and were not simply created in their contemporary form, but he was the first to provide a truly plausible mechanism whereby they could have changed in the way they have as the generations passed. The mechanism he proposed was that of *natural selection*.

Darwin's hypothesis of evolution by means of natural selection has two main components. First, it borrows from the prognostications of the English cleric Thomas Malthus – who, at the turn of the eighteenth and nineteenth centuries argued that the human population was bound to outgrow its ability to produce more food, and so was bound, sooner or later, to run into severe trouble. Darwin applied this principle to all living things. All, he perceived – even slow breeders, like elephants – seemed bound to produce more offspring than their environment could support. So, like it or not, all creatures were bound to be thrown into competition with their fellows. Note indeed that in Darwin's view the main competition was not between a wild horse and the wolf that eats it, but between two wild horses that are both trying to escape from the same wolf: when the race is to the swift.* This has very interesting implications of many kinds as we will see in later chapters.

Second, Darwin perceived that in any one generation, individuals of the same species vary. One wild horse is very much the same as another wild horse of the same age and sex, but they are not identical. Some do run faster than others. The faster one is indeed the one who escapes.

These two components – inevitable *competition*, and *variation* – lead to what Darwin called 'descent with modification'. If there

*The original quote, from Ecclesiastes 9:11, says the precise opposite: 'I returned, and saw under the sun, that the race is not to the swift, nor the battle to the strong'. As we will see in Chapter 3, this is relevant too – for chance has also played an enormous role in evolution.

were no variation, then there would have been no increase in the speed of horses as the generations passed – although such an increase can certainly be inferred from the fossil record. If all wild horses ran at identical speed, then it would be a pure lottery as to which one fell to the wolf. On the other hand, if there were no wolves (or lions or bears, or what you will) to chase the horses, then there would be no particular reason why swift ones should survive at the expense of slow ones. So the competition between horses does not simply lead to a decimation – a random removal of a few, in the way that Roman armies removed a few soldiers *pour encourager les autres*. It ensures that the ones that are best adapted to the circumstances – what Herbert Spencer, following Darwin, called the 'fittest' – are the ones that survive, and they, of course, are the most likely to produce offspring.

Note – which is why this is all so relevant – that Darwin's notion of evolution, the crucial biological insight of the greatest of biologists, rests on two assumptions about the nature of heredity. First, it assumes that like begets like: swift horses are more likely to give birth to more swift horses, than slow horses are. If it were not so – if all horses gave birth to a random selection of more horses – then natural selection would not work, because the horses born in the next generation would turn out in the same way, whoever survived in the generation before. But then, this observation is hardly exceptionable: resemblances do run in families (and Darwin made friends with breeders of livestock and plants, and himself belonged to an enormous extended family).

Second, however, a system of heredity that could support the mechanism of natural selection would have to allow variation to occur. Again, this is a common observation. Siblings tend to resemble each other more than they resemble other people chosen at random, but siblings are never identical unless they are identical twins (and even they diverge through life's vicissitudes).

Darwin was a very thorough thinker, and a tremendous worrier (none more so!). He provided the explanation he set out to provide: a plausible mechanism of evolution. But to be perfectly happy in his own mind (and indeed to satisfy all his critics, many of whom were extremely astute) he also wanted to provide a mechanism of heredity that could, in practice, underpin his mechanism of natural selection: one that would ensure that 'like begets like', but would also produce at least a modest degree of variation in each generation. In this, he failed. As he lamented in one doleful passage in *Origin of Species*:

> . . . no one can say why the same peculiarity in different individuals . . . is sometimes inherited and sometimes not so; why the child often reverts in certain characters to its grandfather, or other more remote ancestor; why a peculiarity is often transmitted from one sex to both sexes, or to one sex alone, more commonly but exclusively of the right sex.

Now, of course, the mechanisms of heredity have been worked out. As this chapter will soon explain, they are in principle very simple, so simple that some biologists have wondered at Darwin's failure to work them out for himself. Some have even gone so far as to suggest that Darwin was, in fact, not particularly bright.

But those who doubt Darwin's intelligence are themselves immensely foolish. It is not pure chauvinism that prompts me to defend Darwin as the greatest genius of biology. We could argue simply that none of Darwin's immediate peers, who included people of unquestioned intellect – Thomas Huxley comes most obviously to mind – was able to infer a plausible mechanism of heredity either.

In fact, if we think about the matter objectively, Darwin's 'failure' to provide a plausible mechanism of heredity is absolutely typical of the history of science in general. Only a small proportion of scientific cogitation leads to insight. The rest leads into blind alleys, and once the scientist is in a blind alley, it is extremely difficult for him or her to get out again. Usually, scientists are rescued from the various gum trees up which they climb only by their peers, who are looking at the problem from a different angle.

Darwin, I believe, was simply the wrong kind of thinker to arrive at the correct mechanism of heredity. He conceived his grand overview of evolution by looking at thousands of different instances, in thousands of different species: beetles, finches, barnacles, orchids, human beings; in other words, through the eyes of a tremendously accomplished naturalist. Nothing short of such a grand sweep could suggest a convincing mechanism that could be seen to apply to all of them.

But when you start to look at the details of heredity on such a grand scale, confusion reigns. As we will see – and as indeed is common experience – even very simple mechanisms of heredity can give rise to complex patterns of inheritance. Besides, there are mechanisms which, though simple in principle, are not particularly simple in detail, and they produce enormously complicated patterns of inheritance. Add to that the problem of sudden random change, sometimes caused by *genetic mutation* (just to anticipate) and sometimes caused by 'recessive' genes that make themselves felt only now and again, and sometimes caused by accidents or diseases in the womb (accidents which, in Darwin's time, could not easily have been distinguished from true genetic changes). Any character that an individual is born with is, by definition, 'congenital', but congenital characters (such as congenital disorders) may result from particular genes, or may be caused by accidents in the womb or, for example, disease organisms passed on by the mother. Characters that are properly called 'hereditary' cannot be assumed to be genetic, either: for example, syphilis may be passed from generation to generation. In short, any line of inheritance occasionally throws up 'sports', or 'monsters', creatures that are quite out of the ordinary. All in all, it is quite impossible to see a coherent pattern of inheritance simply by scanning the whole of life, as a naturalist tends to do. Darwin's *cri de coeur* in *Origin* is all too easy to explain.

However, if Darwin did have a true intellectual fault, it was that he was not numerate. He admired people who were, like his cousin, the pioneer statistician Francis Galton. He fully acknowledged the importance of maths in rigorous analysis. But his own experiments, though beautifully meticulous and inclusive, tended primarily to be qualitative: 'This happens, and this and this'. Statistical analysis was lacking. As will become evident throughout this book, you cannot carry out serious genetic studies – indeed you hardly see the patterns of heredity at all – unless you are a statistician (although in practice there is no maths in this book, largely because the author is roughly as innumerate as Darwin). One important point is that statistical analysis depends on large samples; you simply cannot see the patterns in small samples (or if you do it is only by luck). If Darwin had had several hundred children instead of a mere ten or so, and several wives instead of one, and several thousand cousins instead of a few score, then he might well have been able to infer, for example, 'why the same peculiarity . . . is sometimes inherited and sometimes not so'. But observations even of all the hundreds of people randomly encountered throughout his life could not truly reveal the orderly patterns.

In fact, Darwin did entertain two notions of heredity in particular: both germane to our theme. First, throughout his life he toyed with variations on the notions of the French biologist of the early nineteenth century, Jean-Baptiste Lamarck. Lamarck suggested that offspring inherit characteristics that were 'acquired' by their parents. Suppose, for example, that ancestral giraffes had short necks. Suppose that those ancestral giraffes spent their lives stretching those necks, to reach the higher leaves. As a result of all those efforts, said Lamarck, the offspring would be born with slightly longer necks than their parents. They too would stretch, and their offspring in turn would have even longer necks. And so on.

Such a mechanism was finally scotched in the late nineteenth century by the German biologist August Weismann. He suggested that information from the body cells (such as the muscle cells of a giraffe's neck) did *not* pass to the gametes (the eggs and sperm), so that heredity could not directly be influenced by the activities of the parents. Even so, Lamarck's hypothesis was far from foolish, and the derision he received in his own lifetime was founded in prejudice and ignorance. There should be no disgrace in science in being wrong, only in being dishonest, or dogmatic, or obfuscatory.

Besides, there is a twist to Lamarckism that makes it highly relevant today. After all, we may say – following Darwin's theory of natural selection – that giraffes did not acquire long necks because their parents stretched their own necks; it was just that natural selection favoured the particular individual giraffes who happened (by chance) to have the longest necks.

Ah, we may ask, but *why* did natural selection favour long necks in giraffes? Why did it not favour long necks in okapis or anteaters? The answer is – because the ancestral giraffes were in fact feeding on leaves in high trees, and okapis and anteaters were not. In other words, the stretching of the ancestral giraffe necks did not lead directly to long-necked offspring. But it was only *because* the ancestral giraffes had a *predilection* for feeding in tall trees that natural selection favoured long necks in the first place. In other words, animals do have some (unconscious) 'control' over their own evolution, even though natural selection is the mechanism that finally applies. At least, they tend (albeit unconsciously) to put themselves in a position in which natural selection favours such-and-such a character, rather than another.

Darwin, however (just to hammer this point down), was not looking to Lamarck for an explanation of evolutionary change: natural selection is in general an alternative to Lamarck's 'inheritance of acquired characteristics'. He did, however, entertain the idea that the mechanism of *inheritance* that Lamarck proposed might be correct, and that it could underpin natural selection just as well as it underpinned Lamarck's own theory of evolution. Indeed, *after* Darwin published *Origin of Species*, he wrote a long essay suggesting that body cells (such as giraffe neck cells) in fact produce 'gemmules' or 'propagules', which, he suggested, contained summaries of information about themselves; and that these summaries then passed to the reproductive organs, thence to become part of the hereditary information. He called this proposed mechanism, 'pangenesis'. It was, of course, in direct opposition to the notions of Weismann, which were published a couple of decades later.

Darwin always sought the opinions of his friends, and he asked Thomas Huxley what he thought of pangenesis. Huxley put on his 'sharpest spectacles and best thinking cap' and replied with wonderful diplomacy: 'Somebody rummaging among your papers half a century hence will find *Pangenesis* and say, "See this wonderful anticipation of our modern theories, and that stupid ass Huxley preventing his publishing them".'* But Huxley, as Darwin well knew, was a very wise ass indeed, and he kept his ingenious but crackpot notion to himself.

In general – whatever the details – Darwin supposed that parental characters were combined (more or less) in the offspring by a kind of blending, like a mixing of inks. He must have known that this was unsatisfactory. After all, red flowers crossed with white flowers *may* produce pink flowers (as we will see). But the cross may equally well produce offspring that are all red, or all white, or a mixture of the two.

There was a broader objection, however, one pointed out

**Darwin*, Adrian Desmond and James Moore, Warner Books, New York, 1992, p. 532.

in 1867 by a professor of engineering from Glasgow University, Fleeming Jenkin. For Darwin, in *Origin*, was not seeking simply to explain evolutionary change. He affected – as the title of his seminal book proclaimed – to explain the origin of *species*. The central notion of the species (at least in Darwin's day) was that each species differed *qualitatively* from another. The new species should be able to shake off the qualities of the ones that came before. But Jenkin – albeit with the racialism typical of his time – suggested a scenario that would seem to proscribe such radical change. Thus, he said, a white man cast away on an island of black people might well be acknowledged as their king. If he were, then he would enjoy all the reproductive success he might hope for. This reproductive success would be a measure of his 'fitness', in that particular environment; and hence the deified white man in the island of blacks would be bound to be favoured by natural selection, as envisaged by Darwin.

Yet, said Jenkin, this hypothetical successful white man 'cannot blanch a nation of negroes'.* In short, the 'blending' inheritance envisaged by Darwin could not apparently produce the kind of absolute changes that Darwin envisaged – changes that indeed could lead to the origin of new species.

In summary, Darwin produced the theory that has transformed biology, and indeed has changed the course of modern philosophy more profoundly than any other thinker of the past three centuries**. Yet the mechanism he proposed cannot work unless the process of heredity operates in a particular way: a way that can produce variation from generation to generation even while respecting the general condition that 'like begets like'; and in a way that would allow obsolete characters to be shuffled off completely and absolutely. But what that mechanism might be, Darwin failed absolutely to perceive.

Here we come to yet another irony, in fact to several more. First, the mechanism of inheritance that Darwin sought and needed was worked out and published during his own lifetime – indeed, just a few years after *Origin* appeared – by a scientist/monk in what was then called Moravia. Second, however, this crucial

***Science and Philosophy*, Derek Gjertson, Penguin Books, London, 1989, p. 37.

**For a full treatment of this point, see *Toward a New Philosophy of Biology*, Ernst Mayr, Harvard University Press, Cambridge, Mass., 1988.

insight was overlooked by the scientific community at large, and was in fact forgotten until rediscovered at the beginning of the twentieth century. Third, when the vital mechanism of inheritance was finally rediscovered, it was not at first acknowledged as the key to Darwinism, the means by which natural selection could actually work. By contrast, biologists argued for several decades that if the newly discovered mechanism of inheritance was correct, then natural selection must be wrong.

But I am running ahead of the story. The monk who provided the vital mechanism that Darwin needed was Gregor Mendel.

GREGOR MENDEL

Gregor Mendel (1822–1884) was almost an exact contemporary of Darwin: he was born just thirteen years after Darwin was born, and died two years after Darwin died. By the mid-1860s he had completed experiments which provided precisely the mechanism that Darwin's theory of evolution needed to round it off. Many commentators have said what a pity it was that Darwin never knew of Mendel's work. But then, they sigh, Mendel was a monk, who did his work in the Augustinian monastery of St Thomas at Brunn in Moravia (now Brno in Slovakia) – a distant country of which we knew as little then as Neville Chamberlain apparently did in 1939. Mendel announced his pivotal ideas in two lectures in the Natural History Society of Brunn, and they were published in the society's *Transactions* for 1866.

Yet Mendel was not a country bumpkin. He had studied mathematics, physics and biology in Vienna. Moravia and Brunn were not obscure. There was almost a century to go before the 'Iron Curtain' descended: Moravia in the mid-nineteenth century was very much a part of Europe. Sets of the essential *Transactions* were kept in England both in the Royal and the Linnean Societies. Darwin was extremely widely read, and other biologists brought matters of interest to his attention. Mendel also visited England, and some have rumoured that he actually visited Darwin. The great British population geneticist E.B. Ford avers that he did not: as Ford records in *Understanding Genetics* (Faber and Faber, London, 1979, p. 13) 'I am . . . the last of those who shared their friends with Darwin, and among the last who knew one of his children (Major Leonard) quite well . . . and am confident that no meeting between Darwin and Mendel ever took place'. Yet I remain intrigued by the notion that Darwin may well have known of Mendel's key experiments but – like everyone else at the time – failed to see their significance.

Such a failure does not reflect ill on Darwin. It would be perfectly understandable. After all, Mendel conducted his experiments with a few carefully selected characters in carefully selected plants – that is to say, with garden peas – under highly contrived conditions. Not even he was able to see that he had in fact discovered universal laws. It would indeed have been stretching credibility too far to suggest that the rules he had worked out in peas could also explain the caprices of inheritance in human beings. Here we have the nub of the problem, for although the basic rules of inheritance are simple, and universal, the *realities* of inheritance are such that the existence of those rules could not be inferred except by exploring deliberately simplified cases, in highly contrived circumstances. But Darwin was a broad thinker, and in this, he was hoist on the petard of his own breadth.

Mendel carried out his seminal experiments on the garden pea, *Pisum sativum*. In particular, he explored the patterns of inheritance of eight different characters, which included stature (short or tall), the colour of the unripe pod (yellow or green), the colour of the cotyledons within the seed (yellow or green), and the behaviour of the seed as it dried – whether it remained round, or became wrinkled.

Clearly, his experiments were highly contrived. He knew perfectly well – because he was an accomplished horticulturalist, as indeed were his parents – that the pattern of inheritance in domestic plants is sometimes orderly, and sometimes much less so. Garden peas are *inbreeders*: the seeds are fertilised by pollen from the same plant. Indeed the pollen comes from the same flower: it has to, because the stigma which receives the pollen and the anthers that produce it are completely enveloped by the petals. Inbreeding plants are also *true-breeding*: you do not see the erratic pattern of inheritance that Darwin observed in human beings (and other animals) and which can also be seen (for reasons that will become evident later) in, say, cabbages. Mendel knew, before he began, that peas would give him clear results, if any plant would; and that cabbages (say), would probably not.

Furthermore, Mendel knew perfectly well that only some characters in garden peas are inherited in an orderly pattern. Indeed he recorded the fact:

> The various forms of peas selected for crosses showed differences in length and colour of stem; in size and shape of leaves; in position, colour, and size of flowers; in length of flower stalks; in colour, shape, and size of pods; in shape and size of seeds; and in colouration of seed coats and albumen. However, some of the traits listed do not permit a definite and sharp separation, since the difference rests on a 'more or less' which is often difficult to define. Such traits were not usable for individual experiments; these had to be limited to characteristics which stand out clearly and decisively in plants.

So Mendel very deliberately elected to study the inheritance of a few carefully selected traits, in a well-chosen species. This is good science; it is a well-established principle that complex issues are often best approached through the study of simple cases. We can see, though, why even those who we know were aware of Mendel's work – why even Mendel himself – did not perceive that the rules that applied to highly selected characters in highly selected plants in practice apply to most characters in most animals, plants and fungi.

Yet Mendel's experiments, deliberately contrived to give simple results, also showed why it was so difficult to discern any order in inheritance among creatures at large, for even the simple examples he chose to study led quickly to enormous complexities.

Thus, to begin as simply as possible, round-seeded varieties of garden peas when left to self-pollinate produced round-seeded offspring, and wrinkle-seeded varieties, when self-pollinating, produced wrinkle-seeded offspring. This is what 'true-breeding' implies. But in one of his first experiments, Mendel explored what happened when round-seeded were crossed with wrinkle-seeded. This he did by removing the anthers of one kind, so they could not self-pollinate, and then brushing their stigmas with the pollen of the other kind; the time-honoured technique of the plant breeder.

A cross between two varieties is called a *hybrid*. The first generation following a cross between any two specified parents (whether of the same or different varieties) is called the *F1 generation*; and their offspring are in turn called the *F2 generation*, and so on. In the event, the answer to the first of Mendel's questions is that the F1 hybrid offspring of round-seeded and wrinkle-seeded peas all had round seeds.

What then had happened to the quality of wrinkledness? Mendel now allowed the F1 hybrid plants to self-pollinate. The result of this was that some of the F2 generation had round seeds – but in others, the quality of wrinkledness had mysteriously reappeared. This, of course, is exactly the kind of phenomenon that Charles Darwin had drawn attention to: that a character may miss a generation, and then crop up in a later one.

Mendel was not content merely to observe that some F2 plants were round and some were wrinkled. He counted them. There were 5474 round ones, and 1850 wrinkled ones. The ratio is very nearly three to one.

A short diversion is called for. The great British twentieth-century statistician R.A. Fisher pointed out that Mendel's figures were, in fact, too good to be true. All Mendel's published results show very clearly the kinds of ratios that confirm his ideas. But life isn't like that. In truth, all biological systems are subject to time and chance, and the kinds of figures that are *really* obtained from experiments such as Mendel's only rarely show exactly what they are supposed to show. Some critics have darkly hinted that Mendel, holy man though he was, fiddled his results.

I do not believe for one second that that is the case. Here, rather, is yet another quirk of science history. For it is only in the twentieth century that biologists (like physicists) have routinely begun to work in teams, and it is only now that those teams have come routinely to include statisticians. Indeed, everyone acknowledges nowadays that if complex experiments in biology are truly to be informative then statisticians must be involved at the design stage.

Mendel was not a statistician in the modern sense – for indeed, statistics in his day was still primitive. His statistics was of the commonsense, accountant's variety. He probably did not perceive that rigorous statistics were necessary to *test* hypotheses. Probably, rather, he saw the experimental results primarily as a means of *illustration*, of confirming what common sense already showed was obvious. In the same way, two centuries earlier, Isaac

Newton recorded precise experiments with light that he could not possibly have carried out. But Newton, unimpeachably honest, was not at all contrite. '*Of course* the experiment did not turn out exactly as I recorded,' he replied when challenged (though I paraphrase): 'This is the seventeenth century for Goodness' sake! What do you expect with prisms like these? But it is obvious what *would* have happened, if I had been able to control all the factors precisely.' Mendel doubtless felt the same. He counted peas until (so common sense suggested) he had counted enough to make the point. So what else do you need?

The three-to-one ratio that Mendel observed has several implications. First, it suggests that inheritance can indeed follow simple arithmetical rules. The patterns of inheritance are not invariably messy. Second, the clear ratio clearly militates against the kind of explanation of inheritance that Darwin found himself adhering to: that inheritance is like a mixing of inks. There was no mixing here. The F2 peas were either wrinkled, or they were not. They were not half-wrinkled, or wrinkled in parts. The quality of wrinkledness had not been diluted, and still less had it been extinguished. It had merely been suppressed for a generation. But how?

Mendel, genius that he unquestionably was, provided an explanation that was simple, satisfying and – so all experiments subsequently suggest – correct. The characteristics of a plant ('*characters*') were determined not by vague pervasive philtres that could be mixed like inks. Instead, each character was determined by a discrete 'factor' (Mendel used the German *Anlagen* for 'factors'). These are the factors that we now call *genes*, and for convenience, I will use that excellent term from now on.

Each individual, said Mendel, contains two copies of each gene; one inherited from its mother, and one from its father. And each individual (just to round off the point) passes only *one* copy of each gene on to its offspring. *And this is the essence of classical genetics, from which all else follows. The rest of this book is a footnote.* It is, however, quite an interesting footnote, so I will continue.

Each true-breeding round pea, so Mendel inferred, contained two copies of the gene for roundness, and each true-breeding wrinkled pea contained two copies of the gene for wrinkledness. In fact, of course, the roundness gene and the wrinkledness gene are different versions of the same gene; that is, different versions of the gene that determines seed shape. Different versions of the same gene are called *alleles* of that gene, and 'allele' is an extremely important term.

Each kind of pea passes on only one copy of its pea-shape gene to its offspring. Hence the true-breeding round pea passes on one roundness allele to each offspring, and the true-breeding wrinkled pea passes on one wrinkledness allele to each offspring. They can do nothing else, each one contains only one kind of allele for that particular gene. Hence the F1 offspring of a cross between a true-breeding round and a true-breeding wrinkled all contain one roundness allele and one wrinkledness allele.

Mendel realised the roundness allele is *dominant* over the wrinkledness allele. So long as it is present, the wrinkledness allele remains inoperative. That is, it is *recessive*. So the peas of the F1 cross-bred generation are all round, because in each one, only the roundness gene makes itself felt.

Consider what happens now, however, when the F1 offspring themselves start to breed. Stage one is to produce gametes: eggs or sperm. Each gamete can contain only and copy of each kind of gene. But the F1 offspring each contain two versions of each pea-shape gene; both a wrinkledness allele and a roundness allele. Each one can pass on only one of the two alleles to each gamete: *either* the wrinkledness allele *or* the roundness allele but not both.

Hence, if these F1 offspring are randomly crossed with other F1 offspring the subsequent, F2 generation contains a fine old mixture. Each individual could, in principle, inherit two roundness alleles (one from each parent) – in which case its own seeds would be round. Or it could inherit two wrinkledness alleles – and then its seeds would be wrinkled. Or it could inherit one roundness and one wrinkledness allele – in which case, because roundness is dominant, its seeds would be round, just as if it had inherited two roundness alleles. Common sense immediately allows us to see that there are three ways out of four in which the F2 seeds could finish up being round, and only one way out of four in which they could finish up being wrinkled. If common sense fails, Figure 1.1 makes it all obvious at a glance. In fact, as Mendel found, in this experiment the ratio of round peas to wrinkled peas in the F2 generation is indeed 3:1.

Mendel, as already noted, did not establish the basic vocabu-

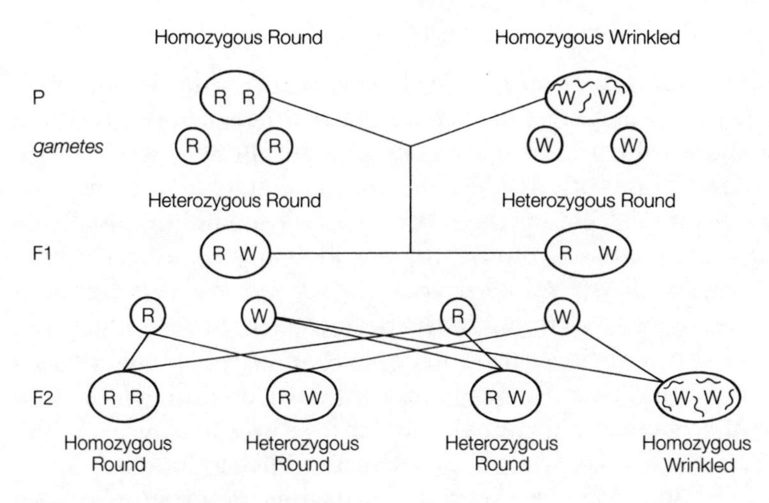

Figure 1.1 Mendel's famous 3:1 ratio
The parents (P) are round homozygous, and wrinkled homozygous. Their F1 offspring are therefore all genetically heterozygous – but round is dominant so they are all phenotypically round. The F2 generation contains one homozygous round and two heterozygotes – all of which are phenotypically round – plus one homozygous wrinkled; the only one that is wrinkled phenotypically.

lary of genetics; indeed, the term 'genetics' was not coined until 1909. But it will make this historical account easier if at this point we introduce a few basic twentieth-century terms that will occur throughout this book.

First, an individual that contains two identical alleles of a particular gene is said to be *homozygous* for that gene; while an individual that contains two different alleles of a particular gene is *heterozygous* for that gene. Clearly, too, we must differentiate between an organism's *phenotype*, which refers to how it looks; and its *genotype*, which alludes to the genes it contains. Thus homozygous round-seeded peas, with two roundness alleles, are phenotypically the same as heterozygous round-seeded peas, containing one roundness and one wrinkledness allele. But genotypically the two are different – as is shown by the phenotypes of their offspring. Finally, the total complement of genes (alleles) within any one individual is called its *genome*, and the total array of alleles within any breeding population of organisms is called the *gene pool*.

Mendel's first experiment with peas is about as simple as heredity can be, yet in principle it illustrates almost all there is to know about its basic rules. So why, in reality, does heredity seem complicated – so complicated, indeed, that it fooled Charles Darwin?

Mendel himself again provided the essence of the answer. It is easy (even by playing simple mind-games, and without carrying out experiments at all) to envisage complications: and once you add a few complications, the patterns that rapidly emerge become very intricate indeed. Furthermore, it soon becomes impossible to see any patterns at all unless the numbers of offspring are very large indeed. In short, once you get beyond the very simplest examples, you have to be a statistician to see what is going on. Mendel was a statistician of a simple kind, while Darwin was a wonderful observer but was not a statistician at all. He had no hope of seeing clear patterns from observations of single lineages of creatures that have only a few offspring, such as human beings.

Specifically, Mendel did not stop with his experiments on wrinkledness and roundness in peas. He went on to show that the yellow colour of seeds dominates green colour. When yellows were crossed with greens all the F1s were yellow, and when the F1s were allowed to self-pollinate, the yellows in the F2s outnumbered the greens three to one, exactly as with roundness and wrinkledness.

He then tested the inheritance of both characters together, and found, first of all, that whether a pea's seeds were wrinkled or round had no bearing on whether they were yellow or green. The two characters were inherited independently – and this independence of inheritance is one of Mendel's fundamental laws. He then found that when homozygous round-yellows were crossed with homozygous wrinkled-greens the resulting phenotypes were in the ratio of nine round-yellows, three round-greens, three wrinkled-yellows, and one wrinkled-green. This is another famous Mendelian ratio, 9:3:3:1; and the reason for it can be seen at a glance in Figure 1.2. But the ratio cannot be seen at all unless there are at least sixteen offspring, and because of the stochastic (chance) variations this ratio is not likely to emerge at all clearly unless the numbers of offspring are very large indeed. Clearly, too, some combinations of characters are rare: only one in sixteen of the pea offspring (on average) is both wrinkled *and* green. A plant breeder would hope to produce such comparative rarities by producing tens of thousands, or millions of offspring. In a human family, such rarities would simply turn up now and again, apparently 'out of the blue'.

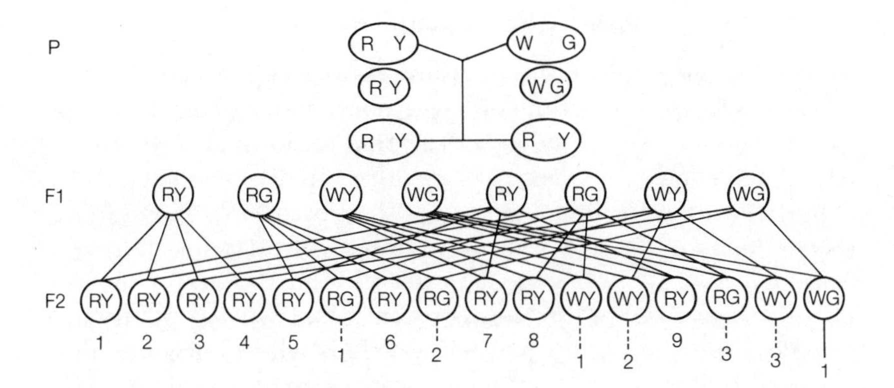

Figure 1.2 Mendel's Law of Segregation
Most characters in most organisms are inherited independently of other characters. Thus round peas can be either yellow or green, and wrinkled peas can also be either yellow or green. But because the round gene dominates the wrinkled gene, and the yellow gene dominates the green, the offspring of doubly heterozygous parents appear – phenotypically – in a ratio of 9 : 3 : 3 : 1.

The important point to note here is the cumulative effect even of 'simple' complications. Just add one complication to another, and the outcome rapidly becomes very confusing indeed, just as Darwin observed. We may draw a parallel with the modern mathematical concept of chaos: a few simple physical rules, piled one on top of another, produce effects which, in detail, are totally bewildering. Some have said that Mendel was lucky to have seized upon examples in which clear rules could be easily discerned. The facts instead suggest that he was simply a genius, who knew that complexities have to be sought by focusing first upon simplicities. The approach to complexity through simplicity is René Descartes's principle of *reductionism*.

Mendel's seminal experiments, duly announced and recorded, were then ignored and effectively forgotten for the rest of the nineteenth century. Mendel himself embarked upon experiments which, I feel, his own native good sense must have warned him against. His senior, Karl Wilhelm von Nageli, suggested to Mendel that he should try to make sense of the heredity of the hawkweed, *Heiracium*. Mendel accordingly worked on *Heiracium* from 1866 to 1871, and got precisely nowhere. The reason is now clear: *Heiracium* reproduces by *parthenogenesis* – that is, the eggs develop into embryos without being fertilised by the males. Parthenogenesis is in fact a form of asexual reproduction, even though it is based upon sex cells. Clearly, Mendelian laws of inheritance do not apply to it. Von Nageli would not have known this, however. His intervention may in practice have held up the development of genetics by several decades – but if it did, it was inadvertent. He was not trying to be mischievous. He was merely being realistic, in a way that Darwin might have approved of.

Perhaps Mendel was frustrated by his lack of recognition: this is suggested by his recorded comment, '*Meine Zeit wird schon kommen*' – 'my time will surely come'. But in practice he was elected Abbé of Brunn in 1868 and became more and more involved in church politics until his death in 1884. Further advances in the science of genetics were left to the twentieth century. The achievements of that science have indeed been spectacular. At the purely theoretical level – we will come to practicalities later – twentieth-century geneticists have advanced the whole process of unification. Thus they have shown that the ideas of Mendel do indeed complement those of Darwin, and have revealed that Mendel's abstract '*Anlagen*' have a precise chemical basis, and act in clearly definable ways, thus unifying the parallel sciences of genetics, biochemistry, and indeed pharmacology. Genetics, in short, as this book will unfold, has effected the grandest of all conceivable syntheses within the overall science of biology.

But it has also revealed a huge range of further complexities in the mechanisms of heredity: complexities that fully vindicate the confusion of Darwin, and emphasise the genius of Mendel, in pinning down the essentials.

'CLASSICAL' GENETICS IN THE TWENTIETH CENTURY: AN INSTANT GUIDE

In Russia, in the middle decades of this century, the ideas both of Darwin and Mendel were rejected on ideological grounds. Soviet 'crop improvement' was consequently based on non-Mendelian principles – a fact from which their agriculture has never recovered.* It is ironical, then, that the first scientist truly to recognise the significance of Mendel's results, soon after their publication, was a Russian botanist, I.F. Schmalhausen. In general, however,

*See I. Michael Lerner's *Heredity, Evolution, and Society*, University of California, Berkeley, 1968.

Mendel's work lay fallow until 1900, when it was discovered by three scientists working independently: the Dutch biologist Hugo de Vries; the German botanist Carl Correns; and the Austrian Erich Tschermak. Within a few years the French zoologist L. Cuenot, the American biologist W.E. Castle and the British biologist William Bateson showed that Mendel's laws apply to animals as well as plants. Bateson also – in 1909 – coined the word 'genetics'.

Since then, to be brutally brief, the central science of genetics – 'classical genetics' – which deals with the *patterns* of inheritance, has been developed by such giants as T.H. Morgan in the United States and R.A. Fisher (the statistician) and C.D. Darlington in Britain. The creatures that have emerged as the principal 'models' for such studies have included various bacteria, including the famous and ubiquitous *Escherichia coli*, which lives in the gut of human beings and of many other animals, usually harmlessly but also causing various forms of travellers' diarrhoea; yeast, which is a fungus-like single-celled organism; true fungi, such as *Neurospora*; maize, among plants; and, most famously of all among animals, the fruit-fly *Drosophila*, favoured first by Thomas Morgan. *Drosophila* has many visible characters that have a simple genetic basis, produces a lot of offspring, and has a generation time of just a few weeks; an ideal subject. There have also been many classic genetic studies on various snails, which are convenient because many of their genes show their presence by affecting the colour or form of the shell; on a vast range of domestic plants and animals; and on various odd but convenient species such as tobacco (*Nicotiana*) and *Primula*, and the small floating water-weed, *Arabidopsis*. Scientists, in short, are opportunists. They make use of whatever is to hand that seems most likely to throw light upon the immediate problem – just as Mendel worked on garden peas, and Darwin played around with pigeons.

These classical genetic studies have revealed that Mendel was right to begin work on characters that he knew, or sensed, had a simple pattern of inheritance. For they have revealed a great many complications, of which the most important are as follows:

First, it is clear that the same character or set of characters – the same phenotype – may be brought about by quite different genetic means. Thus, some strains of wheat are tall (that is, have long stems) and some are short. But some are tall because they produce more of the growth 'hormone' giberellin – while others produce the same amount of giberellin but have cells that are more sensitive to it; and some are tall because they are less responsive to changing day-length and do not so quickly acknowledge that autumn is pending, and so they go on growing for longer. Among dogs, Great Danes are big because they have genes which in a general way promote growth, while St Bernards are big largely because they produce large amounts of the growth hormone from the pituitary gland (and they are in fact 'pituitary giants'). As discussed further on pages 38–9, several snails of the Pacific islands have transparent shells, but appear to be coloured because the mantle beneath is coloured. But shell transparency has a different genetic basis in different kinds of snail. And so on.

Neither can we infer that two phenotypically different creatures are necessarily different genetically. Phenotype in general is 'plastic'; that is, the form, colour, behaviour, and even in some cases the sex of an organism may be determined not only by its genome, but by circumstance. Oak trees that grow in crevices in rocks are very different in form from the tall, straight trees of oak forest, or the luxuriantly spreading behemoths of open fields. Some salamanders sometimes retain a larval form throughout their lives, and sometimes assume an adult form, depending on what life has to offer them. Successive generations of children in the western world have tended to be taller than their parents, in a manner that looks very Lamarckian, but isn't. Tortoises are female if incubated at high temperatures, but male if incubated in the cool. Female naked mole-rats stay small and infertile so long as the queen mole-rat is ruling the nest, but can themselves grow into big, fertile queens if the boss lady dies. In Chapter 3 we will explore the phenomenon of *gene expression*: the fact that genes make their presence felt under some circumstances, but not necessarily under others. Thus, a gene may become operative in a cool environment for example, but not when it is warmer. In general, plasticity of phenotype is largely but not exclusively explained by the effect of environment on gene expression.

Then again, most characters in any one organism are *polygenic*; that is, they are brought about by more than one gene. Skin colour in humans is one such character – as Fleeming Jenkin observed:

so the offspring of one black parent and one white may be any colour from creamy to dark.

On the other hand, most genes are *pleiotropic* in their effects, which means that they affect more than one character, and often affect characters that are not at all related. Thus did the great British geneticist E.B. Ford observe that different colour variants of brown butterflies in fact have different resistance to temperature. The difference in colour does not, in this case, directly affect the temperature response, but the genes that result in colour differences also produce a difference in physiological response.

Easily confused with pleiotropy is the phenomenon of *linkage* – a phenomenon that Mendel himself observed. For genes, as we will explore more fully in the next chapter, are 'carried' on physical structures known as *chromosomes*, within the cell nucleus; and as Thomas Morgan first established, they occur in rows, one behind the other. Each place on the chromosome, where each gene occurs, is called the *locus* of that gene. Genes that are carried on the same chromosome may be inherited together, effectively as a package. If they are very close together on the same chromosome – that is, closely 'linked' – then they are extremely likely to be inherited as a pair.

This phenomenon has several very important connotations. First, it shows that one of Mendel's fundamental rules – the one that says that different genes can be passed on *independently* of all others, so that round peas and wrinkled peas can be either yellow or green – is not invariably true in practice. If the roundness gene was linked to the yellowness gene then the two qualities would tend to be passed on together. Note, however, that whereas two different characters brought about by a single pleiotropic gene will *always* go together, two characters that are caused by different genes that happen to be linked will go together only sometimes. Indeed, geneticists can ascertain where particular genes are positioned on chromosomes – that is, begin to create a chromosome *map* – by measuring the frequency of linkage (or, more precisely, by comparing the frequency with which the two characters turn up together, with what would happen if they were not linked at all).

The phenomenon of linkage – which depends on the physical proximity of genes on any one chromosome – has itself been the subject of vigorous natural selection. For within the cell, genes act in concert with other genes. In general, as we will see in the next chapter, genes produce enzymes; and enzymes work in conjunction with other enzymes, to form 'metabolic pathways'. Enzymes work most efficiently if they are compatible with all the other enzymes in their pathway – for example, if the first enzyme in the pathway turns out its product at exactly the pace, and in the form, that the next enzyme in the pathway is best able to deal with. Hence it is advantageous if genes that produce enzymes in the same metabolic pathway are inherited as a group; and natural selection has accordingly favoured the coming-together of genes that habitually work together. Groups of genes that have thus been selected are called *clusters*, and Darlington called such clusters, passed on and working in unison, *supergenes*.

A special example of linkage is *sex linkage*. In most animals (though not in tortoises and some others) sex is determined by special chromosomes known as sex chromosomes. All mammals, for example, are female if they contain two X chromosomes – that is, are XX; and are male if they possess an X and a Y chromosome – that is, are XY. The allele that causes the human blood disease haemophilia is recessive, and is carried on the X chromosome. If a male inherits an X chromosome with the haemophilia allele, then he has the disease; for there is no equivalent, dominant allele on the Y chromosome to override it. If a female inherits an X chromosome with the defective allele, then she will probably be clear of the disease, because her other X chromosome will probably carry a normal allele that will dominate. Such a female will, however, be a *carrier* of the allele, and a carrier has a 50/50 chance of passing on the chromosome with the defective allele to each of her children. A girl could be a haemophiliac herself, if her father was a haemophiliac, and if her mother was a carrier, though this is a most unlikely state of affairs. Hence, in general, haemophilia is a *sex-linked* disorder: it is far more common in boys than in girls. Haemophilia blighted Queen Victoria's family, and Darwin may well have had this in mind when, in *Origin*, he wondered why 'a peculiarity is often transmitted . . . to one sex alone'. Domestic cats provide a more homely example. Gingers are always toms. The female equivalent of the ginger tom is the tortoiseshell. The one exception – apparently tortoiseshell males – is brought about by a chromosome anomaly known as *Klinefelter's syndrome* (of which more later) in which individuals that look male (but are sterile) contain two XXs or more, and at least one Y chromosome (XXY, XXYY, XXXY or XXXXY).

Dominance, too, need not be the all-or-nothing affair that Mendel observed in peas. The degree to which a gene makes itself felt is its *penetrance*, and penetrance varies, for all kinds of reasons. Sometimes one allele is merely *partially dominant* over its partner. Thus in some cases a red-flowered plant crossed with a white-flowered plant produces a pink-flowered offspring. Then it may seem as if the two characters are simply being mixed, like coloured inks, but this is not so. If, in such a case as this, the pink hybrids are self-pollinated, then the offspring will include some pure reds and some pure whites, as well as some pinks.

In 1908, H. Nilsson-Ehle showed partial dominance and polygenic inheritance working together, to influence the colour of the kernels of wheat. Some wheat strains have dark red kernels, and are true-breeding. Some have white kernels, and they, too, are true-breeding. When the two types are crossed, the F1s are all medium red. But when the F1s reproduce, the F2s show a range of colours. For every sixteen plants (on average), one would be dark red, one pure white, and there would be fourteen intermediates: four pink, four fairly dark-red, and six medium-red. The point here is that colour is determined by two alleles, with the red partially but not completely dominant over the white. A plant that contains four red alleles (that is, is doubly homozygous) is dark red, and one that is doubly homozygous for white alleles is white. Any other combination produces some intermediate colour, depending on the number of white or red alleles.

Then again, sometimes the effect of one of the pair is actually enhanced by the presence of the other, in which case the gene that is enhanced is said to be *superdominant*.

Some complications are positively bizarre. In particular, not all the genes in animals or plants are carried on the chromosomes, tucked away in the nucleus. Some are carried on small structures (*organelles*) within the cell cytoplasm, outside the nucleus – such as the *mitochondria*, which in both animal and plant cells contain the enzymes that carry out cell respiration; some are carried on the *chloroplasts*, which in plants carry out photosynthesis. During fertilisation, the female egg (ovum) combines with the male sperm. In general – and certainly for most of the time in animals – only the ovum provides cytoplasm, for the sperm, effectively, is all nucleus. Thus mitochondrial and chloroplastic genes – 'cytoplasmic genes' – are in general passed on only through the female line.* Clearly, though, the laws of Mendelian genetics cannot possibly apply to organelle genes. As we will see in Chapters 6 and 7, however, plant breeders, opportunist as ever, make use of cytoplasmic genes in various ways. For one thing, some mitochondrial plant genes induce male sterility, which in plant breeding can be useful; and chloroplasts can provide convenient sites in which to position novel genes during genetic engineering.

Finally, genes, every now and again, *mutate*. That is, they simply change, and start to behave differently. Mutations occur randomly in nature, although in practice some kinds of mutation occur more frequently than others, and some genes are prone to mutate in particular ways, and various environmental stimuli – such as X-rays, ultra-violet radiation, and various chemicals – can increase the rate of mutation.

Genetic mutations are random, and most lead to bad effects. By the same token, you would not seriously expect to improve the workings of a television set by giving it a random kick, even though we have all resorted to it. Mutations that occur in body cells can lead to cancers, which is why excessive sun-bathing (over-exposure to ultra-violet) can lead to skin cancer; and why the diminution of the ozone layer (reducing the atmosphere's natural ultra-violet filter) exacerbates the problem. But body-cell mutations are not passed on to the offspring – for the reason that Weismann identified: the gametes stay aloof from such calamities.

However, mutations that occur within the sex cells themselves *can* be passed on to the offspring, and they become a new source of genetic variation within the subsequent generations. As I have said, mutations are random and most of them are bad. *Yet mutation in practice provides the* chief *source of new genes – and is therefore the chief source of the variation upon which evolution by natural selection depends.* Mid-Victorians did not know about genes and genetic mutation. But Darwin made it very clear that the variation which he saw as the essential precondition of natural selection was randomly produced. Mid-Victorians –

*In fact, this is not invariable. Modern studies are providing many examples of mitochondrial inheritance through the male line. In some groups, such as conifers, this is common.

including mid-Victorian scientists – preferred to believe that if evolution happened at all, then it unfolded according to a plan prescribed by God. They found the idea that random events should be the driving force extremely offensive, as indeed do many people today. I believe (and so do many modern clerics) that Darwin's view of evolution by natural selection is perfectly compatible with a broad and satisfying view of religion, even if there is some conflict with particular points of theology. I will discuss at least part of this issue in later chapters. But the nature and implication of genetic mutation certainly belong here.

Incidentally, as we will see in a later chapter, X-rays and other mutagens play a relatively small but still significant part in plant breeding. They produce new variants – not for natural selection to act upon, but for breeders to select between. What can be called *directed* mutations, however – specific and deliberate alterations in genes – are now playing a very large part in genetic engineering.

Finally, as we will see in the next chapter, the genome in general is now known to be far more restless than was hitherto envisaged. It now transpires – extraordinarily – that pieces of DNA, and indeed entire genes, may hop from place to place within the genome – within the chromosome, or from chromosome to chromosome; and that such pieces may (or may not) cause other genes to mutate as they do. Individual genes may also multiply within the genome, sometimes to produce hundreds of copies. In general, it seems that whenever biologists pin down living systems, they turn out to be more fluid than conceived. This innate fluidity and restlessness, as we will see many times throughout this book, has many implications for evolution, and for human attempts to manipulate the genome.

These, then, are the main complications of heredity. Some are caused by innate complexities of genetic inheritance, and some result from the confusion caused by *non* genetic influences upon hereditary. We need not be surprised that any one biologist, even the genius Darwin, failed to see the simple, underlying pattern. We may rather be suitably astonished – and grateful – that another genius of a different kind did.

Put the two species of genius together – the grand overview and the adumbration of precise mechanism – and we surely have the perfect synthesis: the neatest, tidiest and most convincing unifying vision of life's diversity that we could ever have dreamed of; the biological equivalent of the physicist's grand unifying theory. Well, that is in fact how things have turned out. The grand unifying theory of biology – what Julian Huxley in the 1940s called 'The Modern Synthesis' – is known as 'Neodarwinism'*. Now some footnotes have been added to the Synthesis, as discussed later, but it remains none the less the 'paradigm' of modern biology, where 'paradigm' (as defined by the modern philosopher of science Thomas Kuhn**) means, effectively, world view. Neodarwinism is, in short, the essence of modern biology.

NEODARWINISM: THE MODERN PARADIGM

You may feel intuitively that Darwin's and Mendel's ideas do fit together, neatly. You may reasonably suppose, then, that when Mendel's ideas were first rediscovered at the beginning of this century, biologists would have said, 'Eureka! Here (in Mendel) is the perfect body of ideas to complement, flesh out, and generally complete the beautiful vision of Darwin!' In fact biologists in the first decades of this century were split, between those who thought Mendel must be wrong (or that his ideas must apply only to a minority of creatures), and those who thought Darwin must be at fault. (In the same kind of way, as we will see in the next chapter, William Bateson, who contributed so much to the early development of genetics, for some time vigorously opposed the apparently obvious notion that genes are carried on chromosomes.)

The trouble was felt to lie in Darwin's notion that evolutionary change was generally gradual: a step-by-minute-step progression from one state to another. Mendel seemed rather to suggest that an organism either had one character or another

*There is room for confusion here, because the term 'Neodarwinism' is used by different philosophers in different contexts. In particular, Herbert Spencer and others in the decades after *Origin of Species* sought to apply ideas of natural selection to politics and ethics, and this application is sometimes called 'Neodarwinism'. I do not believe that Darwin himself would have approved of this. But in any case, I am *not* using the term in any sense other than the one defined in the text.

**(The seminal work is *The Structure of Scientific Revolutions*, published in 1962).

– yellow flowers or white; wrinkled seeds or smooth – with no infinity of shades between the two. Yet Mendel himself had pointed out that many characters even in his garden peas were of a 'more or less' character. Besides, it soon became clear that most characters are, in fact, polygenic; and if there are ten genes coding for skin colour, say, which could each be present as one of several different alleles, then any shade between is possible. In short, Mendel's genetics are perfectly compatible with Darwin's gradualism, once we acknowledge the *reality* of genetics: that most characters are polygenic, and therefore, within their limits, almost infinitely variable. The intellectual ground-work – the realisation that this is so – was laid in the 1920s and 30s, and the Modern Synthesis (Neodarwinism) finally assumed its proper place in the 1940s.

In practice, the Modern Synthesis – the essence of modern biology – can be expressed as follows. Most creatures on the face of this Earth practise sex, in one form or another: that is, different individuals exchange genetic information. Sometimes, as in most bacteria, one individual simply squirts a seemingly random bundle of genes into another; but plants, animals, and fungi carefully and precisely divide their genomes into two, as we will discuss in the next chapter, and then splice the two half genomes to make a new and qualitatively different whole.

Because organisms practise sex, all the individuals who are in contact with one another – that is, all who are within the same *population* – are potentially able to combine their genes with each and any of the others. To be sure, males cannot mate with other males, but they can mate with each other's daughters, so that any one gene in any one individual could, in some future generation, find itself sharing a genome with any other gene from any other individual. Thus, all individuals in any one population partake of a common *gene pool*. 'Gene pool' is a respectable technical term – and one with a truly evocative metaphoric ring. All of us are members of populations and each of us does indeed dip the common reserve of genetic information.

In the neodarwinian paradigm, evolutionary change is regarded as a shift in a population's gene pool. As the years and generations pass, some genes are gained, and others are lost. Each individual allele in the pool is liable to change in *frequency*; in fact, evolution can simply be envisaged as a shift in gene frequency. In practice, most biologists would not acknowledge that evolution had truly taken place until and unless some alleles had actually been lost altogether, while others had been gained and established. There has, for example, been quite a shift in the frequency of the various alleles in the gene pools of human populations in most of the continents of the world over the past few centuries, as (for example) Moors from North Africa and Mongols from Asia came and mostly went. But such changes would be seen merely as a temporary fluctuation. There has been little or no absolute loss or gain of alleles, not enough to register as an evolutionary shift.

New alleles may appear in the pool for one of two main reasons. The first – and most radical – is mutation. Any gene may simply change, in ways that we will explore in the next chapter. Note, however, that any one gene can be present in more than one copy, so if one allele alters, this does not necessarily mean that the organism has lost all copies of the original. Mutations sometimes result from a simple 'mistake': sometimes because of some assault from outside (such as radiation), and sometimes are induced by the migrations of other genes within the genome.

The second source of new alleles in any one gene pool is *introgression*. This means, simply, that sometimes some creature from a different population wanders in, bearing alleles that are not possessed by the original population, and stays to interbreed and so add its genes. As we will see in Chapter 6, introgression can hold up – and occasionally enhance! – attempts to breed better crops, as wild plants pollinate the domestic ones. Contrariwise, cross-breeding with domestic cats constantly extends – or some would say 'pollutes' – the gene pool of wild cats in Britain, while feral domestic dogs are presently corrupting the few remaining wolves of Italy. Note that introgression – unlike mutation – does not create brand new genes that may not have occurred on Earth before. But introgression does bring in genes that are new to any one gene pool, and it can be very important indeed for evolution. In either case, the first *appearance* of new genes – the initial source of genetic variation – is pure chance. At least this is true in nature. Breeders, and genetic engineers, increase variation in their subjects' gene pools primarily by arranged introgression; that is, *crossing* with individuals from different populations.

Whether or not a new gene increases in frequency once it has appeared, or peters out, is partly a matter of time and chance, and partly of natural selection. In animals, for example, if a new mutation occurs in an ordinary body cell – an *autosomal* cell – then it will not be passed on to the next generation, and so will have no real effect. Changes in muscle cells, as Weismann said, can have no direct effect upon gametes. However, this is *not* true of many plants, which first may reproduce asexually to produce a new plant (as a strawberry sends out runners), which in turn may produce flowers, which do contain sex cells. So a mutation that occurs in a strawberry runner may well be transferred into strawberry gametes, and hence be passed on and shared sexually with other strawberries. This is one profound difference between animals and plants.

If, by whatever means, the mutation does finish up in a gamete, and is passed on, one of several fates befalls it. Most mutations either have very little effect or are harmful (they arise as random changes in a delicate mechanism, so how could it be otherwise?). If the mutation does not affect the survival of the individual that inherits it, then it is said to be 'neutral', and gene pools clearly accumulate a large backlog of 'neutral' mutations. These are important, because variation is variation, and what is neutral in one generation may prove useful in a later one. If mutations are harmful, then the inheritor may suffer, and so the mutant gene is liable to die out. But if the mutation is recessive then it will not immediately make itself felt, and so it is that most populations of organisms contain recessive genes whose effects are indeed harmful (known as 'bad genes' or *deleterious alleles*). These are the source of all the *single gene disorders* in human beings. If by chance the mutation is beneficial, then its possessor is likely to have more offspring than a non-possessor, and so the new, beneficial mutation will increase in frequency from generation to generation. The population may grow as a result of a beneficial mutation, but no population can grow indefinitely. Sooner or later, then, the individuals with the mutated genes find themselves in competition with the non-mutated gene, and if the new gene is better and the competition is fierce, those with the original allele die out. This is natural selection in action: individuals with a beneficial character (brought about by a beneficial gene) ousting those without it. The acquisition of a new gene, coupled with the loss of an old one, truly represents a qualitative change in the gene pool. Once such a qualitative shift has occurred, then evolution can properly be said to have occurred. Genes acquired by introgression similarly peter out, hang around, or increase and contribute to evolutionary change.

Alleles are also lost from the gene pool, partly by natural selection, and partly by time and chance. How natural selection works is obvious: genes that produce uncompetitive characters are simply pushed out. Loss through time and chance is called *genetic drift* – and may result in the loss of very beneficial genes. Here there are two main mechanisms. First, and importantly, each individual passes on only half of its genes to each of its offspring. Unless an individual has a huge number of offspring (as a fly does), then it is statistically unlikely to pass on all of its genes to the next generation. If the population is large, this does not matter, because the genes in any one individual are likely to be represented in another, who very possibly will pass them on. But in small populations of animals that are not fecund (a few score rhinoceroses in a park, for example) there can clearly be a huge loss of alleles, and hence of overall genetic variation, in each generation. There simply are not enough new individuals in the new generation to soak up all the genes of their parents.

The second mechanism of loss by genetic drift is simply that some individuals in any one generation fail to breed at all. Some die young, others fail to find a mate. If the non-breeders are the sole bearers of some rare allele, then that allele dies with them. Again it is clear that such loss is far more likely in small populations, where rare alleles may indeed be represented only in one or a few individuals. Loss is also likely in many modern wild habitats that have become fragmented, because then many animals cannot find suitable mates. Again rhinoceroses come to mind, particularly the few hundred remaining Sumatrans, increasingly cut off from each other as the forest is felled around them. Clearly, these genetic considerations have enormous implications in conservation, which I will discuss in Chapter 8.

This, then, is the neodarwinian view of evolution: a gene pool changing over time, through the interaction of natural selection, and time and chance. Since it was formulated, in the middle decades of this century, there have been one or two modifications of varying significance, which we will come back to at the end of this chapter. But these changes have not burst the paradigm, they have merely enriched it. Neodarwinism remains the essence of all modern biology.

Let us refer again, though, to the *title* of Darwin's seminal work of 1859: *On the Origin of Species by Means of Natural Selection.* It begs a very important question: What exactly is a species?

THE IDEA OF THE SPECIES

The seventeenth- and eighteenth-century biologists who laid the foundations on which Darwin built had two principal cultural roots. They were Christians, and they were Platonists. As orthodox Christians, following Genesis, they believed that God had created all living creatures, and that he had made them in their present forms. As followers of Plato (whether they knew it or not), they believed that each actual creature was the embodiment of an 'idea', an idea in the mind of God, as to what that creature should be like. To them, each 'species' was an ideal, and the individuals in each species were all representatives, approximations, of God's fundamental vision of what their particular type should be like.

As everybody knows, many orthodox Christians – both scientists and non-scientists – objected to Darwin's ideas. What most people don't know, however, is that they objected most strongly not to the idea of evolution *per se* – of 'descent with modification' – or even to his stark contention that the course of evolution was shaped by the vicissitudes of environment, and not, at least directly, by the moulding hand of God. What really offended critics was Darwin's insistence that one species could change into another, or – even worse – that one species could diverge to form several, or even thousands of, new species. Before Darwin, it was possible to reconcile the general idea of evolution with the story of Genesis by supposing that in the beginning God did indeed create a whole array of creatures – each one a type, a 'species' – which subsequently changed by evolution. But Darwin envisaged that any one species could give rise to many, and indeed suggested that present-day living things may *all* have evolved from just one first ancestor. This goes against Genesis – and hence is non-Christian. But it is also non-Platonist, because it suggests that the basic ideals of God – the species – were, in fact, changeable. The idea that God could change his mind seemed somewhat blasphemous.

Modern biologists still have a very strong sense that 'species' are real. Creatures within any one species may be very variable, but there none the less are clear differences between one species and others. Thus all domestic horses, though they range from Shire to Shetland, belong to the same species, called *Equus caballus*. But domestic horses are clearly different from asses (although they are also in many ways similar). They are even more different from rhinoceroses (although the anatomist can still see clear similarities). Human beings also seem variable, but they too are all of one species: in fact, within the human species the total genetic variation – that is, the total variety of alleles – is remarkably small. Black or white, brown or yellow, short or tall, human beings are much of a genetic muchness. But we are clearly different from chimpanzees (though in many ways obviously similar), and even more different from monkeys (though similarities are still discernible).

But modern, post-Darwinian biologists no longer regard species in mystical, Platonic terms, as ideas of God. Instead, their approach is functional. No definition of species can be perfect, so it is foolish to try to frame one. But the kind of rough-and-ready definition that I learnt at university works well enough for most purposes: 'Two creatures (of appropriate age and sex) are of the same species if they are able to breed together sexually to produce fully viable offspring.' This is simple and to the point, but it contains a wealth of notions, which are worth discussing.

First, we can readily see how such a definition fits into the neodarwinian paradigm. We can simply say that all the creatures of any one species share a common gene pool. In fact this might not be literally true, for the gene pool of any one species may be divided into several different gene pools for geographical reasons, for example. Each pool then represents a different 'population', and all the populations together form the whole species. Thus the lions of Asia are now confined to a single reserve, the Gir Forest of north-west India, and they are separated, by several thousand miles of country that now is hostile to lions, from their counterparts in Africa. But when Asian lions and African lions are brought together in zoos, the resulting hybrids are indeed 'fully viable', and in fact, as we will see again in Chapter 8, the first serious attempts to conserve Asian lions in zoos were spoiled because Africans were allowed to cross-breed with them (and the breeding programme had to begin all over again). Thus, all lions are considered to be of the same species, even though in practice, in the wild, Asians do not interbreed with Africans. However, Asians are visibly different from Africans (for example they have smaller manes), and they are generally recognised to form a separate *subspecies*.

The neodarwinian paradigm also shows clearly, at least in general terms, how one species evolves from another, or how one species diverges to become many. In the first case, we have merely to envisage that over many generations, a great many alleles are lost from the pool, and a great many more are incorporated, until the creatures in the population are so different from their ancestors that they would not be able to interbreed with them even if those ancestors still existed. In the second case, we have merely to envisage that a single gene pool becomes divided into two or more; and that each separate pool then evolves in its own way, until the differences are too pronounced to allow for inter-breeding. Thus, for example, tree snails of the family Partulidae inhabit many islands of the Pacific, and there are (or until recently were) about 100 different species, which all probably arose from only one or a few ancestors that once lived on the giant Southern continent of Gondwana.

This, then, is the basic neodarwinian concept of species, and it is obviously very powerful. But in Nature there are always complexities. Why and how is it, for example, that we recognise all domestic horses as being of the same species even though they are very variable, yet can clearly distinguish horses from asses – even though some horses may look more like some asses than they look like other horses? Why is there variation within species, and what is the nature of the distinction between species?

SAME ONLY DIFFERENT

Just to begin this discussion, recall if you would the concept of the recessive gene, a gene that does not make its effects felt unless it is inherited in double dose. As we have observed, genes that are in any way deleterious tend to be recessive, basically because if they were not they would damage every individual that possessed them, and would tend to be rapidly eliminated from the gene pool by natural selection. But recessive deleterious alleles can lurk in the gene pool, cropping up only now and again.

The gene pools of most wild animals (there are laboratory strains of many creatures in which this is not true!) contain at least some deleterious alleles. If these are inherited only in single dose, then their possessor is heterozygous for that allele and is called a 'carrier'; and carriers are unaffected by the recessive allele, even though they may pass it on to their offspring. The recessive 'bad gene' is harmful only in homozygotes: that is, if inherited in double dose. Our attention is naturally drawn to those deleterious alleles that cause frank disease, such as cystic fibrosis in humans. But 'deleterious' is a matter of degree – so that an allele which, for example, slightly reduced the spring of the tendons would be deleterious in an animal that relied upon speed.

Thus *in general* it pays to be heterozygous; and *in general* creatures that are highly homozygous – that is, possess many of their alleles in double dose – are not as 'vigorous' as those that are more heterozygous. Darwin himself noted this: for example, that if two different varieties of the same plant were crossed, the resulting hybrid could be stronger than either parent. He called this phenomenon *hybrid vigour* – although he did not, of course, know the underlying genetic reason for it. The more general term these days for 'hybrid vigour' is *heterosis*. By contrast, if two creatures that are themselves highly homozygous and are similar to each other are crossed, then the resulting offspring would be very homozygous indeed. Most of its alleles would be present in double dose, and any (relatively) deleterious ones would make themselves felt. Hence such 'inbred' creatures are often weak. They are said to suffer from *inbreeding depression*. As we will see in chapter 6, the deliberate creation of hybrid vigour, and the avoidance of inbreeding depression, play a great part in livestock and crop improvement.

Similarly, wild species are less likely to go extinct if they have a wide variety of alleles within their gene pool, largely because they are better equipped to avoid inbreeding depression if the individuals are genetically diverse, a matter we will discuss at much greater length in Chapter 8. Furthermore, wild creatures of many kinds go to great lengths to avoid mating with their own brothers and sisters.

On the other hand, mating between creatures that are too different genetically is also hazardous. Most characters are polygenic, which means they are produced not by one gene, but by concert-parties of genes, and some concert-parties work a lot better than others. Breeders of livestock and crops find that in their quest for hybrid vigour, they can go too far. If they cross individuals that are *too* dissimilar, then the resulting offspring may suffer from *outbreeding depression*. The underlying genetic reason here is that its concert-parties of genes, which in the parents work so well, are broken up. In practice, breeders find from experience that some combinations work very well, and produce extremely vigorous offspring, while others do not. In cattle, the Hereford-Friesian cross – Hereford bull and Friesian cow – produces excellent calves. Among dogs, the crossing of Great Danes and St Bernards is disastrous. The two sets of genes from the two huge but different breeds do not mesh well together at all.

Great Danes and St Bernards show the problems that can arise from bad marriages even within species. If animals of different species mate – which some animals do naturally in the wild, and many more can be induced to do in captivity – then there can be several kinds of outcome. If the two creatures were too different, then nothing at all would result: for example, horses and rhinoceroses are clearly related species, but they are far too different for mating to produce any kind of offspring. If the relationship is closer, however, then hybrid offspring may result, but usually – at least among animals – these would generally be deficient in some way or another. For example, if horses mate with asses the result is a hybrid known as a mule. Mules benefit from heterosis, and so are extremely tough, which is why they are employed as working animals. But they are certainly not 'fully viable' because they are sexually infertile. The reason here is that horses and asses have different chromosomes, and thus – for reasons we will see in the next chapter – their hybrid offspring are unable to produce viable *gametes* (eggs and sperm). The fact that horses and asses can mate to produce offspring shows that they are indeed closely related, and they are both placed in the same genus, *Equus*. But the fact that their hybrid offspring are not fully viable *defines* them as separate species.

Whether humans and chimps could produce hybrid offspring

is unknown; or at least, it is unknown to me, though it is hard to believe that somebody, somewhere, has not tried to impregnate a captive female chimp with human sperm. Whether those offspring would or would not be fertile is also unknown. I suspect not, for again there is a clear chromosome difference between humans and chimps. So again, chimps and humans are clearly related (probably more closely than horses and asses). But again, they are different species. Again, we can assume that attempts to produce hybrids between humans and monkeys would fail. They are indeed related, but, as with horses and rhinos, the relationship is too distant to allow issue.

With all this in mind, we might expect to see different populations of animals and plants in nature constantly tending to divide and to form new species; and might also expect to see every degree of separation. So we do. Red wolves and grey wolves in North America are considered separate species, because if left to themselves they keep to themselves. But if they are forced together (for example in captivity, or because one or other type has become extremely rare in the wild), they do, in fact, produce fertile offspring. Fire-bellied toads and yellow-bellied toads in Europe live separate lives in separate areas – except where their two habitats overlap, in the Carpathian mountains and along the Danube Basin, when they do in fact hybridise. Carrion crows of England and Southern Scotland hybridise with hooded crows of Northern Scotland, where the two meet. In such cases you may say, 'Well, if the two populations *can* mate together to produce sound offspring, they should be considered of the same species!' There is no simple answer to that. In nature, species divisions are *not* as absolute as the most tidy-minded zoologists might like them to be. But our definition of 'species' referred to 'fully viable' offspring, and this does not refer only to reproductive prowess. The hybrid offspring of fire-bellied and yellow-bellied toads remain confined to a 'hybrid zone' between their parents' positions. They fail to spread into their parents' territories because they cannot compete with their pure-bred parents. In short, the hybrids are ecologically disadvantaged, even though they may be reproductively potent. In the context of the real, wild world, an ecologically disadvantaged creature cannot be said to be 'fully viable'.

There is one further complication before we move on, one

that has at times caused enormous confusion (and still does) but is extremely important in many contexts. This is the phenomenon of *polymorphism.*

POLYMORPHISM

As if to prove that in Nature nothing is simple, many *bona fide* species include different forms that are extremely different in appearance. Thus in many creatures the two sexes are very different. Male baboons and elephant seals are very different from the females. In some insects, only one sex has wings. In many species (especially among birds) the male is very showy, while the female is drab (though occasionally, as with the eclectus parrots of New Guinea, the females are brighter). Barnacles and some deep-sea fish such as anglers have 'dwarf' males which remain permanently attached to the matriarchs as parasitic appendages. This general phenomenon is called *sexual dimorphism.* When the males and females are commonly seen together – as are elephant seals in their harems, or drakes with ducks – then it is easy to see who belongs to whom. But when the two forms are usually seen apart, mistakes can be made. Many a male and female butterfly, and not a few birds, have in the past been classified as different species.

In genetic terms, however, sexual dimorphism is easy to explain. It is obvious even in relatively monomorphic species that the sex genes, though perhaps few in number, make a profound difference to their possessor's physiology, and it is not surprising that they can profoundly affect feather-colour or body size. More intriguing, however, is that within some species, *different individuals of the same sex and age can also differ enormously.* This is the general phenomenon of *polymorphism.*

Tropical butterflies provide some striking examples. A classic is the African swallowtail, *Papilio dardanus,* which has several forms differing enormously in wing pattern, wing shape, and overall size. Polymorphism is much more common than it may seem, however, because it becomes increasingly clear that many enzymes and other bodily materials can be polymorphic too. Polymorphism sometimes represents a population caught half-way in making some evolutionary shift from one state to another. But it has much more general and interesting significance than this. The term is most accurately applied when the two or more polymorphic forms co-exist in a more or less constant, or 'stable', ratio.

Stable polymorphism can arise in several ways, but two main mechanisms predominate. The first is *frequency dependent selection*, in which natural selection favours any one form, but only so long as that form remains fairly rare. Thus, most of the different forms of *P. dardanus* are Batesian mimics: most of them resemble other butterflies that are poisonous, though they are not poisonous themselves.* However, poisonous animals succeed not by poisoning the animals that eat them, but by *warning* their attackers that they are poisonous; the idea is not to exact revenge, but to avoid being eaten in the first place. The non-poisonous mimics issue the same message, but with them it is pure bluff. But if the mimics become too common – if, that is, they begin to out-number the 'models' that they imitate – then the predators are unlikely to learn that the advertisement is supposed to be linked to poisonousness. Indeed, if they begin by eating mimics, they will get it in their heads that this particular pattern denotes tastiness.

In short, natural selection favours mimics, who have no weapons of their own but who borrow the charisma of those who have. But if the mimics are too successful and become too numerous, natural selection starts to work against them. Thus the only way in which a mimic population can grow is to adopt more than one model. *Papilio dardanus* becomes commoner than it otherwise could by imitating more than one model, while *each* of the polymorphic forms is subjected to frequency-dependent selection.

Papilio dardanus has long since become a text-book example of polymorphism brought about by frequency-dependent selection. More recent are the studies of Bryan Clarke of Nottingham University on the partulid tree snails of Pacific islands.

Professor Clarke studied in particular the *Partulas* on the island of Moorea, one of the French Polynesian Society Islands, which

*The second form of mimicry is Mullerian mimicry, in which different kinds of animals that are *all* noxious, such as different kinds of wasp, reinforce the message they give out to potential attackers by mimicking each other.

also include Tahiti. There are seven *Partula* species on Moorea.* Two of them, *Partula taeniata* and *P. suturalis*, are generalist feeders, while the other five are more specialist. The two generalists are both highly polymorphic. The various morphs within each of the generalist species differ in size, shell shape, shell pattern, and even in some cases in 'chirality' – whether the shell coils to the right or to the left.

In general, the various specialist species can feed in the same place, because they do not all feed on the same things. The two generalists usually do less well when the specialists are around – but even so, generalists and specialists do co-exist in some places. Now, some of the morphs of the two generalist species closely resemble one or other of the five specialist species. But Clarke found that whenever a generalist overlaps with a specialist, the particular generalist morph that is present is always very *different* from the resident specialist.

The reason here, Clarke suggests, may lie in the psychology of the birds that are presumed to prey upon the snails. For birds feed most efficiently by focusing upon whatever edible component of their diet is most common. Thus if birds are surrounded by a lot of red berries and a few yellow berries, they concentrate on the red ones and leave the yellow ones alone altogether. But if the yellow ones are commoner they eat them, and leave the red ones alone. That way, they do not have to waste time making choices. This is an example of a general tendency among all animals to develop an 'optimum foraging strategy'.

As with berries, so with snails. When generalist overlaps with specialist, the specialists tend to out-number the generalists. The specialists, at least in their own specialist places, feed more efficiently than the generalists. But because of the constant attentions of birds, natural selection will favour the generalists provided they look as different as possible from the more common specialists – and provided they remain less common. As soon as the generalist morph starts to increase, and match the specialists in numbers, then the birds start to concentrate on them instead. Here, then, is an advantage in being rare, and in being different.

*I say there *are* seven species of *Partula* on Moorea but in fact, since Professor Clarke began his studies in the 1960s, all seven Moorean species have been wiped out by a predatory snail of the genus *Euglandina*, which has been introduced to islands throughout the Pacific.

Stable polymorphism can also be sustained by a second mechanism. Some creatures gain enormously – disproportionately – by being heterozygous for particular genes. A classic example is that of sickle cell anaemia among African people, in which sufferers are born with defective blood cells which collapse – become sickle-shaped – as the oxygen tension falls, in which state they are useless. The gene that causes this condition seems disastrous, yet it has persisted among Africans, presumably for thousands of years, at roughly constant frequency. The reason, it seems, is that people who are heterozygous for the sickle gene are less prone to malaria than people who have 'normal' blood, because the malaria parasite attacks red blood cells and thrives better in normal cells. So although the sickle gene is disastrous in double dose, it persists because in malarial regions selection favours the people who are heterozygous for the gene. But if the heterozygous people become too common, then they start giving rise to large numbers of people who are homozygous for the sickling gene – and these people generally die before they reach reproductive age. Thus, anaemia exerts heavy selective pressure against individuals who are homozygous for the malaria gene, and malaria exerts somewhat less, but still heavy pressure against people who are homozygous for the normal gene. But there can be no heterozygotes unless there are also some homozygotes of each type. So the three persist in 'stable' equilibrium.

Appearances – or phenotype in general – can be deceptive in deciding which creature belongs to which species. Ah, you will say wisely, what really counts is the degree of difference between two creatures' genomes – that is, *genotype* is what matters, rather than phenotype.

This is of course the case, but even here there are complications – which again can be nicely illustrated by partulas. Bryan Clarke and his collaborators concluded on various grounds that the seven Moorean species really belong to two groups. *P. taeniata* belongs in one group and has just one other close relative on Moorea; *P. suturalis* belongs to the other group, with four other close relatives. Many kinds of hybridisation are possible between the different Moorean species, and with species from other Society Islands. Interestingly, however, successful matings are not always possible between species which, according to all other evidence, are very closely related. On the other hand, it sometimes transpires that species that do not seem to be very closely related, *can* mate together.

Taken all in all, then, it is perfectly reasonable to say that 'genetic distance' between populations is what causes them to be different species. Partula snails are so different genetically from European garden snails, and human beings are so different genetically from baboons, that there would be no chance whatever of a successful mating. But when the two genotypes are not quite so distinct as this, we cannot simply relate the difficulty of hybridisation to the overall difference. Very closely related species may be unable to hybridise simply because *particular* genes have created some 'reproductive barrier' between them. One single gene could throw a spanner in the works. Yet species that are more distantly related might be able to hybridise, provided the genetic differences are not of the *particular* kind that could create a barrier. It is the case, for example, that the genome of human beings is extremely similar to that of chimpanzees. But it is impossible to say on those grounds alone whether or not the two species could hybridise. It all depends on whether *particular* genes (as yet unidentified) have arisen to prevent the union.

So 'species' is a much more elusive and labile concept than Plato or the pre-Darwinian biologists believed. There are all shades of difference between creatures that are only slightly incompatible – or may be perfectly compatible but are kept apart by geography – and those between whom the differences seem absolute.

Overall, too, we see that it pays to belong to a distinct species, and not to be too promiscuous, because by belonging to a distinct species you maximise your chances of producing offspring that are reasonably well adapted to the prevailing conditions. Thus as we will see again in Chapter 5, animals (and plants) have in practice evolved mechanisms for (a) ensuring that they do not generally mate with creatures from other species but (b) ensuring, contrariwise, that they *do* mate with their own kind. Thus we must conclude that although the concept of species is fluid and flexible, animals (and plants) nevertheless take pains to ensure that their own particular species retains its 'integrity', and does not outbreed with other creatures except perhaps *in extremis*. By the same token, it is clear that although any given population has an innate tendency to change (by mutation and drift, and by natural selection) it also has a built-in conservatism.

Any creature that actually exists must to some extent be on to a winning formula, and natural selection is as liable to operate to keep things the same, as to cause things to change.

However, once we have grasped the neodarwinian paradigm, and once we start applying it to real cases, we can play mental games for ever, and also envisage possible experiments to test ideas that would take millennia to carry out. The neodarwinian paradigm is, in short, very powerful indeed, and I am certain that it is here to stay. None the less, there have been some significant new footnotes in recent decades, which we should look at.

MODERN FOOTNOTES TO THE NEODARWINIAN PARADIGM

The first supremely important footnote to the paradigm concerns the *level* at which natural selection operates. Do individuals compete with individuals, as Darwin proposes? Does gene pool compete with gene pool, as some of his twentieth-century followers supposed? Or does gene do battle with gene? This issue has huge theoretical significance, which extends well beyond the bounds of biology, and I will discuss it at length in Chapter 4. The second footnote is worth discussing here. Is natural selection really the prime force of evolutionary change? Or are we simply the outcome and the victims of time and chance?

NATURAL SELECTION *VERSUS* TIME AND CHANCE

Darwin proposed the idea of 'natural selection', so doubts cast on its importance seem like attacks on its author. But although Darwin was the first advocate of natural selection, he has suffered as all great thinkers have done from the over-enthusiasm of some of his followers. Some, like Herbert Spencer, sought to extrapolate his ideas from biology into moral philosophy and politics. Thus emerged 'social Darwinism' – apparently the notion that society *ought* to be as 'red in tooth and claw' as Alfred, Lord Tennyson supposed nature to be, and that the weakest should go to the wall. The gentle and liberal Darwin, who railed against slavery in an age when it was still considered both proper and necessary, who stressed that the naked Tierra del Fuegans were 'wretched' only through circumstance, who sought to provide financial security for his village neighbours, and who wrestled and grieved throughout his middle years with the multifarious sicknesses of his many children, could hardly have approved.

Others, in other contexts, have sought to find a natural selective 'reason' for absolutely every quirk of nature. This reached a height in the middle years of this century, when one enthusiast suggested that the pinkness of flamingoes had evolved as camouflage, as they took off from their feeding grounds and flew across the evening sky. Presumably this protected them against low-flying lions. Darwin himself stressed that other factors besides natural selection must have helped to shape animals. He classed sexual selection as a separate mechanism, which we will discuss in Chapter 5. He also stressed the importance of chance.

However, the notion that chance operates virtually to the *exclusion* of natural selection has become something of a vogue these past few years. This idea takes various forms, which deserve various degrees of attention. All, I believe, enrich the neodarwinian paradigm, but none actually threatens it, as some biologists apparently believe.

First, and in general, various biologists have promulgated the notion that *neutral* changes in the gene pool are in the end of more significance than those brought about by natural selection. After all, new species would inevitably emerge if gene pools became isolated, even by the perfectly chance processes of mutation and drift. There is an important philosophical principle here. All scientists have to guard against proposing theories that are simply unnecessary. They have to ask the question, 'Supposing the mechanism I propose is not operating, what difference would it make?' More formally, it is incumbent upon all scientists to propose a 'null hypothesis', one which says – 'The world would turn out much as we see it now, even if nothing much was going on except time and chance.' The idea that 'neutral' changes could have brought about much of what we see in nature is salutary, as a 'null hypothesis'. It would guard us against pink flamingoes, which seem to be pink simply because they eat pink plankton, and presumably would be blue if they ate blue plankton. But most biologists, while accepting that the null hypothesis of neutralism ought to be given as much scope as possible, find it impossible to conceive of a world that contains so many astonishingly well adapted creatures, unless natural selection was moulding them. As Hamlet said in a slightly different context, 'There is a divinity that shapes our ends, rough-hew them how we will.' For 'divinity' read 'natural selection', and you have it.

But even the most committed Darwinians must now agree that many of the major evolutionary events in the history of the world have indeed happened entirely by chance, that they are 'contingent', as the expression is, on extraneous and uncontrollable events. Thus nineteenth-century biologists liked to believe that dinosaurs had become extinct because they had 'outgrown their strength', or were too stupid, and that as soon as the perkier mammals got into their stride, they swept those dullards aside – through good old Darwinian competition.

But this does not accord with the facts. First, it should have been obvious at least by the late nineteenth century that (a) the dinosaurs in their many forms dominated the entire world for 140 million years – from *c.* 200 million years ago (mya) to around 60 mya; that (b) the mammals first appeared on Earth at about the same time as the dinosaurs – 200 mya; but (c) the mammals stayed small and boring for all the time the dinosaurs were around, presumably because they were comprehensively out-gunned. Modern studies suggest that the dinosaurs disappeared because of a massive climatic change around 65 mya, which may possibly have been precipitated, or at least exacerbated, by collision with a meteor, or perhaps by huge volcanic eruptions which, it now seems, created much of the surface of modern India.

Thus it seems clear that the mammals did not out-smart the plodding dinosaurs. They simply hung around until *chance* pushed the dinosaurs aside. But for that meteor (or volcano) mammals would still be tree-shrews, and human beings would not have evolved at all. It is now clear, too, that in the history of the world there have been at least five 'mass extinctions' that were at least as dramatic as the wipe-out of the dinosaurs, and that each one enabled quite new groups to seize the stage that hitherto had been fairly insignificant – and which presumably would have remained so. But again, whatever survived each mass extinction was subject, as ever, to the shaping force of natural selection. Natural selection must now be seen to operate on a shifting stage, and with precipitate changes of cast. But it operates none the less.

Yet we should make some adjustments to the pristine neodarwinian model. Notably, we should no longer look upon the gene pool simply as a bag of beads, to which some beads (genes) are added by mutation, and others thrown out. As already noted, the genomes of all creatures are far more restless than has been appreciated. A surprising number of genes move around the genome, sometimes causing mutations as they go, while they and others replicate and multiply within the genome. Gabriel Dover of Cambridge University calls these mechanisms 'molecular drive', and suggests that they collectively cause a huge amount of the change that we perceive as 'evolutionary'. The genome is not so much a bag of beans, as a bag of snakes.

Yet I still do not feel (as some do) that such an observation represents a *fundamental* attack on the neodarwinian paradigm. The variants produced by molecular drive are surely subject to natural selection, just as are the variants produced by what we may now call 'classical' mutation. The source of variation does not seem to me (or to many other biologists) to affect the issue. To be sure, we should see natural selection operating on a moving stairway, rather than upon a static stage. But in the end, natural selection decides what lives and what dies, and the ones that live, as Darwin said, are those that are best adapted to the circumstance.

That, then, takes us to the limits of 'classical' genetics: genetics, that is, in which the genes themselves, Mendel's 'factors', are conceived simply as abstractions – or as beads, threaded on chromosomes. Clearly, classical genetics is immensely powerful and, when combined with the evolutionary ideas of Darwin, provides a wonderfully unifying picture of the endless variety of life. But in genetics there is another, quite different line of inquiry to pursue: 'What exactly *are* genes, and how do they work?' This is the subject of the next chapter.

CHAPTER 2

The Reality of the Gene

It is a terrible pity that science has so often found itself in conflict with religion. Both, I believe, have essential things to say, and most of the conflicts between them have been mainly concerned with details of theology, which do not seem to me (or to most modern clerics, at least in Britain) to be of fundamental significance. For example, it does not compromise the Jewish or the Christian faith to acknowledge that the Earth *in fact* took longer than seven days to come into being and to acquire its complement of living things. The Genesis account of the Creation loses none of its power by not being literal.

Was the conflict inevitable? In one sense no, but in one important sense yes. There was a fight to be fought between science and religion; the pity is only that it still continues so vituperatively. That the fight was not quite inevitable is suggested by the history of the seventeenth century, in which modern science properly began. In that century to be sure, Galileo fell out very publicly with the Roman Catholic Church, although the conflict was far from simple, and the Jesuits of the Church initially behaved with more intellectual subtlety than Galileo did. More striking, however, is that most of the seventeenth-century pioneers of science were devout. The view of Isaac Newton was typical. He felt that his intellect was a gift from God, and that its purpose was to enable him, a mere mortal, to explore God's Creation, the better to appreciate its magnificence. To Newton, as I believe to the mainstream of his contemporaries, scientific research was itself an expression of reverence.

Yet the seventeenth century nurtured the seeds of conflict. Devout though Newton was, his studies – and those of others – seemed to reveal a Universe that ran like clockwork. Some – notably the Reverend William Paley in the eighteenth century – argued that where you had clockwork, you must have a clock-maker. But others felt that the 'laws' that Newton and others had revealed provided the complete explanation of the workings of the Universe; in fact, that the discovery of laws squeezed God out all together. This is one expression of 'materialism': the notion that if we once describe the properties of the materials of which the Universe is composed, then we can provide a complete explanation of the whole.

Though the eighteenth century was also in general devout (by the standards of late twentieth-century Britain), this general conflict deepened. The 'laws' of science and maths that continued to unfold in that most 'rational' of centuries seemed to leave no room for God. Hence, when Napoleon asked the philosopher Pierre Simon de Laplace where God fitted into his view of the Universe, Laplace replied, '*Je n'ai pas besoin de cet hypothèse*' – 'I have no need of that hypothesis.'

Yet the processes of life are not so easily contained by the laws of Newtonian physics, and besides, in the eighteenth century, biology was still, by and large, little more than descriptive natural history. A remarkable number of eighteenth- and nineteenth-century biologists were, in fact, clerics, frankly admiring the works of God; the Reverend Gilbert White of Selborne is a classic example. Nevertheless, the eighteenth century saw ideas of evolution first rising seriously to the fore, not least through Charles Darwin's own grandfather, Erasmus, and these ideas do present orthodox religion with another kind of threat, as we will see.

Biology did not get truly into its modern stride until the nineteenth century, and as it did so, it seemed to threaten the established Christian Church yet again, on two main fronts. One of those fronts was the idea largely pioneered by Erasmus Darwin – that of evolution: triumphantly consolidated by Charles in *Origin of Species* in 1859.

The point again, though, is not that Charles himself was anti-religious. He trained initially to be a clergyman; he was at pains throughout *Origin* and other public writings not to offend the established Church; and he was finally buried in Westminster Abbey. Indeed, the notion of evolution *per se* was not necessarily considered blasphemous, for we may argue (as many scientists and clerics did) that God was perfectly free if He chose to create life first in some primordial form, and then shape it as He chose as the generations passed.

Yet Darwin's particular idea of evolution *by means of natural selection* does contain one extremely troublesome idea. For natural selection needs variations to select amongst. And Darwin argued that those variations arise *by chance*. They are not prescribed. It is hard to see how chance variations can be part of a master plan. Yet God is traditionally perceived as the Master Planner.

This particular conflict has deepened in the twentieth century, for modern biologists point out that there was nothing inevitable even in the big evolutionary changes that are clear to see in the fossil record. If the dinosaurs had not been wiped out by a change of climate – possibly brought about by a chance collision between the Earth and some meteor – then the mammals would have remained the small and insignificant creatures that they had been for the previous 100 million years. If the mammals had not been given their opportunity to diversify, then apes would not have evolved. If Africa had not undergone the climatic changes that it did, around five million years ago, then apes would not have taken to living in open savannah, and so there would have been no humankind. Yet Man is supposed to occupy a special place in the Universe, and in the mind and plans of God. What price that idea, if Man himself came into being only by chance?

On the second front, nineteenth-century biology presented established religion with another manifestation of materialism. Until the nineteenth century, most biologists were content with the notion that living things differ absolutely from non-living things: that they partake of some special quality, some aether, called 'life'.* They were content to accept, too – indeed there seemed no other explanation – that 'life' was 'breathed into' living things by God. We cannot simply dismiss this idea. After all, it may literally be the case (a matter to be discussed in Chapter 10). It certainly is not easy to explain even now exactly what happens to people in the second (or millisecond) in which, lying peacefully in bed, they pass from life to death. Of course they stop breathing and their heart stops beating – but why? The cells of the lungs and the heart remain 'alive' for some time after their ostensible 'death', and so apparently do the nerves that drive those organs. It certainly looks *as if* some extraneous 'force', or aether, has departed.

But as biology developed in the nineteenth century, so it took in the notion that first reached fruition in physics in the seventeenth century: the materialist notion, that the behaviour of the whole

*Although some aboriginal religions contain the quite opposite idea: that everything – stones, rivers, stars – is possessed of the spirit of 'life'.

could be explained by the properties of the component parts, and the out-working of innate 'laws'. Mary Shelley's *Frankenstein* was an early nineteenth-century, Romantic ('Gothic') exploration of this idea, albeit a parody. A 'monster', cobbled together from bits of corpses, was restored to life by electricity, a 'force' that was still mysterious – and in many ways still is – but belonged none the less in the material world rather than in that of magic. At the time of the English 'Romantics' electricity was already the subject of *bona fide* scientific research.

More formally, and in the end more profoundly, the early nineteenth century saw the first stirrings of biochemistry, and in particular the realisation that there was no absolute distinction between the materials produced by living bodies, and those that could be analysed and synthesised in the laboratory. Urea, the standard form in which humans and other mammals excrete surplus nitrogen, was one of the first 'organic' products to be chemically analysed. In general, then, the suspicion began to grow that biology – the science of living things – could perhaps be 'reduced'* to chemistry; and indeed that chemistry, in its turn, could in the end be explained by the simple and irrefutable laws of physics. To many – clerics and scientists alike – there seemed to be no room at all for God in this sweeping vision.

By now, this kind of conflict ought to have been resolved – and indeed, there are still plenty of devout scientists of all religions, and plenty of science-rooted priests. But of course it has not. Many scientists see religion as 'primitive' and destructive, while many non-scientists still see science as a general threat. Many more who are not overtly religious see all attempts to explain the material basis of the Universe as an offence to 'human values'. I prefer the idea (essentially that of Isaac Newton) that science is a way of appreciating the Universe, a Universe that could well be seen as a 'Creation'.

The rising science of genetics is, perhaps, the ultimate challenge that materialism can throw down. 'Genes' certainly do not literally 'determine' every aspect of our lives; at least, we are

*'Reduce' in this context does not mean 'diminish'. It is a reference to a scientific approach: that of 'reducing' problems that are too complex to solve into smaller problems that are soluble, and then synthesising the grand explanation of the whole from the individual explanations of each of the parts. As Sir Peter Medawar commented, 'Science is the art of the soluble.'

not clockwork toys, in the way that some seventeenth-century physicists conceived of a clockwork Universe. But genes do set ultimate physical limits on what each of us can achieve. This, of course, is a dangerous and often misused argument when applied to human beings, as we will explore in Chapters 9 and 10, but at least at a commonsense level it is undeniable. High among the reasons why we cannot fly is our lack of genes for wings. I could have trained like a maniac since the age of three and would never have run as fast as Carl Lewis; studied mathematics with equal zeal and never have matched Albert Einstein; and practised the harpsichord day and night yet never rivalled Mozart. I just don't have the genes for it. Even our thoughts are not quite as boundless as Romantic poets would have us believe; it is hard for any of us to think of a nine or an eleven-dimension Universe, for example, which modern physics suggests is 'correct', primarily because we ourselves are built for life in four dimensions only; and one of those dimensions – time – is very difficult for us to think about. All of us, to be sure, might achieve far more than we (or others) might care to suppose, but in the end we can do – and even think about – only those things that our genes allow us to.

Yet genes, so close it seems to the 'essence' of life, are themselves 'just' chemical agents – or so the nineteenth century began to show, and the twentieth century has so far ratified. Chemically, genes are more complicated than urea, to be sure. Yet they are compounded from exactly the same elements as urea – carbon, oxygen, hydrogen and nitrogen (with additional phosphorus in the case of genes). Genetics and biochemistry have long since met in the middle, to spawn the enormously powerful science of 'molecular biology'.

We will return to the philosophical issues – essentially, 'what is life?' and 'does it actually *matter* what life is?' – in a later chapter. Here, we will address the facts of the case: the fact that genes are molecules, albeit of a versatility that is stunning, and ever more surprising.

GENES AS MOLECULES: THE RISE OF MODERN UNDERSTANDING

In retrospective spirit, we can identify three lines of thought that have led to modern ideas about 'genes as molecules'. All of them date from the nineteenth century. The first is the central idea of *biochemistry* – which is that biological systems in general can be explained, at least in part, in terms of chemistry. The second is the work of Mendel himself: the discovery that the features of a living thing are 'determined' not by vague essences, like Darwin's 'inks', but by discrete entities, that interact in patterns that can be described arithmetically and indeed can be predicted. The third was the rise, in the nineteenth century, of the techniques of microscopy, and hence of the science of cell biology, otherwise known as *cytology*. We discussed Mendel's fundamental contribution in Chapter 1. Here we should explore the two other threads: the growth of biochemistry and of cytology. We might begin with the latter.

MICROSCOPES AND CHROMOSOMES

The fact that living things are compounded from collections of cells was first revealed in the seventeenth century by the Dutchman Anton van Leeuwenhoek. In the decades and centuries that followed – as lenses improved, and also techniques for lining them up to increase the magnification and reduce aberration – so the general structure of cells became apparent; in particular, that the cells of animals, plants, and fungi generally contain a central, dark-staining body called the *nucleus*. Then in the 1860s came a crucial discovery: that when cells are dividing, the nucleus temporarily seems to disappear; and that about the time of cell division, discrete 'bodies' appear where the nucleus had been. These 'bodies' could be seen because they took up coloured dyes, and hence were called *chromosomes* (meaning 'coloured bodies').

No-one at first knew what chromosomes were. How could they? But it soon became apparent that they were at least highly intriguing, and possibly very important. First, it became clear that the number of chromosomes was characteristic of each species; and that there was always an even number of chromosomes (or at least, if there was not, then something odd was happening). Second, it became clear that the chromosomes in fact existed in distinct *pairs*: each one with a matching, *homologous* partner. But then in the early 1900s, three American biologists working independently –

Clarence McClung, Nettie Stevens, and Edmund Wilson – found that in grasshoppers and related insects, the different sexes had different chromosomes. Females had a particular matching pair which Nettie Stevens called X chromosomes, while males had only one X chromosome. McClung, Stevens and Wilson went on to propose that this pattern of chromosomes was not simply correlated with sex, but actually *determined* sex. Nowadays we accept that this is the case: that many (but not all!) animals indeed have a special pair of 'sex' chromosomes that are different in males and females (as we will explore further in Chapter 4). So then of course the question arose: if chromosomes determine sex, why not other qualities as well?

Most exciting of all, however, was the behaviour of the chromosomes during cell division, as first observed in the late 1880s. For the way in which the chromosomes themselves divided was seen to be amazingly – almost militarily – orderly. The way this works is shown in Figure 2.1. In essence it runs as follows.

The first sign of cell division is the disappearance of the nuclear membrane, while at the same time the chromosomes, which are not visible during most of the life of the cell, appear as little sticks and ciphers. In the absence of the nuclear membrane, they seem simply to be suspended in cytoplasm.

Then, during 'ordinary' cell division of ordinary cells, each chromosome splits along its length, to form two half-chromosomes, or *chromatids*. So where in a human, say, we began with 46 chromosomes (23 pairs), we now have 92 chromatids.

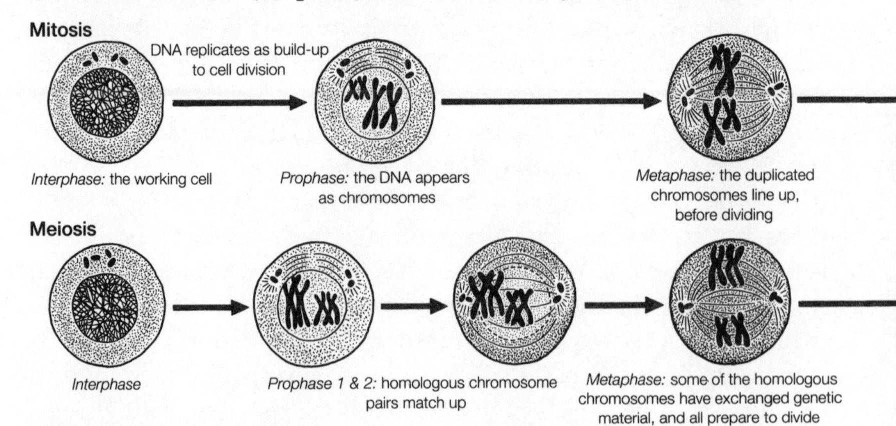

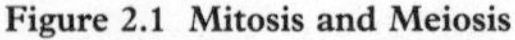

Figure 2.1 Mitosis and Meiosis
The existence of chromosomes was first discovered in the late nineteenth century – and so too was the fact that they divide with military precision. Mitosis produces two daughter cells, identical with their parent. Meiosis produces four gametes, each with half the chromosome number of the parent.

Then one complete set of 46 chromatids (one from each chromosome) migrates to one end of the cell, and the other set migrates to the other. (Though it is now known that they do not simply 'migrate'. They are pulled, bodily, by contractile protein threads.)

Then a new cell membrane grows between the two complete sets of chromatids, to give two cells, each containing a complete complement of chromatids. These chromatids – half-chromosomes – then duplicate themselves again in the manner that will become obvious later. So now we have two daughter cells, each containing a complete set of complete chromosomes, each, therefore, a facsimile of the parent. This 'ordinary' cell division is called *mitosis*.

The production of sex cells, or *gametes*, is similar, but differs in several outstanding respects. It begins like mitosis: the nuclear membrane disappears, and the chromosomes become visible. But then – a crucial difference – each chromosome seeks out its homologous partner, and the two lie side by side, so that the 46 chromosomes of the human cell appear as 23 matched pairs (or rather, as 22 matched pairs, plus the two sex chromosomes).

Then, as in mitosis, each chromosome splits along its length to produce two chromatids. So now the original pairs of homologous chromosomes appear as sets of four: two sets of two chromatids, lying side by side.

Now occurs the essence of sex. The two chromatids lying on the inside of each quartet exchange lengths of material within

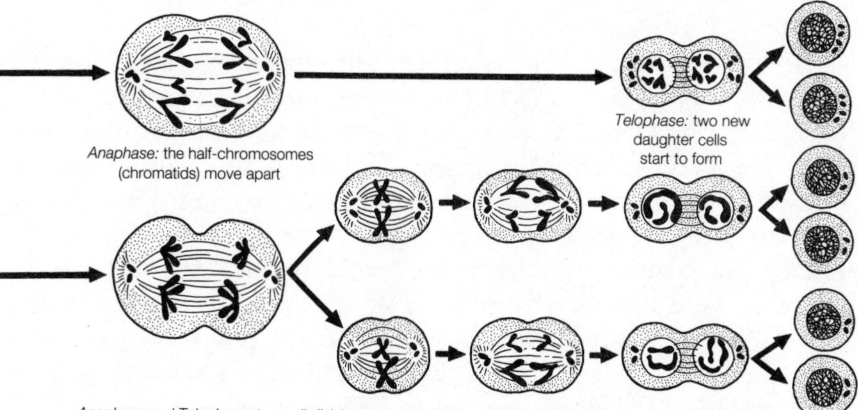

Anaphase: the half-chromosomes (chromatids) move apart

Anaphase and Telophase: two cell divisions now take place, giving four daughters – each one a gamete

them: the key process known as *crossing over*. After this, each set of four chromatids consists of two that are unchanged, and two that are essentially hybrids.

Then the process of chromatid separation and cell division continues essentially as in mitosis – only twice. The net result, then, is to produce *four* cells – gametes – each of which contains only *half* the number of chromosomes of the parent cell. So a human gamete contains only 23 chromosomes. Some of these (a random collection) will be exactly the same as the parent chromosome and some (those that have undergone crossing over), will contain material from each of the parent's homologous pair of chromosomes, and so will be qualitatively different from any parent chromosome.

It was August Weismann, in the 1880s, who first suggested that this division of chromosomes was related to the passing on of hereditary information. Clearly, he suggested this during the forty years – between the 1860s and the 1900s – when Mendel's work was still in abeyance, sequestered and forgotten within the Brunn Natural History Society *Transactions*. With hindsight, we may wonder and be grateful that the manoeuvrings of the chromosomes correspond *precisely* to what might be predicted from Mendel's laws – assuming, that is, that the chromosomes did indeed carry hereditary information. Thus, according to Mendel, the 'factors' that underlie the organism's characters are discrete, and are passed on according to simple rules. These rules suggest that each daughter cell in the normal course of events is simply a facsimile of its parent, but that the gametes contain only half the information contained in the parent cells.

We might assume, indeed, that when Mendel's ideas were rediscovered in the early twentieth century, biologists would shout 'Eureka! Here is perfect coalescence of theory and observation!' But they did not. For reasons I cannot quite fathom, some refused to accept that the behaviour of chromosomes did in fact have anything to do with the passage of hereditary information. Oddly (the history of science constantly proves the Yorkshire adage, 'There's nowt so queer as folk') one of the principal doubters was William Bateson, who did so much to establish the science of genetics. The great population geneticist E.B. Ford records that when Bateson did finally concede that chromosomes carry Mendelian factors, he exclaimed – ' – and all my life's work has gone for nothing!' (*Understanding Genetics*, Faber and Faber, London, 1979, p. 35.) Ford suggests that Bateson was over-reacting.

Never mind. The fact that chromosomes exist, and their possible role as bearers of Mendelian factors, were established in the decades at the end of the nineteenth century and the beginning of the twentieth. Three kinds of questions follow logically. First, 'How do the activities of a gene affect the overall character of the organism?' Second, 'What exactly *are* genes?' Third, 'How do genes actually work?' We will take these questions in that order.

HOW DOES A GENE BRING ABOUT A 'CHARACTER'?

We can usefully begin this story in the middle (acknowledging that strict chronology does not work). Francis Crick (of whom more later) says that he vaguely remembers remarking in the 1940s: 'The most useful thing a gene could do would be to direct the synthesis of a protein.' (*What Mad Pursuit*, Penguin Books, London, 1988.) In practice, this is precisely what genes *do* do: they 'direct the synthesis' of proteins. Clearly, this fact was not finally established by the 1940s; but, equally clearly, Crick by that time had strong reasons for believing it. So, how had Crick, *circa* the 1940s, reached this position?

Much of that question could be answered just by reviewing what proteins do, which is plenty. To begin with, they form much of the fabric of the body. Some, such as insulin, serve as hormones. Others act as 'receptors' on the surfaces of cells, seizing passing molecules of hormones (or drugs) and conveying their messages to the cell; or seizing (and hence recognising) material that is foreign to the body, and hence initiating an immune response. They also provide a large part of that immune response, because the crucial molecules of immunity, the 'antibodies', are also proteins.

Most of all, though, proteins serve as *enzymes* – the natural catalysts of the cell, which regulate the cell's chemistry (that is, its *metabolism*). Thus they help to create all the other components of the body – the fats, the bones, and so on – because they largely control the metabolic pathways through which those components are synthesised. It transpires that *most* proteins serve as enzymes, and that by far the majority of enzymes are proteins (but not quite all, intriguingly, as we will see!). Whatever controls proteins, then – and in particular, controls enzymes – seems to control the whole body. To be sure, if genes make proteins, they are powerful indeed.

The notion that genes do indeed make proteins in general and enzymes in particular can be traced back through several lines of thought. Crucial, obviously, first of all, was the idea that there *are* things called enzymes, an idea that is properly ascribed to the great German chemist Edouard Buchner in the late 1800s. In 1897 Buchner set out to explore the juices of yeast; yeast, after all, was of crucial concern in the brewing, distilling and baking industries, which are three of the biggest industries in the world. He crushed the yeast to extract the juices, and decided to store those juices by the method that has long been favoured by cooks for the storage of perishables – in sugar, which generally preserves biodegradable material by 'sucking' the water out of them (by osmosis) and making it impossible for microbes of decay to act. However, Buchner's yeast juice fermented the sugar, to produce ethyl alcohol. This was the first revelation that the agents which effected fermentation – hitherto known as 'ferments' – could work *in vitro*; outside the environment of the cell. Buchner renamed those ferments *enzymes*, which literally means, 'in the yeast'.

The next conceptual line – which in practice could be established only over many years – has shown that (a) there are many many thousands of different enzymes in nature and that (b) most of those enzymes are proteins, with just a few exceptions. The third point to be established was that the presence (or absence) of particular enzymes is indeed associated with the presence (or absence) of particular genes. Again, a huge body of experiment has now confirmed this. But again, just a few observations can be considered 'classic'.

Chronologically the first of these 'classics' was the work of the English physician Archibald Garrod in 1902, who studied the inherited disease *alkaptonuria*. The urine of sufferers turns black when exposed to the air, which is bizarre, but does not seem to matter too much; but they also tend to suffer from arthritis in later life, which certainly does. Garrod and William Bateson (again) confirmed that the disease runs in families, and was commoner in marriages between cousins; and indeed they effectively established a Mendelian pattern of inheritance (insofar as is possible with small samples). Garrod also showed that sufferers from alkaptonuria excrete homogentisic acid (HA), which indeed turns black on exposure to air. 'Normal' people break down homogentisic acid (which is ingested in food) and then excrete the break-down products. He therefore suggested that alkaptonuria was an 'inborn error of metabolism'. Later he identified three more such 'inborn errors'. From this and later investigations of other inherited diseases he (a) suggested that 'inborn errors' are caused by defects in particular enzymes, and (b) tentatively proposed the now famous adage, 'one gene, one enzyme'.

Seminal, too, was an experiment by George Beadle and Edward Tatum in the United States. They worked on orange bread-mould, *Neurospora crassa* – a fungus that can live on a very simple diet of inorganic salts, glucose or sucrose to provide carbon, and a single vitamin, biotin. They first employed X-rays to mutate samples of the mould, and then crossed these mutants with normal, 'wild-type' *N. crassa*, to show that any effects in the subsequent generations were indeed caused by altered genes, and did not result simply because the irradiated moulds were sick.

The offspring of the mutants that had been created by X-rays could not live on the normal, simple diet. Some of them evidently required one or more extra vitamins if they were to grow. Others could not synthesise one or more of the amino acids needed to construct proteins (of which more later). Each mutant fell short in its own particular way because it lacked a particular enzyme. By further cross-breeding, Beadle and Tatum confirmed that each such deficiency indeed had a simple genetic basis. 'One gene, one enzyme' was thus firmly supported.

In the 1950s (and not until then! It is remarkable how recent so many crucial discoveries have been) the great American chemist-turned-biologist Linus Pauling broadened this adage, effectively to take its modern form: 'one gene, one protein'. He showed that the African form of inherited anaemia, known as 'sickle cell', had a simple genetic basis (more of this, too, later). He also showed that the offending gene caused a specific defect in haemoglobin, the protein-based pigment that carries oxygen in the blood. Hence, defective genes lead to defective proteins other than enzymes.

In practice, we now know that even the 'one gene, one protein' adage needs some refinement. Many proteins, at least as they appear in their finished form in the body, are synthesised with the help of several genes, each contributing a part. Many are folded into their final, functional, three-dimensional form with the aid of subsidiary proteins, known as 'chaperonins', which themselves must be produced by more genes. And most proteins in practice are not fully functional until various 'oligosaccharides' (sugars) have been attached to them, which is again done with the aid of enzymes, each of which is produced by one or more genes. But as a general rule, 'one gene, one protein' serves well enough.

But what exactly is a gene?

WHAT ARE GENES?

Francis Crick thought in the 1940s that genes *ought* to produce proteins, because proteins were so very obviously important. Furthermore, proteins had all the trappings of importance: they were known to be virtually infinitely variable in their structure, and commensurately versatile in their qualities. In fact, proteins were so obviously variable, versatile and important that for the first four decades of this century – in fact, from the time they first started thinking seriously about the problem – most biologists more or less assumed that genes must not only make proteins, they must themselves *be* proteins. This is certainly what Erwin Schroedinger argued in the 1940s, in his highly acclaimed book *What is Life?*

The idea is far from foolish. After all, whatever it was that could create nature's infinite variety of enzymes had itself to be at least equally variable: and what other molecule in nature could fit the bill? Besides, there was a popular and commonsensical idea that genes made proteins by acting as templates, to which the new protein was moulded. What kind of molecule, other than a protein, could provide such a template? We would, of course, expect to find the material of which genes are made in chromosomes, but there is no problem here, for there is plenty of protein in chromosomes. Of course, the idea that genes not only *make* proteins but *are* proteins begs the question, 'What makes genes?', which is the kind of theological point with which bright seven-year-olds discomfort teachers of Sunday school: 'Who made God?'

But there was always another, unlikely candidate: an acid, first discovered in 1871 in the sperm of trout from the River Rhine. Sperm is mostly nucleus, so this could be called a nucleic acid. Biologists did not at first realise the significance of this acid. But they did not ignore it. Analysis quickly showed that it contained a sugar which, in the 1920s, turned out to be deoxyribose. Hence the name 'deoxyribose nucleic acid', or DNA for short. In the 1920s, too, Robert Feulgen of Germany showed by staining that DNA was confined to the chromosomes; it was not simply a general component of the nucleus. Another 'nucleic' acid also turned up, which in fact was found in the cytoplasm at least as much as in the nucleus. This second acid contained a similar sugar, called ribose. Hence it was called ribosenucleic acid (or ribonucleic acid), *alias* RNA.

Yet as a candidate to be the stuff of genes, DNA simply did not seem to have what it takes. Let us assume, first of all, that genes do indeed operate by making proteins. For every protein – and there are many millions of types – there has to be a gene. Can DNA possibly be as versatile as protein?

Well consider, first, how proteins are constructed, and how they achieve their versatility of form and function. Their general structure became apparent in the early 1900s: Emil Fischer of Germany showed that they consist of chains of amino acids. Since there are many thousands of different proteins, it seemed reasonable to suggest that there might be thousands of different amino acids, but in fact it became clear – though not, finally, until the 1940s – that the true number in nature is nearer to twenty. The rich variety of proteins is thus analogous to the rich variety of words in the English language: all 15,000 of them in the *Oxford English Dictionary* are formulated from twenty-six letters, and those same letters, with a little manipulation, can be made at least roughly to represent a potential infinity of other languages as well. Nature, like language, is economical, which is why its extraordinary variety is amenable to study.

It has long been clear too that the behaviour and properties of a protein depend upon its overall, three-dimensional geometry – that is, its *tertiary structure* – and where, within that overall shape, particular amino acids are positioned. To some extent amino acid chains fold 'spontaneously' into their final three-dimensional shape (though as mentioned, many are now known to require 'chaperonins' to help them to fold correctly). But in either case, the tertiary structure ultimately depends upon the order of amino acids in the protein chain, and upon its length: that is, on its *primary structure*. Nowadays 'biotechnologists' seek to design proteins of precisely the right shape, which will do precisely what they want. But more of this in Chapter 7.

Can DNA possibly match such versatility? The chemical evidence, at first sight, suggests not. Thus, it became clear in the first few decades of this century that the two nucleic acids also consist of chains of smaller molecules; not amino acids in this case, but *nucleotides*. But whereas there are twenty different kinds of amino acid within the protein chains, there are only *four* kinds of nucleotide within DNA or RNA. Furthermore, the different nucleotides seem much of a muchness. Each nucleotide has three components: a sugar (ribose or deoxyribose); a phosphate radical; and a 'base', either a purine or a pyrimidine. The only source of variation lies in the bases. Each nucleotide of DNA either contains one of two purines: adenine, 'A', or guanine, 'G'; or else contains one of two pyrimidines: cytosine, 'C', or thymine, 'T'. RNA is much the same, but the bases differ slightly: uracil, 'U', a pyrimidine, is substituted for thymine. So, how can a DNA chain, with only four different kinds of component, provide an adequate code for a protein chain that has twenty different kinds of component?

It doesn't seem to make sense, and yet the twentieth century produced a succession of experiments that first hinted and finally demonstrated beyond reasonable doubt that DNA is indeed the stuff of genes, and that proteins are merely the products. In the 1920s an Englishman, Frederick Griffith, set the ball rolling with his work on the bacterium *Pneumococcus*. The normal, 'virulent' form of pneumococcus causes diseases in mice, but it loses its virulence if it is first killed by heat. There are also 'non-virulent' strains that do not cause disease at all. But Griffith showed that if non-virulent living pneumococci were injected into mice together with killed virulent pneumococci, then the mice died, *even though neither form alone was capable of causing disease*. In other words, the two kinds of bacterium, though each deficient in some way, could combine their strengths and between them re-acquire the quality of virulence. There was thus an exchange of 'information' between the two kinds, an exchange which, as is now evident, is analogous to sex in animals and plants.

Then, in the 1940s, Oswald T. Avery and his colleagues at the Rockefeller Institute found that specific material was transferred from the virulent pneumococcus to the non-virulent type which could 'transform' the non-virulent into a virulent type. So what was this transforming material? Well, it did not lose its transforming ability when treated with an enzyme that destroys protein (a protease). But it did lose its ability to transform when exposed to a DNA-ase. In other words, the transforming material was not protein. It was, in fact, DNA. Avery announced these results in 1944.

The second significant line of inquiry that effectively proved the role of DNA in conveying hereditary information involved phage viruses of the kind known as *bacteriophage*, usually shortened to *phage*. 'Phage' comes from the Greek for 'eat', and phage viruses do in fact make their living as parasites of bacteria (proving Jonathan Swift's adage that 'a flea hath smaller fleas that on him prey; and these have smaller fleas to bite 'em, and so proceed *ad infinitum*'). In particular, biologists studied phages that multiply within the bacterium *Escherichia coli*, which lives in the guts of animals. Inside an *E. coli* host, a single phage can produce several hundred replicates of itself within twenty minutes, but by the 1940s, microbiologists already knew that they can do a great deal more besides. They can also exchange genetic information. Thus, if different *mutant* phages are allowed to infect a single *E. coli*, it isn't long before *normal* types begin to appear: individuals that evidently have combined the 'good' qualities from two or more otherwise defective phages. The parallels with Griffith's and Avery's pneumococcus experiments are obvious.

Viruses are structurally much simpler than bacteria. By the middle of this century, phages were known to consist of a central core of DNA, and an outer coat of protein. In 1952, at Cold Spring Harbor, Long Island, Alfred Hershey and Martha Chase showed that only the DNA of phages entered the bacteria that they attack. The protein coat remained on the outside. Thus the subsequent reproduction of entire phages within the bacterial host was orchestrated by DNA alone. Presumably, the exchange of information between phages was also enacted through DNA.

In short, more and more evidence showed that hereditary

information – 'genes' – was in fact encapsulated in DNA. So now the question is: 'How can a molecule that apparently has a rather boring basic structure – being made of only four basic components – make, or code for, molecules that have a much more complicated structure?' This can be answered only by describing the structure of DNA in far more detail. Some biologists, though by no means all, realised at the start of the 1950s that this was one of the key issues of biology. The answer did not come easily.

THE STRUCTURE OF DNA

Again, to begin the story in the middle, the general model of DNA that is now accepted was first proposed in 1953 by the English physicist-turned-biologist Francis Crick, and the young American chemist James Watson, who worked together at Cambridge (England). Crick and Watson were the great synthesisers: their strength lay not in performing key experiments, or in compiling crucial data, but in making sense of the observations of others.

The data from which they synthesised their model came first from chemistry – analytical and physical – and then from attempts to describe and measure directly the physical structure of the DNA molecule, both with the electron microscope and, most important of all, by X-ray crystallography.

We have already looked at the early chemical findings: that DNA consists of chains of nucleotides. Two outstanding pieces of chemistry after World War II enriched the picture enormously. First, by 1953 the British chemist Alexander Todd had shown that the nucleotides are joined together *via* the phosphates. In fact, the phosphates linked deoxyriboses in adjacent nucleotides – always joining to the number 3 carbon in one base, to the number 5 carbon in the next base (see Figure 2.2). Such an observation is important largely because it imposes restraints: it shows what kinds of shapes the DNA molecule *cannot* assume.

Observations elsewhere revealed subtleties in the proportions of the four bases in DNA. Biochemists of an earlier age assumed that all four types – A, G, C, and T – were present in equal amounts: in other words, that DNA was just a boring polymer, like polythene; long chains of identical units. But after World War II it became

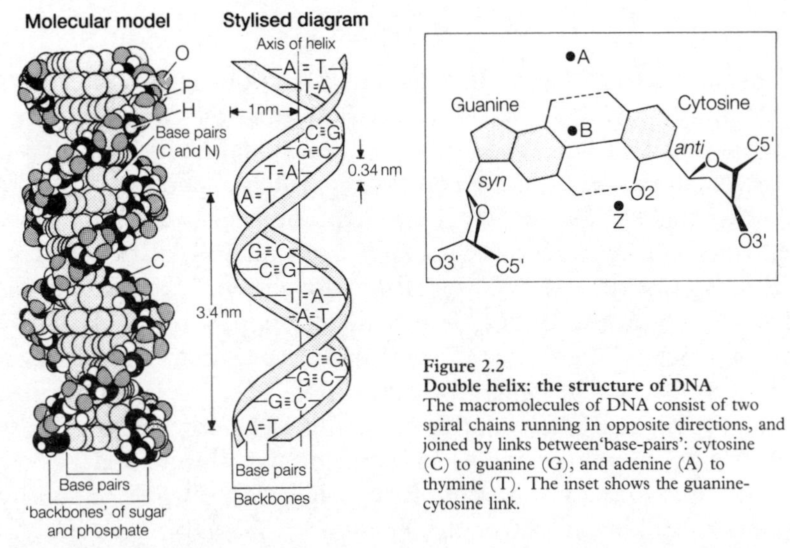

Figure 2.2
Double helix: the structure of DNA
The macromolecules of DNA consist of two spiral chains running in opposite directions, and joined by links between'base-pairs': cytosine (C) to guanine (G), and adenine (A) to thymine (T). The inset shows the guanine-cytosine link.

clear that the four bases were *not* present in equal amounts. So DNA was not like polythene. Furthermore – a highly intriguing finding – the proportions of the four nucleotides differed in DNA from different species.

Then in the 1940s an Austrian biochemist, Erwin Chargaff, working at Columbia University in the United States, added one further and highly pertinent insight. He showed that whatever the total amounts of the four bases, the amount of A always equalled the amount of T; and the amount of G always equalled the amount of C. There are several totally uninteresting reasons why this might be so. But in fact the reason is very interesting indeed – as became evident in the early 1950s, and will be revealed in a few paragraphs' time.

In addition, twentieth-century scientists have been able increasingly to explore the structure of complex molecules directly. The electron microscope, in general use after World War II, showed that DNA molecules were thread-like: many thousands of angstroms long, but only twenty angstroms thick (an angstrom is one ten-millionth of a millimetre). Since individual nucleotides were only 3 angstroms thick, DNA was obviously compounded from many thousands of them.

But the crucial technique was X-ray crystallography, developed in England from the start of World War I first by William Bragg and then by his son, Lawrence, for which they both received the Nobel Prize. The general idea is to deduce the arrangement of atoms in crystals by firing X-rays at them. The atoms scatter the

X-rays, and from the pattern of the diffractions (as revealed on a photographic plate), the position of those atoms can be inferred (the necessary calculations having been made). Early work was carried out on very simple compounds (such as sodium chloride) but even before World War II, W.T. Astbury of Leeds University felt confident enough to begin such work on DNA. It is remarkably easy to extract DNA, by fairly conventional chemical means that have become routine, and it manifests in a form like fluffy cotton. But when wetted, it becomes tacky, and can be drawn into long threads in which the macromolecules line up in parallel; a regular arrangement that is crystal-like. However, to prepare crystals good enough for X-ray is no mean technical feat.

The final phase began after 1949 at King's College London by Maurice Wilkins and Rosalind Franklin, using DNA prepared in Bern by Rudolf Signer. Crucially, they showed that the DNA chains are arranged in a helix, with the purines and pyrimidines on the inside, and the sugar-phosphate backbone on the outside. They showed, further, that each purine or pyrimidine occupied 3.4 angstroms, and that each turn of the helix was 34 angstroms deep, so there were 10 nucleotide molecules per coil. But the diameter of the helix (as was already known) was 20 angstroms – which was too large, if the DNA helix contained only one chain. Perhaps, then, it contained two chains; or perhaps (as Linus Pauling hypothesised at one point) it contained three.

There was one further, highly intriguing line of thought: not a piece of evidence, but an idea, put forward in 1940 by Linus Pauling and Max Delbrück at Caltech. Perhaps, they said, the replication of genes involved a splitting, with each half acting as a template for the re-creation of a complementary half.

Finally, at Cambridge in 1953, James Watson and Francis Crick put everyone out of their misery by proposing that the DNA macromolecule in fact consisted of two helical chains that ran in opposite directions, 'head to tail'. This was the famous 'double helix' – now the favourite symbol of every institution that has anything whatever to do with molecular biology.

The two DNA chains in the double helix are held together by weak chemical bonds ('hydrogen bonds') between the bases that run along the centre: adenine in one chain joins to thymine in the other; cytosine in one links with guanine in the other. This explains the ratio observed by Chargaff.

One chain thus complements the other precisely: if there is a sequence of bases running CATTG in one chain, then it will be matched by GTAAC in the other. In principle, though not of course in detail, this is more or less the same kind of mechanism suggested more than a decade earlier by Pauling and Delbrück.

Watson and Crick described their wonderful vision of DNA in the international science journal *Nature* in 1953. Their paper occupied a mere 800 words; tight as a sonnet. It ended with the wonderfully disarming sentence: 'It has not escaped our notice that the specific pairing we have postulated [between A and T, G and C] immediately suggests a possible copying mechanism for the genetic material'.*

This was the turning point. Molecular biology and all that comes from it (including 'genetic engineering', DNA fingerprinting, and all the rest) began here. Crick, Watson and Wilkins shared the Nobel Prize for Medicine in 1962 (Rosalind Franklin having tragically died at a very young age, in 1958).

But science never stands still. With the structure of DNA established (although final 'proof' did not come until some time later!), the next question is: 'How does DNA *work*?'

HOW DOES DNA WORK?

Here there are several sub-questions. The first, perhaps, is 'How does DNA replicate itself?' The second: 'How in practice are the protein chains put together, on the basis of the DNA instructions?' And the third is the one we have posed before: 'How do the four different nucleotides in DNA code for the 20 different amino acids in protein?'

In the event, Pauling and Delbrück had already guessed the principle of DNA duplication, and Watson and Crick hint at it in the last line of their *Nature* paper. In practice, one helical strand of DNA splits away from its partner, and from the moment the splitting starts, nucleotides within the surrounding medium

*J.D. Watson and F.H.C. Crick, 'Molecular structure of nucleic acids: A structure for deoxyribose nucleic acid', *Nature*, vol. 171, pp. 737–8 (1953). Later came: 'Genetical implications of the structure of deoxyribonucleic acid', *Nature*, vol. 171, pp. 964–7 (1953).

begin to line up against their complementary opposite numbers. As they do so they are joined one to another by an enzyme, DNA polymerase (Figure 2.3). Note – a point crucial to the later development of genetic engineering – the extent to which the assembly and dis-assembly of DNA depends upon teams of enzymes, of which DNA polymerase is one.

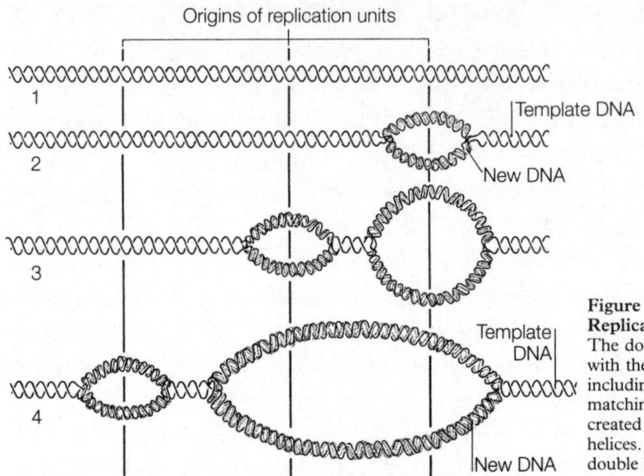

Figure 2.3
Replication of DNA
The double helix divides; and with the aid of enzymes, including DNA polymerase, a matching partner is immediately created for each of the single helices. Then we have two double helices.

The fact that DNA really does duplicate in this way was first shown experimentally in a highly ingenious experiment by Matthew Meselson and Franklin Stahl at Caltech in 1958. They grew cultures of *E. coli* in a medium that was highly enriched with the heavy isotopes ^{13}C and ^{15}N. (The common isotopes of carbon and nitrogen have atomic weights of only 12 and 14 respectively, so DNA built from the heavier forms is significantly more weighty.) Meselson and Stahl then transferred the *E. coli* with their heavy DNA to a medium containing only light isotopes of C and N, and allowed them to divide for one generation. After one generation they found only one kind of DNA in the new *E. coli* cells, and that had a molecular weight that was intermediate between that of heavy *E. coli* DNA and normal, 'light' *E. coli* DNA. So the newly formed DNA clearly consisted of one strand of heavy DNA, and one strand of light. Clearly, then, duplication was achieved first by a splitting of the original double-stranded DNA into two single strands, followed by the replication of each one. (The fact that the replication begins as soon as the splitting begins, and does not wait for a complete separation, was established only later.)

Next question, then: how does DNA make protein? It was clear from the start that DNA could not itself act as the manufacturer of protein. DNA is (mostly) confined to chromosomes within the nucleus; but it can be seen that proteins are synthesised in the cell cytoplasm, outside the nucleus. Clearly, there had to be some intermediary, ferrying the DNA instructions to the sites of protein manufacture. This is where the second nucleic acid, RNA, enters the scene.

That RNA was in fact the intermediary between DNA in the nucleus and the sites of protein manufacture in the cytoplasm seemed likely from before the time that Watson and Crick proposed their DNA model. For example, cells that make a lot of protein always have a lot of RNA in their cytoplasm; and the sugar-phosphate 'backbones' of DNA and RNA are similar, which suggests some kind of liaison between them. (Though in fact, as we have seen, RNA differs from DNA not only in the nature of its sugar – ribose instead of deoxyribose – but also, slightly, in its nucleosides. It contains uracil, instead of thymine; but uracil forms a 'base-pair' with adenine, just as thymine does.)

Even as early as 1953, then, the picture was envisaged: a strand of DNA, once separated from its partner, can *either* begin to make a complementary copy of itself, and so replicate, *or* can begin to make a complementary strand of RNA, which then leaves the nucleus and supervises the manufacture of appropriate protein in the cytoplasm. The creation of RNA complementary to a piece of DNA is called *transcription*. In the cytoplasm, the code now carried upon this RNA is made manifest in protein by the process of *translation*.

In practice, the manufacture of protein is somewhat complicated. It involves three different kinds of RNA, each of which is created by different parts of the DNA. The RNA that ferries the message out of the nucleus is called *messenger* RNA, alias mRNA. But messenger RNA does not make protein directly. Instead it co-operates with RNA that resides permanently in the cytoplasm (though it is first made in the nucleus) in bodies known as *ribosomes*, and thus is called *ribosomal* RNA or rRNA. In addition, each amino acid that is lined up for incorporation into protein is chaperoned into the ribosome by a short length of RNA known as *transfer* RNA, or tRNA. Each tRNA chaperones a specific amino acid, and links up to a particular place on the mRNA, with rRNA acting as a kind of work-bench for this linking to take place. Enzymes again are needed, to push this process along. This is shown in Figure 2.4, but the above description is all that is really needed.

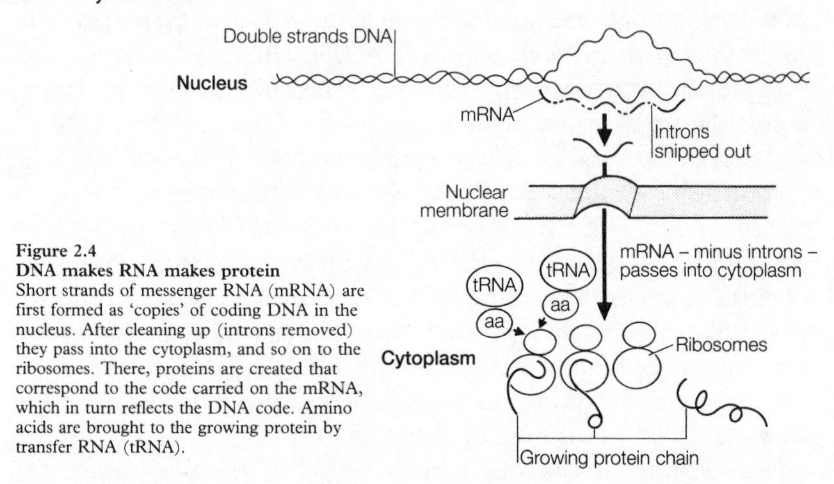

Figure 2.4
DNA makes RNA makes protein
Short strands of messenger RNA (mRNA) are first formed as 'copies' of coding DNA in the nucleus. After cleaning up (introns removed) they pass into the cytoplasm, and so on to the ribosomes. There, proteins are created that correspond to the code carried on the mRNA, which in turn reflects the DNA code. Amino acids are brought to the growing protein by transfer RNA (tRNA).

The various kinds of RNA act simply as executors. The central administrator, which determines what kind of proteins are produced, is the DNA. So our third question must be: How does a particular piece of DNA – a gene – code for a particular protein?

One obvious and known fact was that protein consists of linear chains of amino acids (even if the final shape of a protein is far from being a linear chain), and DNA consists of linear chains of nucleotides. Somehow the sequence of nucleotides has to determine the sequence of amino acids. This raises the question we posed before. There are roughly twenty kinds of amino acid in protein, but there are only four kinds of nucleotide in DNA. So how does four code for twenty?

In combination is, of course, the answer. If nucleotides lying side by side acted in pairs, then four different kinds could produce sixteen different combinations: 4×4. If they operated in threes, they could produce sixty-four different combinations: $4 \times 4 \times 4$. If they operated in fours, they could produce 256 different combinations: $4 \times 4 \times 4 \times 4$. Clearly they cannot operate in pairs, because sixteen is too few. Two hundred and fifty-six is far too many. It seemed most likely, then, that they operated in groups of three. Sixty-four is more than seems to be needed; but it is not absurdly too many.

Sydney Brenner and Francis Crick it was, at Cambridge in 1961, who showed that the genetic code did indeed operate through triplets of nucleotides in DNA. They worked with mutant forms of the T4 phage, which infects *E. coli*. They found that if just one nucleotide was added to a T4 gene, the protein that then resulted was non-functional. If two nucleotides were added, non-functional proteins resulted again. But if three nucleotides were added to a T4 gene, the resulting protein *was* functional. The addition of one or two nucleotides simply messed up the reading of the code, threw the whole reading out of synch. But the addition of three nucleotides added a whole new amino acid, which changed the resulting protein somewhat, but not enough, necessarily, to disrupt its activity significantly. Each trio of nucleotides that codes for a specific amino acid is now known as a *codon*.

Brenner's and Crick's work, too, illustrates immediately the nature of mutation. If a gene operates – as it does – by providing successive trios of nucleotides, then a mutation could clearly be of one of three kinds. If just a single nucleotide was added to the sequence, then this would horribly disrupt the reading of the code. If a single nucleotide was taken away, the code would be disrupted again. If a single nucleotide was simply changed – one substituted for another – then this may or may not cause a different amino acid to be substituted in the corresponding protein, and this may or may not disrupt the function of that protein. In any case, we see that mutations in chemical terms can be tiny changes, but that tiny changes can have huge effects.

Exactly which sequence of three nucleotides coded for which amino acids was worked out in the early 1960s, beginning with the work of Marshall Nirenberg and Heinrich Matthai who showed that artificial mRNA consisting exclusively of chains of uracil produced a protein that contained chains of the amino acid phenylalanine. So UUU codes for phenylalanine. Similar experiments in other laboratories followed and by June 1966, it was known that *all* codons contain three successive nucleotides, and that 61 of the 64 codons code for amino acids (some amino acids clearly have more than one corresponding codon), while the other three (UAA, UAG, and UGA) serve as 'stop' signals, which signal, literally, that the protein should be brought to a close. In short, the genetic code was 'cracked'.

Thus did genetics take on the additional discipline of molecular biology, and thus has 'molecular genetics' emerged as a discipline in its own right. For the rest of this book we will sometimes talk of genes as if they were just abstractions, like beads on a string, and sometimes as specific chemical entities which can be manipulated in specific ways.

Not surprisingly, a lot of molecular biologists felt rather smug by the middle of the 1960s. The structure of the gene was solved, and so, in fair detail, it seemed, was its *modus operandi*. One large question which we have not touched on yet (it is the subject of the next chapter) had also been breached: that is, how it is that genes are switched on and switched off, in the course of an organism's life. When molecular biology was put together with Mendelian genetics, and when Mendelian genetics was put together with Darwinian ideas of evolution as outlined in Chapter 1 – well, the result looked very powerful. Indeed, it seemed practically unimprovable. There was little to do, some felt, except dot the 'i's and cross the 't's. Francis Crick himself sought fresh challenges in a quite new field: in the study of nerves.

There was no call to be smug. Nature is not so easily pinned down. The past thirty years have yielded – and continue to yield – insights that are not mere adornments but are fundamental: powerful enough, in some cases, to put a quite new spin on biology as a whole. We will look briefly at some of these insights in the next few pages, and explore some of their implications in the chapters that follow.

THE RESTLESS GENOME

It was fortunate that the world's first molecular biologists, from the 1950s onwards, worked primarily with bacteria. They did so largely for practical reasons: bacteria are relatively easy to deal with in the laboratory, and a great deal is known about them. This was lucky, however, because bacteria are prokaryotes (their cells lack nuclei) and their genomes are a lot tidier, and easier to understand, than those of eukaryotes (that is, animals, plants and fungi, whose cells do have nuclei). Practically all the DNA in a bacterial genome is there for a clear purpose. Each group of three nucleotides either acts as a codon, to prescribe some particular amino acid, or else serves as punctuation (to show that one gene has ended, and the next begins) or else is part of a control sequence, regulating the behaviour of other genes. It is all very logical: the genes are arranged neatly, like books on a library shelf.

The genomes of eukaryotes, by contrast, are like the shelves of public libraries after a busy Saturday. (I speak as one who used to work in a public library.) The genes, with their orderly series of codons, are interrupted by stretches of DNA that have sometimes been called 'nonsense' and sometimes 'junk'. In any case, they look at first sight like random stretches, interrupting the gene. These interrupting sequences are called *introns*; and the bits of the gene that actually provide the code for protein are called *exons*. The messenger RNA that is first made in the image of the DNA includes a 'replica' of the introns as well as of the exons, but the intron equivalents are snipped out (by enzymes) before the mRNA is transported to the ribosomes for translation. The terms 'intron' and 'exon' are somewhat confusing, since you might reasonably assume that 'exons' are the things that should be left out and the introns left in, whereas in truth it is the other way around. But there you go. There is also 'junk' DNA between the genes. Indeed, it seems that at least 90 per cent of the DNA in the average vertebrate genome is 'junk'.

In fact, it transpires that many of the introns are probably far from superfluous, and clearly are not 'junk'. Their *modus operandi* is not understood in detail, but among other things they are clearly involved in modifying the extent to which any particular gene makes itself felt: that is, its *expression*. We will return to this in the next chapter, and in discussions of 'genetic engineering'. On the other hand, there is no reason to assume that every feature of a living organism has, in fact, been finely honed by natural selection to the point where it is strictly functional. Any whole organism has plenty of superfluous features, including nipples in male mammals and the vestige of a pelvis in some whales and snakes. To some extent, then, introns might indeed be 'junk'. But so long as they do no harm, natural selection will not militate against them.

Even more intriguing than introns are a whole group of DNA sequences of various kinds known generically as *mobile elements*. Some of these simply move from place to place within the genome; unstitching themselves from one chain of DNA, and reinserting themselves somewhere else, in the same chromosome or in another. Often these mobile elements land right in the middle of a functional gene, and when they do they might cause that gene to mutate, and/or to cease to function. Thus it was that the American biologist Barbara McClintock first observed examples of mobile elements in maize, in the 1940s (before, of course, the molecular structure of these elements was understood). Some seeds in some corn cobs had different coloured spots, and she concluded that this was because particular genes in particular cells were being switched on (or off) because of elements moving around within the genome. In practice there are various kinds of mobile element, but she called this particular kind *transposons*. Mobile elements, causing 'spontaneous' changes within the genome that can be quite radical, clearly can have a significant effect on an animal's evolution. We will return to this matter, too, in Chapter 4.

A particular class of mobile element is the *minisatellite*. Minisatellites contain repeating sequences of nucleotides. Each sequence is not in fact identical to the others, but, characteristically, each repeated sequence in a minisatellite contains a 'core sequence' that is identical in all of them. How and why minisatellites evolved and what 'purpose' if any they serve in the organism as a whole is a matter for speculation. They have attracted attention primarily because they are so immensely variable – so much so that every individual organism (by and large) has a unique pattern of minisatellites. On minisatellites, then, is built the modern technique of *DNA fingerprinting*, which is now of such tremendous significance both for forensic purposes, and in animal conservation, as further discussed in Chapter 8.

Viruses are acknowledged as the supreme agents of disease. As organisms, they are as stripped down as it is possible to be. They consist of nothing more than a packet of genes: a central core of DNA or RNA, wrapped in a protective coat of protein. Their survival tactic is to invade the cells of their hosts – which may be bacteria in the case of phages, or any other organism – and latch on to the host's own DNA, either in the nucleus or within the mitochondria or chloroplasts. Then they utilise the host's cell machinery to effect replication of themselves. *Oncogenes* – genes that predispose cells to cancer – could be erstwhile viruses that have integrated themselves completely into the host genome. Viruses themselves almost certainly should not be considered 'primitive' just because they are simple in design. Presumably they arose as fragments of genomes of more complex organisms which (in a manner analogous to that of transposable elements) found they could earn a living as mavericks.

One highly intriguing observation, however, is that viruses (or even fragments of bacterial genome) might integrate pieces of host genome into themselves, and then transmit those pieces to other hosts. Thus they could carry genes from organism to organism, athletically crossing species barriers. Such mechanisms may (or may not) have played a significant role in the evolution of plants in particular. They would help to explain why it is that nodules within the roots of leguminous plants such as clover, which contain the bacteria responsible for 'nitrogen fixation' (Chapter 6), also contain and produce haemoglobin: the blood pigment that carries oxygen within the blood of vertebrates and some insects. Did legumes (or their ancestors) acquire the necessary genes directly from animals? Nothing, as we enter these realms, should be written off *a priori*. Nothing that does not offend what Peter Medawar called 'the bedrock laws of physics' need be thought impossible. In nature – given the immensity of time available, and the enormous number of organisms alive at any one time – events that happen only with extreme rarity may in the end prove important, because an event that occurs just once in a billion organisms in a million years is likely, statistically, to occur sooner or later, and when it does may change the world for ever.

This, however, is speculation; the fact that mobile elements exist within the genome, and that viruses flit between organisms, certainly is not. It would be fantasy to suggest that you or I might acquire and incorporate a gene from an ant, say, or a cabbage. But with viruses behaving the way they do, this should not be considered impossible.

One last, intriguing matter to speculate before we move on. How did it all begin?

IN THE BEGINNING . . .

Molecular biologists have presented a picture of life as a dialogue between DNA and proteins – with RNAs of various kinds acting as the go-betweens. Without DNA to provide the code, there would be no proteins; or none, at least, of a complicated and replicable kind. Without proteins in the form of enzymes, to fetch and carry, stitch and mend, there could be no DNA, for DNA does not and cannot form itself.

This is fine; the picture is perfectly easy to understand. It is also encouraging in a general way, because dialogue implies co-operation. The notion that DNA and protein arose simultaneously in the history of the world seems implausible. Presumably they first arose separately, and first developed the 'dialogue' as a form of symbiosis, in the way that algae and fungi co-operate to form lichens, to their mutual benefit. It is pleasant to contemplate that this early, crucial step in the development of life involved a symbiosis. Even in a world dominated by natural selection, nature is not necessarily as red in tooth and claw as is often perceived.

There is a snag, however, with such a view. Neither DNA nor protein can form themselves. Chemists can show how nucleotide-like molecules could form spontaneously in nature, but for DNA to form spontaneously is a thermodynamic impossibility. Synthesis must be aided by enzymes that can bring energy to bear from elsewhere. Amino acids can form spontaneously, but they cannot link in ways that would create worthwhile biological catalysts. So the dialogue itself is easy enough to understand. But it is very difficult to see how the actors in that dialogue first came into existence.

Recent research suggests that the answer may lie with the molecule that now seems to play the errand-boy role: RNA. For it is now clear that at least some molecules of RNA can act not simply as a code, but also as a catalyst. RNA can in fact play the part of enzyme; and it does to some extent, in some contexts, even in modern organisms. To be sure, RNA in its modern form could not have appeared spontaneously in nature; like proteins and DNA, it is thermodynamically unlikely. But the molecule we see today is highly evolved, and we do not have to assume that it first appeared in its modern form. We have only to envisage that the first 'living' molecules were like modern RNA, but were much simpler, and were able to form spontaneously. So perhaps some primitive RNA was the first molecule of organic (carbon-based) life. Presumably it replicated itself in its own image, in a way that DNA and protein cannot. Perhaps it helped to bring proteins into being through its catalytic ability to link amino acids, and formed those chains in sequences that reflected the sequence of units within the RNA itself. DNA, itself initially a product of RNA, would have evolved into a storehouse of information, containing the structure of the RNA that formed it in more permanent form. From that would evolve the modern pattern: DNA makes RNA makes protein.

This of course is speculation, and the details will probably be for ever uncertain. But the general scenario is at least plausible. I will return to this issue later in this book, however (Chapter 10). There is more to say about the origin of life.

Clearly, biology has come a very long way since Darwin first complained in *The Origin* that 'no-one can explain' the elementary facts of heredity. The elementary facts can be explained in fine detail, sometimes down to the last molecule. Yet even so, the more it is explained, the more elusive and mysterious it all seems. The seventeenth-century vision of a clockwork universe seems to apply reasonably well to the arithmetical certainties of Mendelian genetics, or even to the 1960s vision of the genome, with its orderly ranks of genes. But the vision we have now, of a restless genome, innately anarchic, always prone to change, linked (perhaps) to other genomes by viruses, seems to re-awaken the ancient and poetic vision of life as a candle-flame: never the same for two instances of time, definable only in general terms.

There is one more huge and obvious area of mystery. A liver cell in an animal's body has the same genes in it as a muscle cell – for the genomes in both have been multiplied up from the same egg. So how are the two cells different? More: a dog, say, is different from a puppy, in looks, in behaviour, in physiology. A puppy is even more profoundly different from an embryo, and an embryo is nothing like an egg. How can the same genome give rise to a succession of creatures that are changing from one second to the next? These two kinds of question are related, and are the subject of the next chapter.

CHAPTER 3

And After Many a Summer

The Greeks and the Romans shared their Gods and agreed that those Gods were fickle. Mortals traded with them at their peril. If ever mere humans dared to emulate the Gods, or ventured to suppose that any success they met with derived from their own endeavours, then they would surely be struck down. The Greeks had a word for such presumption: *hubris*.

The Romans had a figure of myth to illustrate the point: a young man called Tithonus. Aurora, Goddess of dawn, fell in love with him, and she granted him immortality to keep him by her side. But in this she conspired in his hubris, for immortality is the prerogative of the Gods. The other deities did not override her, but they did find a loophole in her contract. They left Tithonus with his immortality, but they did not grant him eternal youth. As he grew older he became more infirm, until in the end he cried for death. But Greek and Roman Gods and Goddesses, sly though they may be, must keep their promises. Sick, crippled, self-disgusted though he was, Tithonus could not die.

Alfred, Lord Tennyson, commemorates the fate of Tithonus in a poem that contains one of the finest of all his stanzas:

> The woods decay, the woods decay and fall,
> The vapours weep their burthen to the ground,
> Man comes and tills the field and lies beneath,
> And after many a summer dies the swan . . .

'Me only', the poem continues, 'cruel immortality consumes'.

Aldous Huxley was one of the twentieth century's 'Renaissance men': a polymath. He was the brother of Julian, and a grandson of Thomas, evolutionary biologists both, who featured in Chapter 1. Biology, then, was in his blood. But Aldous was primarily a writer: of essays, poems and novels. Appropriately, in 1939 he published a marvellous piece of science fiction based largely on the biological thinking of his day. The title, *After Many a Summer*, is taken from Tennyson's fine poem.

Huxley's novel concerns an eighteenth-century English aristocrat, the Fifth Earl of Gonister, who seeks, like Tithonus, to live for ever. The Earl observes that carp may live for hundreds of years (for example in the ponds of abandoned monasteries) and concludes that they must have the secret of longevity. He decides that the answer must lie in their nutrition; at the time that Huxley created the Fifth Earl, after all, the full power and range of vitamins was just beginning to be appreciated – including the ability of some creatures to synthesise a wide range of vitamins that human beings cannot make. So the Earl decides to eat the gut contents of carp, which he refers to in his diary as 'stercoraceous pap'.

The approach succeeds. The Earl does live for ever: or at least, at the time he is re-discovered, lurking in the dungeon of his ancestral home, he is known to be at least two hundred years old. His housekeeper – born in the early nineteenth century – is still with him, for she too has partaken of the pap.

Like Tithonus, the Earl has lost his youth, but unlike Tithonus, he has not simply become decrepit. Instead, decade by decade, he has become more and more like an ape. By the time the heroes of the novel come upon him, he and his erstwhile housekeeper are out-and-out simians.

I have not related this story just through whimsy, charming though it is. For in the Roman myth, Tennyson's re-telling, and Huxley's re-working, there lies most of what we need to know about the way that genes operate, the way that living things acquire their complexity, the way that they unfold from egg to ancient, and the processes of senility and death; and in these stories, too, lies a very important insight into the nature of evolutionary change.

So now for the serious science. We can usefully begin with more discussion on how genes actually work.

MORE ABOUT HOW GENES WORK: GENE EXPRESSION

There are two essential unifying concepts. The first is that genes emphatically are *not* passive bystanders in the life of the cell. The cliché that compares DNA to a set of blueprints is another of science's misleading metaphors. Blueprints are drawn up before a job starts, then put in a drawer (or indeed lost. It usually makes very little difference). But DNA is an active administrator of the modern kind: sleeves rolled up, frenetically and intimately involved in the second-by-second running of the cell from conception to the grave, and extremely sensitive to events all around it. What should *not* generally change in the course of a gene's lifetime is its overall structure: if that changes, then that is a mutation, which in the short term is usually a disaster, even though (as we saw in Chapter 1) mutations in the long term are an essential source of variation. What *can and does* change in response to immediate circumstance is the activity of the gene: whether or not it actually functions at any one time, whether or not, as geneticists say, it is *expressed*.

Gene expression is the second important concept, one that serves to unify great swathes of biology. A gene that is 'dominant' is one that is expressed. Recessive genes, by contrast, are shut off. Partial dominance implies some expression by both homologous alleles; 'superdominance' implies extra-vigorous expression of the dominant allele in the presence of some particular partner; 'penetrance' alludes to the degree of expression. Many hormones operate, directly or indirectly, by switching on or shutting off particular genes in particular cells, so that they start or stop directing the production of the particular protein for which they are responsible. Many drugs operate in the same way. I say 'directly or indirectly' because some chemical agents do in fact operate directly on particular components of the protein-producing mechanism, while others remain on the outside of the cell and their messages are relayed to the protein-producing mechanisms *via* 'secondary messengers', inside the cell. But the outcome is the same.

The mechanisms by which genes are switched on or shut off were first revealed by the French biologists François Jacob and Jacques Monod in the late 1950s, for which they received a Nobel Prize in 1965. In essence, working with lactobacillus bacteria, they showed that each stretch of DNA that serves as a gene has another stretch 'upstream' of it which acts as a control region. This upstream control region has various components, including a 'promoter', a 'regulatory gene', and a 'terminator'. In response to a suitable outside stimulus, the promoter instructs the regulatory gene to produce a protein which in fact acts as 'repressor': repressing the activity of the gene it is supposed to regulate, positioned downstream along the DNA. In general, then, in the absence of instruction to the contrary, genes express. But when their regulator genes assail them with repressor protein, they are shut off.

Since these pioneer experiments, many different mechanisms for the control of gene expression have been revealed, especially in eukaryotes (that is, in organisms whose cells have nuclei: animals, plants and fungi). In eukaryotes, indeed, these control mechanisms may apparently operate almost anywhere along the chain of command that in the absence of a control allows a gene to express by producing protein. All kinds of intriguing oddities are being revealed: that the mysterious introns may, in fact, play a serious role regulating the genes in which they find themselves (a fact of great significance in genetic engineering, which we will look at again in later chapters) or that the topographical position of a particular stretch of DNA within the nucleus may also affect its expression – for example, whether the piece of chromosome that harbours the gene is towards the periphery of the nucleus, or near the centre. We mentioned in Chapter 2 that the position of a particular stretch of DNA within the genome is far from immutable. Since gene position influences gene expression, we might expect that natural selection would favour some positions rather than others.

The fact that genes can be regulated in various ways, and that regulation of gene expression is the principal means of controlling the cell as a whole, clearly means that in eukaryotes in particular there are many different ways of fine-tuning the activities of any one cell. All multi-celled organisms make exquisite use of this fact, and each body cell in an animal, plant or fungus is perpetually restrained and stimulated by a host of hormones or other 'factors' (such as growth factors) produced by other cells in the body. Thus do the cells of a multi-cellular organism work in what seems like miraculous harmony. Thus, too, we move away from the classical notion of a body regulated by some central executive, such as the heart or the soul. We can of course identify hierarchies within the body organisation; for example, in the way the brain influences the hypothalamus, and the hypothalamus influences the pituitary gland at the base of the brain, and the pituitary in turn influences just about every other gland in the body. The overall picture, though, is of the body as a giant cocktail party, with every cell talking to every other cell. Most of the cells, most of the time, ignore most of the chatter of other cells, but within the cacophony are clear threads of information which do pass cogently from the speaker to those that have ears to hear. Basically, or at least much of the time, the cells in the body's perpetual cocktail party are telling each other to switch on, or switch off, particular genes. Cancerous cells, in general, can be perceived as rogues that for one reason or another ignore or override the signals that come from other cells: particular genes within them that control cell replication, and which in healthy tissue would be suppressed by outside or inside influence, instead find free expression. On the other hand, pharmacologists are now learning some of the body's control tricks, and are emulating them in new generations of drugs. Again, the more control mechanisms there are within the cell, the more opportunities there are for intervention, and for fine-tuning.

So much for the general principles of gene activity and regulation. Once we start to apply those principles, we see how very powerful they are. First, they help us to explore what has often been seen as the greatest mystery of all: not how an organism *is*, but how it *becomes*: how it contrives to change, constantly, from the moment of conception to the time of death. This embraces an entire suite of phenomena: cell differentiation; embryogenesis and maturation; ageing and death. We should look at these matters.

HOW ORGANISMS WAX AND WANE: THE PROGRAM OF THE GENES

Consider, first, an animal body as it is; we can look at how it becomes, a little later. It contains a multiplicity of different tissues: liver, muscle, nerves, lymphocytes, what you will. Within those tissues there are many different kinds of cell; indeed, it has been estimated that the human body as a whole has 100 million different *kinds* of cell; and of course, there are billions of cells of each kind.

Yet all the billions of cells derive from a single fertilised egg, or *zygote*. They may be only a few generations removed from that zygote (it does not take many generations to multiply up from one to billions) or they may be many generations down the line. But whichever is the case, the body cells (as opposed to the sex cells, or gametes) will have been produced by the process of mitosis, and in mitosis, as we saw in Chapter 2, all the genetic material in the mother cell is very carefully replicated and conserved in the daughters. Thus, apart from a few odd cells such as the red blood cells of mammals, which contain no nuclei at all, and apart from the gametes, all the 100 million different kinds of cell in the body contain exactly the same kinds of genes as all the others; that is, exactly the same genes as the zygote from which they descended.*

From the few remarks so far, however, it is obvious how such variety can be achieved. Liver cells do indeed contain the same genes as muscle cells. But in liver cells the genes appropriate to liver are turned on, and those that could produce the contractile proteins appropriate to muscle are turned off; in muscle, the genes for contractile proteins are turned on, and those that would produce special liver enzymes are turned off. There is a class of cancers known as 'teratomas', in which the genes of particular cells escape the local controls, and start to produce the wrong kind of tissue in the wrong place.

In practice, the process of *differentiation* – by which the cells of an animal embryo turn into liver cells, or muscle cells, or what you will – occurs in stages, but generally begins very early in embryogenesis (embryo development). The single cell of the zygote divides to become two; two becomes four, four becomes eight, and so on. Soon there is a small ball of a few hundred cells known as a *blastocyst* and by that time differentiation is already in train. In 'advanced' animals – vertebrates, arthropods, echinoderms such as starfish, molluscs, segmented worms, and many others – the different cells first become committed to one of three camps: ectoderm, mesoderm or endoderm. 'Ecto' become tissues associated with the outside, such as the epidermis and retina; 'meso' become muscles and the like; 'endo' become gut linings and associated tissues such as liver. In yolky eggs the destiny of any one group of blastocyst cells is clearly influenced by their position relative to the yolk. In non-yolky eggs, such as those of placental mammals, it is harder to see what it is that first sets the trend. As embryogenesis progresses, each group of

*In Chapter 6, however, we discuss the phenomenon of *somaclonal variation*; mutations that give rise to genetic novelties within the body, and which are pressed into service by plant breeders.

cells apparently produces various 'factors' which influence the fate of cells around it; and in turn is influenced by 'factors' that those surrounding cells produce. These factors in general diffuse from the places they are generated to their targets and thus form gradients of concentration; and the effect of any particular factor upon its target clearly depends to a great extent upon concentration. This is demonstrable. There have been many experiments in which bits of embryo that are moved to new positions change their destiny – unless they have already become too committed to their original course, in which case they continue to grow as they were before. What is less demonstrable (in general, so far) has been the precise nature of the controlling factors, and their *modus operandi*.

It is clear that as embryogenesis proceeds, as the cell generations pass and the descendent cells become more specialised, so individual cells lose their versatility. If the two cells of a two-cell embryo actually separate in the womb, then each daughter cell can grow into a complete organism and identical twins result. The two twins together, both descended from the same cell and each genetically identical to the other, form a *clone*. If a two-cell embryo is split artificially then each one similarly develops into a clone of identical twins. This has long since been achieved with animals such as frogs whose large eggs develop outside the body, and has more recently been achieved in a range of mammals, including sheep. Indeed, Dr Chris Polge and his colleagues at ABC (Biotechnology Company) in Cambridge have produced identical quads from four-cell sheep embryos, and even produced identical quintuplets from five of the eight cells of an eight-cell sheep embryo. Such work is primarily intended for agricultural purposes – to multiply 'elite' (outstanding) animals quickly and reliably. But it is also of great fundamental significance, and has some potential value in conservation where it can be important simply to produce as many offspring as possible in the shortest possible time.

Once we get past the eight-cell stage it becomes impossible to pull this particular trick. One reason is that the young embryo does not increase in total size for the first few divisions, so the individual cells at the eight-cell stage are about as small as it is possible to be if they are to survive alone. But once the embryo does begin to increase in overall size, then differentiation also begins. And once animal cells begin to differentiate then, in general, they lose their ability to grow into a complete organism, because they lose their ability to produce all the different tissues that are characteristic of a complete organism. A cell committed to a life as endoderm cannot start generating mesoderm cells.

Cells that retain the potential to produce all the cell types of the organism that they represent, are said to be *totipotent*. Those that are able to produce some of the cell types but not all, are called *pluripotent*. The individual cells of eight-cell animal embryos are totipotent; and the 'stem' cells of the bone marrow that give rise to all the different cell types of the blood are pluripotent. Some extremely differentiated cell types, notably those of mammalian nerves, cannot divide at all once they are first laid down.

From all that has been said above about control of gene expression, you will doubtless have inferred what is happening. In the totipotent cells of the very young embryo, a great variety of genes are switched on, or retain the capacity to be switched on, but as cell division and differentiation proceed, so successive groups of the genes in each cell type are switched off until only the characteristic liver genes are operating in the liver cells, and the characteristic muscle genes in the muscle cells. In practice, though, life is not quite so simple as that. In the zygote itself, it seems that *none* of the genes is actually switched on. The first few divisions are evidently under the control of 'factors' within the zygote cytoplasm, 'factors' that clearly were bequeathed by the female, since sperm contains very little cytoplasm. The totipotent animal cell, then, does not contain a full suite of active genes. Rather it contains a full set of effectively quiescent genes that have not yet been turned off. Some of the genes evidently start to be turned off more or less as soon as others are switched on. Overall, though, I like the general notion that growth and development to a large extent represent a *loss* of potential. I often think that the same applies to the psychological development of children. Rapidly they lose the tremendous facility that young children have for assimilating language. Rapidly, in so many, the wondrous flights of childish imagination are channelled into dull practicality. Those who escape this fate may become geniuses. As Einstein said of himself, 'I never stopped thinking like a child.'

Obvious from this description is the awesome vulnerability

of the embryo. Its correct development depends upon the beautifully timed switching on and switching off of genes in successive generations of cells, and that switching on and switching off in turn depends upon events in previous generations of cells, or in cells elsewhere in the embryo, which each send out their own particular signals. Yet we noted above that the expression of genes is enormously sensitive to the presence or absence of chemical agents in the surrounding medium, including drugs of many kinds. Imagine the consequences for an embryo if a particular gene in a particular group of cells is switched on or off at the wrong time. Actually, you don't have to imagine. 'Accidents of birth' are a fact of life. Chemical agents acting upon gene expression are only one of several causes. But their potential significance is obvious.

Embryogenesis in mammals is followed by birth, and birth by growth, maturation, and decline. As William Shakespeare's Jaques put the matter in *As You Like It* (Act II Scene vii): first, 'the infant, mewling and puking in the nurse's arms'; then 'the soldier, full of strange oaths'; followed by 'the lean and slipper'd pantaloon'; and finally, 'Sans teeth, sans eyes, sans everything'. In all, Jaques discerned seven ages, but we could as soon acknowledge seventy times seven, for the change is continuous. Life can be seen as an ever-changing candle flame which, as Jonathan Miller once put it, 'performs' its shape.

From all that I have said, the general mechanism of change is obvious. As the organism grows and matures, different suites of genes become active, or are switched off. The soldier with his strange oaths has the same genome as the mewling infant, but different genes are active within that genome. The analogy between genes and blueprints may be misguided, but it is tempting to compare the genome to a computer program, which issues a *series* of instructions, each waiting on the one before. No analogy works exactly, however. There is nothing quite like life except life itself.

Yet there remains one oddity in that overall picture. We can see why natural selection was bound to favour the progression from infant to adult. Adults cannot be born as adults, because they obviously cannot be born at the same size as their parents, so they have to go through a growth stage, which is the infant. But the infant cannot reproduce. If infants did reproduce when they were still small, then their offspring would presumably be even smaller, and there are severe disadvantages in being *too* small. So infants have to grow into adults, if the organism is to reproduce.

But what evolutionary advantage can there be in old age? Why does the soldier degenerate, to the 'lean and slipper'd pantaloon'? And why does the decay continue, until the pantaloon is 'sans everything'? What advantage is there in death?

THE SLIPPER'D PANTALOON

Many have speculated on the reasons for decay and death. Some have suggested that there is advantage to the species as a whole if the old individuals crumble and die, to make way for 'fresh blood'. The idea that individuals should die to make way for others is not entirely implausible, but it needs to be severely hedged. Thus in Chapter 4 I will argue that natural selection often favours parents who take particular care of their children. The 'aim' of reproduction, after all, is to pass on the genes, which child-care obviously abets. Child-care will be favoured provided the risks do not out-weigh the benefit. For example, natural selection would not favour child-care in animals who tried to look after their babies but were habitually killed before their offspring were self-reliant. It would, rather, favour animals who produced offspring that were self-reliant from the outset. In the latter case, however, natural selection could well favour parents who kept out of their offsprings' way. It could, in fact, in extreme cases, favour the actual death of the adult, because dead animals do not compete. There are indeed spiders who die after laying eggs, and whose bodies are later consumed by their own emerging infants. Many other animals (such as octopuses) die precipitately after reproducing, in a way that suggests a genetically prescribed *hara-kiri*.

Such mechanisms apply only in particular cases. It is quite wrong to suggest that natural selection could in general favour the death of individuals 'for the good of the species'. The generalisation is that if an individual animal had a set of genes that enabled it to go on living and reproducing, then natural selection would favour that individual. Individuals compete *within* species. As will be discussed in Chapter 4, animals do not risk their own lives *except* in some cases to protect their own very close relatives, including, of course, their own children. 'Laying down the life for the benefit of others' does apply in some cases, then. But it does not explain the *general* phenomenon of death.

Others, then, have suggested more mechanistic explanations; for example, that the genome contains some kind of time clock, which ticks away, and ensures that the genetic program has a limited span. If so, why? And why is it, if that is so, that some individual animal cells – particularly fetal cells – may be effectively immortal when raised in culture? And how is it that some organisms simply do not seem to age? Carp may not be quite so long-lived as the Earl of Gonister supposed, but so far as I know they show no particular signs of senility when they do die. Cedar of Lebanon trees certainly live for thousands of years. If them, as the Earl asked, why not all of us?

Sir Peter Medawar proposed what seems like a fine evolutionary explanation. He pointed out first of all that the chances of dying within a long period of time are greater than within a short time. Thus, even if there were no ageing process at all – even if we all stayed indefinitely at the physical peak of youth – there would none the less be more fit people at forty years of age, than at one hundred years of age. Within one hundred years, even with no physical decay, there is a greater chance of coming to grief than there is in forty years. The same applies to laboratory test-tubes (says Medawar). They do not decay, yet old test-tubes are rarer than new test-tubes. In the wild (and the human genome underwent most of its evolution before we enjoyed the cosseting of civilisation) the rate of accident is high, just as it is with test-tubes.

Medawar pointed out, too, that genes which affect an animal after it has reproduced cannot be weeded out by natural selection. This ought to be modified; it is possible to envisage mechanisms by which post-reproductive genes could have some effects. I would argue, for example, that the wisdom of old elephants contributes to the survival of young elephants, and thus a gene that conferred post-reproductive survival in an elephant would be more likely to be represented in future generations, because individuals that possessed that gene (having inherited it from their mothers and grandmothers) would be more likely to survive. But that is a relatively minor quibble. Undeniably, genes that affect the life of the animal before it has reproduced are far more open to selective pressure than those that do not operate until afterwards. You may ask, 'But why should an animal stop reproducing, if it is fit?' Here again there are many possible answers. One is that some animals are built that way, so that, for example, female mammals are born with all the eggs they will ever possess and they undergo menopause (if they live long enough) at the time their ovaries run dry.

Add to such general observations the fact that all genes in all organisms may at some time mutate, and *most* mutations are harmful, at least in the short term. All lineages – successions of generations – are bound to experience mutations as the generations pass.

But, says Medawar, deleterious mutations that occur in genes that are not expressed until late in life will be less likely to be weeded out by selection than those that mutate early in life. Mutations that occur after reproduction will be subject only to the weakest selective pressure, or to no pressure at all.

Moreover, natural selection would favour organisms that developed methods of mitigating the effects of harmful mutations. One obvious point (to which we will return) is that harmful mutations cause less trouble if they are recessive – because then they appear only in the homozygotes, who in turn are produced only through the marriage of two heterozygous carriers. Thus in practice most deleterious genes that we know about in the human species are recessive. Recessiveness is a matter of expression. Natural selection would therefore favour not simply the elimination of 'bad' genes, but also the spread of genes that suppressed those deleterious genes. By the same token, natural selection would also favour mechanisms that *delayed* the expression of recessive genes, for the later they were expressed in life, the less effect they would have – given, after all, that 'after many a summer' most individuals are liable to be killed by accident in any case.

Hence, said Medawar, old organisms are liable to accumulate deleterious mutations – for two reasons. First, natural selection will not operate strongly upon deleterious mutants that do not make themselves felt until the organism is old. Second, natural selection will *favour* mechanisms that delay expression of harmful genes. Thus, as we age, the program of our genes contains more and more mistakes. Our skin loses its bounce. Our eyes become rheumy. You know the rest, or will do, in the fullness of time.

There are loose ends, of course. It bothers me, for example, that many song-birds can clearly live healthily for twenty years or more in protected environments, whereas few in the wild may survive more than a couple of seasons. But as a broad series of generalisations Medawar's hypothesis works none the less.

Finally, there is of course scope for things to go wrong with the timing of the program. In one inherited disorder children seem to run through the program very rapidly, and may become physically senile before they are ten years old. In contrast, genes that should be shut off after infancy may persist too long. Before birth, for example, mammals have a special form of haemoglobin which seizes oxygen more tightly than the kind they acquire after birth. This enables the fetus to grab oxygen from the haemoglobin of its mother, where the blood supplies of the two are brought close together within the placenta. But poor timing causes some babies to retain fetal haemoglobin for longer than they should after birth, and since this highly acquisitive haemoglobin is also more reluctant to release oxygen in the tissues, some respiratory distress results. Medawar suggested that some cancers at least may result from the switching on (or persistence) of fetal genes, with an excessive propensity for growth. Retinoblastoma, a cancer that grows in the retina of children, and if untreated invades the optic nerve, may be an example. In general, the range of plausible pathologies is clearly vast.

Let us return, though, to the early embryo, and the notion that growth and differentiation move into gear together, so that as development gets under way, totipotency gives way at best to pluripotency, and finally (in many cases) to lineages of cells that either can divide only to produce more of their own kind, or else are unable to divide at all. Here there is a sharp contrast with plants, for most plant cell types that I know of (apart from those that lose their cell nuclei altogether, like the vessel cells of flowering plants) effectively retain totipotency throughout their lives. Thus a gardener may cut a stem from a pelargonium, perhaps but not necessarily dust the end with 'hormone powder', and stick it in suitable compost, to produce a whole new plant. More generally, as we will see again in Chapter 6, almost any plant cell can be cultured under suitable laboratory conditions and will divide to form undifferentiated tissue called 'callus', from which entire plants may re-develop.* The new plants, produced from cuttings or from single cells within a callus, are all genetically identical and hence are members of a clone.

We may suggest an arm-waving explanation for this difference between animals and plants (an 'arm-waving' theory being one that seems plausible but for which there is no evidence worthy of the name). Animals in general survive by being mobile – because after all, they must pursue their food. Mobility in turn imposes severe restraints on shape, and demands that the different parts of the body should coordinate closely. Plants, by contrast, stay in the same place, with their roots deeply embedded in the soil that supplies their nutrient. They have no need to move, but they do have to endure the slings and arrows of the environment to a degree that animals do not, because most animals can move out of trouble. It seems to me to follow that the different parts of the animal need to be more specialised (in general) than the different parts of the plant, or agility will be compromised, but that the different parts of the plant need to retain greater powers of regeneration, because they are more likely to be damaged. In short, the plant needs to be able to adjust its shape as it grows, just as an oak growing in a crack in a rock is a very different shape from one with the sky to itself, and rooted in deep rich soil in a park.

How fundamental is this difference between animals and plants? On the face of things, it is very profound indeed. Regeneration of entire plants from almost any cell is in some cases simple, and it is therefore possible to produce very large clones of plants from single masses of callus. But it is quite impossible to generate an entire dog, say, from a dog somatic cell, and quite impossible, therefore, to produce large clones of dogs. Only small clones of mammals can be produced – by splitting early embryos, as we have already described. Furthermore, as noted in Chapter 1, *any* mutation in a plant cell may theoretically find itself in future generations, because any plant cell could theoretically give rise to an entire plant which would subsequently develop sex cells.

*In practice this cannot yet be done for *every* tissue. But it can be done for so many that the failures can be ascribed not to some deep biological principle, but a simple lack of appropriate technique.

But in animals, as August Weismann noted, changes in ordinary body cells are not represented in the gametes.

Yet this apparently profound difference may simply be one of degree. After all, some plant species are still very difficult to regenerate from single cells. On the other hand, some specialist groups of animal cells have retained (or secondarily re-evolved) a high degree of pluripotency. Worms and starfish, for example, can regenerate enormous chunks of themselves. Some lizards practise 'autotomy' ('self-cutting'): they deliberately shed the end of their tail if this is grabbed by a predator. But then they grow a new tail. There seems no reason, either, why genes in differentiated animal tissue should be *irreversibly* suppressed. Unless the suppressed genes are actually destroyed (which they are not) there seems no absolute reason why they should not be turned on again.

So, could we in fact produce large clones of mammals from ordinary body cells? Well, it certainly is not possible at present. But there is no reason to suppose that this is theoretically impossible. To achieve this, the switched-off genes of the differentiated cell would have to be switched on again. The nucleus with its re-awakened genes would then have to be placed in the cytoplasm of a zygote of the same species (whose own nucleus would first be removed), and the re-nucleated zygote would then be placed in the womb of a suitable foster-mother, to come to term. The last part of this – transferring the nucleus and inserting the new-formed zygote – is possible with present-day technology. The re-awakening of all the switched-off genes in differentiated animal body cells may or may not be achieved in the decades to come. It will be surprising if it does not prove possible. The consequences – the possibility, for example, of producing effectively immortal clones of particular human beings – we could leave to the science fiction writers, though we will in fact discuss this matter again in Chapter 10.

In the more immediate term, there is already promise of producing clones from cultured embryo cells, which are not yet turned off. In order to see whether a particular animal was worth cloning or not, it would be necessary to take some cells from it at the early embryo stage, culture them (or freeze them for culturing later), and then let the rest of the embryo develop, to produce a whole animal that could be put to the test. But at present this remains science fantasy.

From all that has been said so far, two points are very clear. The first is that the timing of a gene's expression is, in effect, just as important as the thing the gene actually produces: a 'good' gene acting at the wrong time is a menace; a 'bad' gene whose effects are postponed is preferable to a bad gene that exerts itself from the start; any one gene that comes on line – or fails to come on line – early in the genetic program can profoundly affect the form and destiny of the organism throughout its life; and overall, the smooth running of the whole organism from conception to grave depends upon a harmonious and amicable switching on and switching off of suites of genes as time goes by.

The second key point is that the timing of gene expression is itself under genetic control, because the genome includes genes whose specific job it is to control the activities of other genes.

Since timing is so important, it is clear that natural selection must lean very heavily indeed upon those genes whose function it is to control the switching on and switching off of other genes. It is clear, too, that huge evolutionary changes could be brought about by alterations in the genes that control the expression of other genes. In practice, ever since the nineteenth century biologists have been finding more and more examples of evolutionary change that has indeed been brought about primarily by changes in the timing of the program (even though the nature of the program is only now being revealed).

EVOLUTION AND TIMING

Again we can conveniently begin this discussion in recent times and work back. In 1984 a group of genes was discovered within the famous fruit fly *Drosophila* that apparently controlled the entire body plan of the fly – throughout life. These genes were called *homeobox genes*. Each homeobox gene oversees the organisation of a segment of the fly – given that flies (like crustaceans and worms) are obviously highly segmented creatures. In practice, each homeobox gene organises the back half of one segment, and the front half of the segment behind; in other words, genes have a different view from us of where one segment ends and the next begins. The entire group of homeobox genes were arranged along their chromosome, side by side, in the sequence in which they act upon the organism; showing once again, very graphically, that the positioning of genes on chromosomes is important, and is itself a key target of natural selection.

Things began to get really interesting, however, when comparable homeobox genes were found in vertebrates. In truth, the number and lay-out of the genes was more complex than in insects, presumably because vertebrates have a more complex overall body plan. Thus the homeobox genes of mice – known as *Hox* genes – are arranged in several more or less replicated sets, each on a different chromosome, but again, the vertebrate homeobox genes are arranged along their respective chromosomes in the order in which they will act. Stephen Gaunt at the Agricultural and Food Research Council's Institute of Animal Physiology and Genetics Research at Cambridge has shown in a brilliant series of radiographs how *Hox* genes come into operation. Basically, genes can be seen expressing when they start to produce mRNA: the first step in protein synthesis. Radiographs are produced by attaching radioactive probes to the mRNA. Successive radiographs of mouse embryos show successive regions of the mouse (equivalent to the segments of a fly) 'lighting up' as time passes, and the embryo develops, from the head down to the tail. These genes, which establish the overall body plan, clearly come on line very early in the embryo's life. What they do (or fail to do) affects all that comes afterwards. They are among the first genes to exert themselves in the genetic program.

What is truly astonishing, however, is that the *Hox* homeobox genes of the mouse do not simply operate in the same way as those of the fruit fly. The mouse's and the fly's homeobox genes are very clearly homologous. In other words, the mouse and the fly inherited their overall controlling genes from a common ancestor. The grand group (phylum) to which mice belong is the Chordata (of which the vertebrates are a sub-phylum). The fly, an insect, belongs to the super-phylum known as the Arthropoda. The chordates and the arthropods both appear first in the fossil record in the Cambrian period, which began about 550 million years ago. In other words, 550 million years ago, the fly-group and the mouse-group were already very different from each other. The ancestor from which they both apparently descended must have lived nearly one billion – 1,000 million – years ago. It is intriguing to have such proof that flies, after all, are our extremely distant cousins, more distant by far even than fish, but still related. Equally intriguing is to contemplate that the homeobox genes themselves have survived, not entirely but still remarkably unchanged, for perhaps a billion years.

When a gene survives for many generations more or less unchanged it is said to be highly *conserved*. The point is that natural selection can and does act just as vigorously to keep things the same as it does to bring about change. After all, if a system works, it is obviously worth hanging on to. Any variation from a system that works is liable to be worse, and natural selection will therefore tend to eliminate all deviants. Those genes (or bits of DNA that do not function as genes, such as introns) that do change rapidly as the generations pass are of two kinds. Either it simply does not matter if they change, so that mutations go unpunished (which may well be the case with most mutations within the DNA of introns), or else natural selection for some reason *favours* variation (which it does in the case of genes that produce antibodies to protect against disease). This will be explored further in Chapter 5; the point here is that homeobox genes are clearly among the most highly conserved of all. The importance of genes that lay down the overall plan of the body could not be more dramatically illustrated.

Clearly, too – as actually revealed by Stephen Gaunt's radiographs – their working is an exercise in timing. They *must* be among the first in the genome to come on line, and they must come into operation in precisely the right order, at precisely the right times.

Homeobox genes have been conserved because they are of enormous value. But there is another, quite opposite reason why genes might be conserved. Genes sometimes remain in the genome, passed slavishly from generation to generation, if they are simply of a kind that does no particular harm. It is the matter of expression again: if a gene is not expressed, or is expressed only in a harmless way, then natural selection will not single out the individuals that possess it (because they will not be phenotypically distinct from those that do not). Thus it is that all modern animals retain genes from their ancestors that in practice are useless to them – as well as some which (like homeobox genes) are worth hanging on to. The useless features produced by redundant genes that natural selection has had no particular cause to get rid of, are called *vestiges*.

Some of these ancestral genes turn up in the mammalian – including the human – embryo. Human embryos have tails, which they lose long before birth. Tails are an ancient vertebrate feature. More dramatically, human embryos have gill pouches: a reminder of our very different ancestors, 400 million years or so ago, among the fish. There has been no particular need for natural selection to eliminate gill slits in human embryos, so we have kept the genes that produce them. Since genes are not actually lost as development progresses, we must assume that each of us, though adult, still possesses genes for gills. In fact, of course, it may be that natural selection even serves to conserve gill-slit genes: for they may in practice act pleiotropically, and produce characters in the embryo that are useful, apart from the gill slits that seem so bizarre; or they may simply be linked to useful genes, so that elimination of gill slits would also eliminate worthwhile characters. The fact is, however, that gill-slit genes remain with us.

It remains the case, too, that so long as an organism retains the genes for some out-moded feature, then evolution can, in theory, begin at some later stage to make use of those genes again; and indeed there are many examples in nature of genes that previously served some function that has become redundant, but are pressed into service later. The inner ear-bones of mammals are clearly derived from bones that once helped to articulate the jaws of ancient fish. Mammals simplified the jaw, which left spare bones, which the ear commandeered. This is another way of saying that the genes that had become redundant as jaw-bone genes were granted a new function as ear-bone genes. It can pay, in short, to keep old genes in store.

The late nineteenth-century German scientist and philosopher Ernst Heinrich Haeckel was greatly impressed by the fact that mammalian embryos had gill slits. It struck him, indeed, that embryos of different vertebrate types are all fairly similar at first – early human embryos are not too different from those of lizards or fish – but as they grow older, they get more and more dissimilar. He also perceived that young animals of different species tend to resemble each other more than the adults do: so that baby humans and baby gorillas have much in common, whereas adult humans are very different from adult gorillas, particularly adult male gorillas. These observations led him to formulate his famous adage: 'Ontogeny recapitulates phylogeny.' Ontogeny refers to the whole process of growth and development; phylogeny refers to ancestry. In other words, as an animal develops, from embryo through child to adult, it re-enacts its own ancestry. A human is first like a fish, then like a generalised tetrapod, then like a generalised primate, and only finally like a human.

Nowadays, with the knowledge of gene expression, we can see that Haeckel's adage, though interesting, is not literally true. To be sure, young creatures tend to resemble each other more than older creatures, and very young creatures can produce ancestral features. But as a *rule* of biology, Haeckel's notion does not stand up. If ontogeny does seem sometimes to recapitulate phylogeny, that is simply the way things look. There is no particular mechanism to ensure that this should happen in any orderly fashion.

Yet Haeckel came very close, in his grand general theory, to recognising a phenomenon that is of enormous importance: *neoteny*. One animal may evolve from another by altering particular features, in the general way that Darwin envisaged. But it can also evolve in theory simply by altering the timing of the appearance of features. In particular, an animal may retain the features of the child into adulthood, or it may (which is the mirror-image of this) bring forward sexual potency, so that the young form becomes sexually adult. This happens. There is an enormous range of examples in nature of animals that seem to have evolved in precisely this way. This process – effectively turning a child directly into an adult – is neoteny.*

Perhaps the most famous and classic example is the axolotl, a kind of salamander that now lives in just one lake in Mexico. Amphibians in general begin life as aquatic larvae, complete with gills, and then metamorphose into adults: salamanders, newts, frogs, toads, or legless caecilians. Axolotls do not metamorphose. They remain larval in shape and way of life – complete with big, feathery 'external' gills – even after they are sexually mature. This failure to metamorphose has been ascribed to a lack of iodine in the water, but this seems simply to be untrue. More

***Neoteny* is the acquisition of sexual maturity by juveniles. *Paedomorphosis* is the retention of juvenile features in the adult. In practice, it can be difficult to say which is which, and 'neoteny' serves as a general term for all such phenomena.

likely it is an ecological adaptation. Life in Mexico is often easier in the water than out of it, for the land is often desert. So why not stay in the water? Whatever the reason, this is neoteny writ large. Axolotls presumably retain the genes that would enable them to assume the characteristic adult form of salamanders, but they do not switch them on. Some close relatives of the axolotl *do* metamorphose, and other axolotl relatives metamorphose sometimes, but not always.

Yet neoteny may be a much more general phenomenon than this. It seems that the ancestors of the insects may have had a great many body segments, each with a pair of walking legs, somewhat like modern-day centipedes. These ancestors (like centipedes) probably acquired more segments, and pairs of legs, as they grew older. Yet to have only three pairs of legs is very efficient. It seems likely that insects evolved from centipede-like ancestors by neoteny. They became sexually mature after they had acquired three pairs of legs.

Many present-day insects, of course, including flies, bees, beetles and butterflies, have taken the phenomenon of metamorphosis to extreme limits. Or, rather, they have divided their genetic program into very distinct phases: larva (the growing stage, which is sometimes but not necessarily legless); pupa (the quiescent, often over-wintering, transitional stage); and adult (the reproductive stage, in which insects are usually winged). Note, then, that natural selection *can act on each of these phases separately*. Thus we may find wonderfully elaborate larvae giving rise to fairly drab adults, and *vice versa*; and very similar larvae giving rise to very different adults. More to the point, we find that the larvae and the adults may have completely different lifestyles, so that one may be aquatic and the other aerial; one carnivorous and the other vegetarian; and so on. Here, in effect, we have three creatures in one, with three different (but linked!) genetic programs. There is scope, here, for insect larvae to become neotenous in their turn, and thus give rise to a whole new class of creature. But I know of no examples where this has happened.

Human beings exploit the phenomenon of neoteny in their attempts to 'improve' domestic animals by breeding. Haeckel was right to observe that baby animals resemble each other. Childish facial features that seem common to all mammals include a shortness of muzzle, a high forehead, and large eyes.

Physiologists can explain this in mechanistic terms. Mammalian eye development stops early in life, so the eye is relatively larger in the young animal. The volume of an animal increases according to the cube of its linear dimensions, so a big animal needs to eat more, relative to its linear dimensions, than a small animal does. Hence big animals need jaws that are not only larger than those of small animals, but are also *relatively* larger. Psychologists point out in addition, however, that the childish face has also come to serve a vital social function. Mammalian adults have evolved gentle and parental responses to childish features. We do not usually attack children when they cry, which is as well for them and as well for us, for they carry our genes. Instead, their big baby eyes move us to offer them assistance.

So it is that breeders have produced neotenous dogs that retain the short muzzle and high forehead of puppyhood throughout life and effectively serve as child substitutes. 'Toy' dogs in general, like poms and Skyes, are supreme examples. They do not simply resemble puppies. They resemble human babies in a general way, and evoke the same maternal sympathies.

Behaviour is similarly influenced by age. Wolves are the ancestors of dogs, and wolf puppies are very dependent on the adults. They are sociable and constantly seek leadership. Breeders of all kinds of dogs – even macho gun-dogs – have favoured retention of such qualities. Domestic dogs should feel dependent on their owners. Adult wolves may be friendly towards humans but they are not 'tamable'. That is, their psychological development passes beyond the dependent puppy stage, and they become true adults, competing for pack leadership, and recognising the authority only of the true pack leader. This theme is explored further in Stephen Budiansky's *The Covenant of the Wild* (William Morrow, New York, 1992).

Human beings, too, may have evolved neotenously, from chimp-like ancestors. Baby humans are indeed like baby chimps, with high foreheads and short jaws. But adult humans are also somewhat like baby humans; at least, they are much more like baby humans than they are like adult chimps. We never acquire the jutting jaw and brows of the ape. Yet presumably we have not lost the genes that give the chimp its jaw and brows; indeed, as discussed later, the genome of the human is uncannily similar to that of chimps. We have simply suppressed the genes of jaw and brow. Or, we might say, we have postponed them.

This brings us back to Aldous Huxley and *After Many a Summer*. If we lived long enough, would we in fact begin to manifest the genes of ape-dom, which we may still carry, and perhaps have simply postponed? In practice, presumably not. Huxley's scenario would work only if we had inherited the genes that cause apes to mature *as a complete program*, complete with its own controls, so that, once given time to come on line, this program would indeed add another phase to human development. We should rather envisage – following Medawar – that in old age, genetic programs are jumbled. The point, after all, is simply that the expression of deleterious or unnecessary genes is postponed, but there is no suggestion that the postponement is orderly. To invoke for once the metaphor of genes-as-blueprints, it is as if the blueprints are simply thrown higgledy-piggledy into the hold, perhaps with a label around them, 'Not wanted on voyage'. If we just went on getting older and older, we would not transmogrify into some new creature, like the Fifth Earl. Rather, like Tithonus, we would simply crumble away, as we worked our way through the increasingly disordered rejects of our genome.

Finally, just to hammer the point: although we may retain the genes of ape-dom, which could give us big jaws or hairy arms, there is no good reason to suppose that these genes could *ever* be expressed, no matter how long we lived. After all, we see no obviously simian stigmata in centenarians. The genes of ape-dom may not simply have been postponed. They may have been shut off irrecoverably.

Thus, Huxley's novel remains a fantasy, but a pleasing and salutary one, none the less. It is instructive as a morality tale; it reflects (albeit with distortions) the evolutionary biology of its day; and it anticipates, by more than twenty years, the first explanations of gene expression.

This, then, is how genes operate within individuals. In the next chapter we will see how they compete within gene pools; and how, in particular, they affect behaviour.

CHAPTER 4

The Games Animals Play

Science is so vast and deep that it must be broken down into many different specialities – but all the time the fragments are rushing together like imploding planets, to form a quite new view of life: what may properly be called a new *paradigm*. The fragmentation phase, though necessary, has a dour and dogged quality. It creates specialists who, as they say, know more and more about less and less. But when the disciplines coalesce there is true excitement: everything is re-illuminated; everything is suddenly seen in the light of everything else; long-since divorced ideas are juxtaposed with all the delight of long lost friends.

The coalescence of ideas in biology during the past few decades has been cataclysmic. We have seen nothing less than a grand coalition of ecology, evolutionary biology, genetics and behavioural science. For good measure, this consortium has disinterred an idea that most philosophers thought they had long since buried: the notion that biology has interesting contributions to make to moral philosophy. If this coalition has a name, it can reasonably be called *sociobiology*: the term first coined by the Harvard biologist E.O. ('Ed') Wilson in the 1970s to describe inquiries into the evolution – and hence into the root 'causes' – of social behaviour.* The sociobiological approach may indeed be seen as a new paradigm: extending and enriching the neodarwinian paradigm of the mid-twentieth century. The word 'Science' is fine and resonant. But the new biology shows once again that we have lost a great deal by abandoning the ancient expression 'Natural Philosophy'.

The new sociobiological synthesis rests upon three essentially separate ideas: that particular genes may underpin particular aspects of behaviour; that natural selection operates at the 'level' of the gene; and the notion that the 'life strategies' of organisms can usefully be analysed mathematically, and in particular by game theory. These ideas are indeed different, but it is in each other's company that the full power and significance of each can be best

*E.O. Wilson, *Sociobiology, the New Synthesis*, Belknap Press, Harvard, 1975.

appreciated. The biology of the 1990s is still firmly rooted in the synthesis of Darwin's ideas with Mendel's, as achieved a few decades ago. Yet the modern 'sociobiological' approach, and even the modes of speech of the biologists, are very different now from that of the mid-twentieth century.

We should look at each of the three fundamental notions in turn.

THE FIRST IDEA: GENES FOR BEHAVIOUR

The first of the ideas – that an animal's behaviour is shaped largely by heredity – is far from new, in fact it dates back at least to Darwin. Even today, however, there are many who object to it. Some, understandably but in the end unimportantly, are troubled because they find the idea implausible. Others, for reasons that have worthy roots but are taken to foolish extremes, find its implications distasteful.

As for the first objection – well: it is indeed weird and wonderful that genes shape behaviour because genes just make proteins, and how can proteins tell us what to do? Yet there is a vast mass of evidence, both commonsensical and rigorously experimental, to show that they can and do. At an elementary level we might simply observe that general behaviour is obviously affected by hormones, and hormones are either proteins and so are direct gene products, or else they are steroids, which are shaped by enzymes. Thus, the attitude of most female animals towards males is profoundly altered in those brief periods when they are on heat, a period accompanied by a surge in steroid sex hormones. Male elephants in 'musth' seem to have the strength of several elephants, and become extremely moody and dangerous, and this mood-change is correlated with, and largely caused by, surges in testosterone. In all animals, adrenaline induces 'fight or flight'. The daily – 'circadian' – rhythms of vertebrates are effectively determined by pulses of the proteinous hormone melatonin, from the pineal gland; and so, too, are the yearly cycles of appetite in high-latitude animals such as red deer. Doses of melatonin help human beings to cure jet-lag, and also to avoid some forms of winter depression. So yes, of course: gene products can be seen to have a profound effect upon general behaviour.

More difficult to accept is that genes should affect the *details* of an animal's – or a person's – behaviour. Thus we may ask, how can gene products – proteins – tell a young carnivore how to hunt (which many of them do successfully even *without* parental training)? How can these products tell birds how to build nests – in detail! – and how to raise their young? Again, early learning and practice are now known to play a far greater part in shaping these skills than was appreciated in the past, but again, the 'instinctive' – which in practice means 'genetic' – basis of such behaviours cannot be denied. Birds brought up by foster parents of other species do not learn how to build their foster parents' style of nest. Instead, they are likely simply to become confused. The mating and other rituals of hybrid birds are often a hopeless mish-mash of behavioural tics, half inherited from one parent, and half from the other, and inadequately cobbled together.

But at least these days we have a 'model' to show us how such behavioural patterns *could* be built into the animal's nervous system. The behaviour of computers is controlled by a program, and the program has a physical structure, which is held on a disk. Presumably our brains develop – or are born with – arrays of proteins that are at least analogous to the iron oxide crystals on the disk surface. We do not know what those hypothetical brain proteins are, or how they are arranged, but at least we can now see how the 'instinctive' behaviour of animals (or people!) *might* be built into the system. However, I am emphatically *not* arguing that animals' brains in general operate in the same way as computers, or in a general way are like computers. Indeed, the more that psychologists study the behaviour of animals such as dogs or chimps – or humans – the more they become aware of the contrasts. I want only to suggest that this particular aspect of computers – the ability to store instructions for behaviour in a simple physical form – shows what *can* be achieved, by relatively simple means. Finally, on a quite different philosophical tack, we should not reject the notion that genes influence behaviour just because we cannot explain the mechanism in detail, any more than we should reject the concept of gravity, which remains elusive after 300 years.

In short, we cannot now reasonably deny that genes shape behaviour both in general and in specific ways, and it is reasonable to posit that *particular* genes shape *particular* aspects of behaviour. The implications of this, as we shall see, are truly wonderful. They have spawned the new science, which Ed Wilson called sociobiology. Indeed, they have transformed the whole of modern biology.

But here we come to the second objection. Many people find those implications distasteful, and indeed threatening: so much so, that sociobiologists have at times been abused, banished from campuses, and even physically attacked.

Behind much of this antipathy lies a misunderstanding. Sociobiologists posit that genes influence behaviour, and indeed that individual genes are responsible for individual facets of behaviour, and then explore the ramifications of these hypotheses. But many critics have misconstrued the basic claims. They fear that 'influence' means 'determine', and hate the notion that the behaviour of human beings should be prescribed by their genes from conception to demise. Such 'genetic determinism' threatens the notion of free will – and what makes human beings human, if not their freedom to choose how to act? By the same token, it seems to take away personal responsibility, for if a man commits rape – well, he might argue that his genes told him to do it. If women are tied for life to the kitchen sink – well, that is their genetic destiny.

Then again, if our lives are shaped entirely by our genes, then we must be unchangeable, and incapable of 'improvement'. This notion harks back to the nineteenth century, when a highly respected profession of highly paid charlatans known as phrenologists solemnly passed judgement upon their subjects' personality and brain-power from the shape of their heads, and declared that whereas some were creative geniuses 'by nature', others belonged to the 'criminal type', a destiny from which they could not escape. Men and women were hastened to the gallows by such nonsense. The same kind of thinking has reinforced racism, in which people of other physical types are held to be less intelligent, more rapacious, and less trustworthy than the person making the assessment. Here there are two kinds of genetic assumption: that the genes that indicate race (e.g. those for skin colour) are linked to genes that influence behaviour or brain-power; and that behaviour and brain-power are in fact 'determined', and not merely influenced, by the underlying genes.

Since arguments based on 'genetic determinism' are still employed in intellectual backwaters of the western world to condone racism, biologists have been right to point out its shortcomings. But they are simply wrong – or mischievous – to suggest that competent sociobiologists are 'genetic determinists'; and by railing against all and sundry, the critics have committed numerous libels (which probably does not matter) and muddied the waters of an exciting area of intellectual endeavour (which does). In practice, the best way to deal with 'genetic determinism' is simply to expose it for the nonsense it is. There are two main lines of argument.

First, as already suggested, genes can influence behaviour (and thought processes) to very varying degrees. At one extreme, they can simply influence general mood – perhaps through the intermediary of some hormone. At the other extreme, genes can actually provide a detailed program of behaviour, like a computer program. Commonsense, everyday observation suggests that different animals differ greatly in the *extent* to which their behaviour is programmed, as opposed merely to being influenced. In general, animals with simple nervous systems are more heavily programmed than more complex creatures. Different worms of the same species tend to behave in similar ways in similar circumstances, and the same worm behaves in much the same ways at different times. The behaviour of mammals is in general far more flexible. Their behaviour is said to be 'contingent' on all kinds of influences, some of which are obvious, but many of which are not. The behaviour of a chimpanzee at any one time is influenced by its immediate surroundings (whether for example there is noise or quietude); by immediate past experiences (whether the animal has just had a fight, or has been carefully groomed by its favourite female); and by more distant past experiences (including those of childhood). In bright animals, too – including all primates – the behaviour of individuals is influenced by its 'culture': by the collective past experiences of the particular society to which it belongs. Thus, a chimp's own childhood experiences are influenced strongly by the personality and behaviour of its mother, and her personality and behaviour in turn were partly shaped by her own lifetime's experiences, and so on and so on, back to the beginning of chimp-dom.

Thus we can properly say that the behaviour of the chimp is certainly *influenced* by its genes. Genes are responsible for laying down the ground-plan of its brain, and its general responses are characteristically chimp-like, and in many ways quite different from those of sloths or anteaters. Clearly, too, a chimp's responses are *limited* by its genes. Insult a chimp and it may respond in various ways, but you can be sure it will not issue a libel writ. But – the second part of the argument – 'influence' and 'limit' do not mean 'determine'. Modern studies now show that the very anatomy of the brain – at least at the micro-level – is influenced by early experiences; at an elementary level, for example, the way that the neurones join up in the visual cortex of a young kitten is influenced by their early exposure to light. The mental development of children – including their performance in IQ tests – is obviously influenced by upbringing, with emotions affecting the issue as much as mental stimulus. Who denies this? Where is the 'determinism'?

To look at things the other way, it is technically possible now to analyse the chimp's genome atom by atom. In a few centuries' time we might understand in minute detail what every stretch of chimp DNA actually does, and how it all interacts. Yet we will *never* be able to predict exactly how a particular chimp will respond to a particular set of circumstances. That will depend on a whole number of things that go on in the chimp's head – and in the heads of its companions and its ancestors – which have nothing to do with its genes, except insofar as its genes were involved in constructing its head in the first place. To be sure, the behavioural psychologists of thirty years ago were effectively suggesting that the behaviour of chimps – and dogs and pigeons – could be 'predicted'. But their claims were based on experiments in which the 'stimuli' were extremely simplified (bells, lights, coloured discs) and the animals were bored out of their skulls, often very frightened, and anxious above all to avoid punishment. If an animal is given no room for manoeuvre then its behaviour is predictable by definition.

Finally, all the comments that apply to the chimpanzee obviously apply to human beings several times over. Our awareness is extraordinary; our ability to adjust to those around us, and to learn from previous generations, is obviously unparalleled. The notion that we are none the less limited by our genes should not be threatening. It certainly does not bother me that I cannot do arithmetic as well as a £5.00 pocket calculator – indeed that nobody can – although such calculation is a simple trick which we could all of us do if our brains were wired up differently. Neither should we be threatened by the general notion that we are equipped with 'programs' that we can call upon. Anthropologists have long been struck by the similarity in basic gestures, smiles and frowns and general body language, between different peoples who have never had contact with each other. The idea that these might be genetically derived should excite feelings of human unity, rather than of threat. Neither the notion that we are limited by our genes, nor the idea that we have a stock of behavioural programs, amounts to anything like 'genetic determinism', or to any degree undermines the supremacy of culture or of individual choice. It is evil to use the idea of 'genetic determinism' as a weapon, and an agent of social control, but it is paranoid to presume that all reference to the (possible) genetic roots of behaviour amounts to a defence of genetic determinism, and still less to advocacy of such an absurdity.

But I set out only to establish that genes do influence behaviour in general, and that particular genes do program particular patterns of behaviour. Just how important this notion is can be realised when we consider the next idea in the late twentieth-century paradigm: the notion that natural selection acts not on individuals, or on groups, but at the 'level' of the gene.

NATURAL SELECTION AND THE SOLIPSIST GENE

The second of our three component ideas grew up in the 1960s – when several quite opposite ideas were in vogue! – and has been beautifully described by the Oxford zoologist Richard Dawkins in *The Selfish Gene* (Oxford University Press, Oxford, 1976). Dawkins himself has contributed much to the theory.

What does it mean to say that natural selection operates at the 'level of the gene'? Darwin envisaged that natural selection operated upon *individuals*. For example (though this is my example, not Darwin's), through much of Asia, until recent centuries, lions and tigers lived side by side (though now only a few lions are left in Asia, in the Gir Forest of north-west India). Since they both preyed upon large herbivores, the two cats were to some extent in competition. But the point is not that lions as a whole competed with tigers as a whole. In practice, *individual* lions competed with *individual* tigers – and, equally to the point, individual lions competed with *other* individual lions. This remains true (as will become clearer later in this chapter) even though lions habitually hunt co-operatively. Lions might hunt together, but (as Darwin would have said) this is for their own individual convenience. Nature remains red in tooth and claw.

As often, however, Darwin was ill-served by some of his followers. To be sure, his ideas benefited from their twentieth-century coalition with Mendelism, to produce neodarwinism. But the neodarwinian paradigm has at times led people astray. As we saw in Chapter 1, neodarwinism invokes ideas of gene pools and populations. An individual at a given time may out-strip its fellows, but this temporary advantage is not consolidated as an evolutionary change until the upstart's genes have spread through the population and – to press home the point – until other genes have been pushed out as a result. Only then has there truly been a qualitative and effectively irreversible change in the gene pool as a whole.

Probably because they had the idea of populations thus fixed in their minds, many biologists in the 1960s in particular began to invoke various notions of 'group selection'. The idea here was that nature was not quite as red in tooth and claw as Tennyson suggested. Instead, animals could often be seen to sacrifice their own interests – their right to breed or even their own lives – 'for the good of the group', or even 'for the good of the species', which in either case might be formulated into 'the good of the gene pool'.

The Nobel prize-winner Konrad Lorenz argued this case in *On Aggression* (Methuen, London, 1966). Consider, for example – although again this is my example, rather than Lorenz's – two stags fighting for a hind. They do not generally fight to the death, and if possible, they avoid fighting at all. They roar and flaunt their antlers, and the one who feels he has the feebler antlers and roars less vigorously backs down.* If they do engage in combat, one will soon give up; but the winner

*Tim Clutton-Brock and Steve Albon showed that the strongest stags are the ones that roar at the highest rate – for roaring requires strength! Others acknowledge this strength, and know when to back off. (*Behaviour*, vol. 68, pp. 136–69. Quoted by John Krebs in *The Tinbergen Legacy*, p. 68.)

does not generally go in for the kill. Once the loser submits, the game's over.

All this seems very difficult to explain in terms of individual advantage. Lose though he might, it is surely in the weaker stag's interests to battle for a mate, for otherwise he seems to have no opportunity to mate at all. Surely, too, it is not in the winner's interests to show mercy. If he kills his rival, he quashes that particular threat for ever more. The only explanation for this and other apparent examples of gentlemanly behaviour, Lorenz argued, was that natural selection acted to favour the group as a whole. After all, if stags did fight to the death at every opportunity, then the entire breeding season would be mayhem, the winner would probably be too beaten up to mate in any case, the wolves would have a field day feeding on wounded deer, and the entire group would be wiped out. Natural selection surely would favour behaviour that stopped short of mass self-destruction.

In similar vein, V.C. Wynne-Edwards, who wrote the extremely influential *Animal Dispersion in relation to Social Behaviour* in 1962 (Oliver and Boyd, Edinburgh) gave many apparently very convincing examples of group selection. For example, birds (and other animals) sometimes produce fewer offspring when the population is high than when it is low. Wynne-Edwards argued that this served to reduce the competition within the group as a whole, and thus benefited the group as a whole. Individuals might suffer, by curtailing their own reproduction, but this, again, was better than mass suicide. Robert Ardrey, in *The Social Contract* (Collins, London, 1970), argued that humans alone had chosen to override the gentlemanly codes of their animal ancestors, which made them beastlier than the beasts.

From such examples we can easily see that group selectionist ideas do have a subtle and commonsensical feel, and are also emotionally appealing, which is a powerful combination. They seemed indeed to advance upon Darwin's notion that selection acts upon individuals, in the same kind of way that the neodarwinian paradigm advanced the general ideas of *Origin of Species*.

Yet ideas that were first developed through the 1960s and 70s showed that the subtle and appealing group selectionists were profoundly wrong, and that Darwin, though not quite right, was actually nearer to the truth. The new idea is that natural selection acts not upon the group, nor upon the individual, but upon the irreducible unit, the gene itself. To be sure, genes make themselves felt by their effect upon phenotypes: that is, by their effect on the individuals who possess them. But that is not the point. The point is, if a new mutant allele arises that confers some advantage on the individual, and that individual has offspring, then it is the *gene* that is multiplied, and spreads through subsequent generations.*

How, then, can we explain the behaviour of the stag in 'selfish gene' terms? Why doesn't the dominant stag press home his advantage in a fight? Why doesn't he kill his opponent, and remove the threat once and for all? Why is the weaker animal so often content to accept the stronger one's bluff, and back down without a fight? Surely he has no chance at all of mating *unless* he fights for mates, so he should at least 'have a go'? Surely there can be no possible explanation either for the mercy of the strong or for the prudence of the weak, except in terms of group selection. The group benefits from their common sense, but they themselves – and the allegedly selfish genes they contain – seem to miss out.

Point one is that the original observation is flawed. When you look closely at animal conflicts, you see that the winner is in fact quite indifferent to the fate of the loser. If the loser dies – well, that's a bonus. And in practice, losers often die. Old bulls die from battles with young bulls; old dogs with young dogs; old lions with young lions. If there is any restraint on the winner's part, it is entirely for his own benefit. All fights involve risk. It takes energy to deliver the *coup de grâce*. If the losing stag runs away (as he is generally anxious to do, once the game is up) the winner cannot afford to waste time and energy giving chase. By the time he has done so, some other stag will have seduced his hinds. So all animals are *minimalists* when it comes to a fight, but only for their own benefit. They do not fight for exercise. They cannot afford to do more than is necessary.

By the same token, the weakling who backs off rather than fight is in practice maximising his chances of reproducing. He lives to fight another day. Discretion is the better part of valour. He can use the coming winter to rest up, and the following spring

*As Robert Trivers of Harvard University comments in the Foreword to *The Selfish Gene*, the pioneers of the new way of thinking were R.A. Fisher – dating from the 1920s and 30s, and then, from the 1960s and early 1970s, William Hamilton, George Williams and John Maynard Smith. Trivers himself should also be numbered as a pioneer, and Dawkins is certainly among those who followed hard on the pioneering heels.

to build up his muscle, while the stag that now is dominant will squander his winter's energy and spend the next spring trying to recover. Besides, stags do not secure the services of their hinds quite as efficiently as they might care to believe. While the stag's back is turned, the young bloods are liable to sneak in and mate: *sneak mating* is now indeed a recognised technical term. The hind benefits from this. She has the benefit of sperm from a young fit male who at least has the courage and wit to do the sneaking.

In practice, of course, this explanation of the deer's behaviour seemed simply to revert to Darwin's point – that natural selection works on the level of the *individual*, rather than the group. There is nothing in the above to suggest it acting on the *gene*. But in fact a modern sociobiologist would go further than I have done. He would argue that there are genes which underpin aggressiveness, and underpin degrees of caution. Genes that resulted in a sensible degree of aggression would spread through the gene pool, because their possessors would leave more offspring, whereas genes that underpinned out-and-out aggression, or unqualified cowardice, would die out. To be sure, it is the possessor of the gene who does the fighting, or the running away, not the gene itself, and it is the possessor who breeds, or dies without issue. But as Dawkins says, from an evolutionary standpoint, individuals should be envisaged not as stags, or ants, or human beings, but simply as the vehicles of genes.

Yet still you may argue that although we can provide alternative explanations of the deer's behaviour, we still have not shown that the alternatives are better than the group selection idea. Still less, from what I have said above, can we presume that the 'gene level' idea of natural selection is better than the 'individual level' idea. That is true. My above example does not show that individual selection improves on group selection and that gene selection improves on individual selection. However, when you apply game theory (as we will discuss shortly) you can see that although selection might in theory act at just about any level, 'higher' level selection (e.g. at the level of the group) is bound to be slower and therefore less efficient than selection at the level of the gene. A gene that conferred a short-term advantage to the individual that possessed it would spread, even if it brought long-term disadvantage to the group as a whole. In short, group selection probably does operate, but too slowly and therefore too feebly to make any difference.

Second, different field studies of different animals show different things. Sometimes group selectionist ideas seem appropriate, but sometimes they do not. Gene-selection explanations *always* work. Scientists – like every other kind of philosopher – have in the end simply to decide what kind of ideas they feel are most worthwhile. In general, it is in the nature of science to prefer ideas that apply universally, rather than those that apply only sometimes, or require endless *ad hoc* tinkering.

We might point out, indeed, that naturalists over the years have tended to see in nature what they wanted to see. In a perhaps more romantic age they liked to believe that stags were noble, just as Edwin Landseer portrayed them. Thousands of hours of carefully annotated observations by twentieth-century biologists have bred a higher degree of scepticism. A herpetologist recently told me of a fight he witnessed between two Madagascan tortoises. One left the other lying on his back, to die of sunstroke. But after the winner had gone a few yards down the track he turned back, and put his rival on his feet again. Old-fashioned naturalists would have seen nobility in this. The modern, hard-headed herpetologist opined rather that the victorious tortoise's blood was up, and he had turned back to continue the battle. Tortoises are extremely bellicose, but they are not bright.

To return to our thread: Wynne-Edwards's example of the birds that regulate their family size can also be explained very simply in terms of gene advantage. Thus birds have to expend a huge amount of time and energy in raising their young, and they take enormous risks in doing so. If they try to raise more than the surroundings can sustain, they flog themselves to death and the youngsters languish anyway. If they raise fewer than they could, then other birds out-breed them, and so replicate their genes more effectively. So many birds in practice effectively calculate how many youngsters they can hope to raise in a given year. If they perceive a lot of competition, they lay fewer eggs. Better to raise a few safely than waste energy and risk the lot. If there is plenty of space, they take advantage and lay more. The mechanism that underpins this behaviour may be very simple: it may simply be that in crowded conditions, birds feel stressed, or find that food is hard to come by, and either of these could suppress reproduction. But mechanism is not the issue, the point is simply that many birds do indeed seem to gear their breeding to suit the social circumstances. But the decision to exercise constraint is self-centred, nevertheless. More to the point, it is *gene* centred; it promulgates the gene that led to the behaviour in the first place.

It transpires, indeed, that sneak mating, cuckoldry, and good old-fashioned adultery are extremely common in nature. In fact they are probably the rule, rather than the exception. Hedge sparrows are a classic example. At first they seem models of nuclear family propriety, with the cock helping the hen to feed the nest of chicks. Closer inspection reveals that in any one group, some males are polygynous (many wives), some females are polyandrous (many husbands) and some nests are full of chicks that in fact have several fathers. Both the males and the females do reasonably well out of this, because each attaches its own genes to genes from several different partners, which must maximise their chances of survival. Yet each group of chicks is none the less looked after by two devoted parents, which maximises the chicks' chances too. It is a somewhat extreme strategy, and would surely break down if too many individuals finished up looking after too many of their rivals' children. It could of course be regarded as good socialism: a kind of avian kibbutz. But at the level of the gene, nature is as red in tooth and claw as ever.

One last aside. Richard Dawkins spoke of 'selfish' genes. But the word 'solipsist' would be more accurate. After all, selfish people on the whole are perfectly aware that other human beings exist, but they just don't give a damn. Solipsists are sublimely unaware that there is anyone at all in the world except themselves. Genes are not aware even of their own existence, but they are nearer to being 'solipsist' than 'selfish'. But then, metaphors are meant only to be heuristic (to make ideas easier to understand). Besides, *The Solipsist Gene* would not have sold many copies.

The final beauty of the solipsist gene hypothesis is that it lends itself so neatly to the third great innovation of modern biology: the shift, to centre stage, of mathematics.

MATHS

In science, maths is part of the thinking process. It is the arbiter. Maths has been in biology at least since the nineteenth century; though Darwin was not a mathematician, he felt the lack of it, and greatly admired the pioneering statistical studies of his cousin Francis Galton. Pioneers like Sir Ronald Fisher and J.B.S. Haldane in the middle years of the twentieth century brought various kinds of maths – not only statistics – closer and closer to the centre stage. Crucially, in the middle decades of this century, John von Neumann developed game theory, of which more later. Now there are more and more branches of biology that proceed as physics does, with a dialogue between the field observers and the maths-based theoreticians. Outstanding among the new mathematical biologists are William Hamilton and John Maynard Smith.

These last two have applied rigorous maths to the idea that particular genes underpin particular aspects of behaviour, and the notion that natural selection acts at the level of the gene, and thus have produced two of the most startling and powerful insights of late twentieth-century biology. The first, due to Hamilton, is that 'selfish' genes, mindlessly pursuing their own ends, concerned only with their own survival and replication, do not necessarily produce selfish individuals: that indeed, animals for reasons purely of survival can and do behave with *altruism*. The second, due to Maynard Smith, is the notion of *evolutionarily stable strategy*: the notion that natural selection will inevitably produce societies in which more than one behavioural strategy will operate at any one time, a kind of behavioural polymorphism.

ALTRUISM

It is the role of genes to replicate themselves. Replication is what they do. It is their thing. They do it for the reason that Aristotle would have approved of: the talent for replication is built into them. They do it blindly and selfishly, but those that do it best spread throughout the gene pool (or indeed throughout the individual genome through the mechanisms of molecular drive!) and those that do it less well are out-competed by those that do it better. This is a simple idea but an astonishing amount follows from it.

Incidentally, it is of course literally wrong to say that genes *want* to replicate because they do not literally *want* anything. But biologists these days tend to say that genes 'want' to do things (or that the creatures that contain those genes 'want' to do things, whether woodlice or grass) as a metaphor and a piece of shorthand.* It is after all the case that if genes did literally 'want' to multiply they could not arrange things better. I am not wont to buck trends and will use the 'want' metaphor.

In practice, a gene that seeks to replicate and so to spread has two courses open to it. First, it can encourage the organism that contains it to reproduce – or at least enhance that organism's chances of doing so – and second, it can help to ensure the survival of other organisms in the population which also contain copies of itself.** Sir Ronald Fisher and J.B.S. Haldane were the first to draw attention to the latter of these two options: that, for example, an animal could benefit its own genes by saving the lives of its siblings, or even cousins, who shared those genes. But it was William Hamilton in the 1960s who turned those comments into a rigorous theory which, as Dawkins proclaims, has proved to be one of the most significant conceptual advances of the late twentieth century.

Hamilton's theory begins simply, and in essence is simple, although the underpinning maths is not. It is obvious, first of all, that each of us contains a selection of our parents' genes: in fact, you (say) contain (copies of) 50 per cent of the genes contained within each of your parents. By the same token, each

*It was not always so. When I first started studying biology, behaviourism reigned: the essentially positivist idea that the psychology of animals could be studied only by the things that they could be directly observed to do, and not by supposing that they did anything off-stage, such as 'thinking'. The behaviourists developed a purity of expression that became a neurosis and must have repelled entire cohorts of potential biologists, who fled to 'the arts' instead. I remember one biology teacher – not one of mine, I am delighted to say! – solemnly averring that it was wrong to say that a cuckoo lays eggs. What the cuckoo does, she said, is 'to exhibit egg-depositing behaviour'. But biologists nowadays tend to say that even ferns or mushrooms 'want' to do things. This is only a metaphor, let me emphasise again, but a very refreshing one!

**There is of course a third way for genes – being selfish – to replicate, which is simply to multiply within the genome, by molecular drive. Within-genome multiplication has serious consequences in evolution, as we discussed briefly in Chapter 1. But it has no obvious social consequences, and does not therefore belong in this chapter.

of your children contains 50 per cent of your genes. A different way to put this – and conceptually much more useful – is to say that *each* gene in *each* of your parents has a 50 per cent chance of being in you as well; and each of your genes has a 50 per cent chance of being in each of your own children.

It takes time and effort to look after offspring, and it can involve considerable risk. As biologists say, it requires *investment*. Yet parental care to some degree is usual in nature. Even those animals that apparently abandon their young take very good care to ensure that the eggs are laid in safe places, surrounded by good food. To be sure, codfish seem a little cavalier, abandoning millions of eggs to the plankton. But butterflies, for example, though they often die before their offspring hatch, assiduously stick each egg in some sheltered spot on a favourite food plant. Mammals, birds – some fish, some insects, spiders – are prepared to give their all to look after their offspring, and often do.

Indeed, parental care seems such an obvious fact of nature, that we take it for granted. If you were asked why you look after your own children you would be mildly affronted. The first answer, perhaps, would have to do with love, and duty. But if pushed to think about the matter biologically, you would probably – very reasonably – wind up with a tautology, a circular argument. Animals that do not take some steps to get their offspring off to a good start are likely to leave no offspring at all. Therefore the non-caring lineages die out. Therefore those that survive must be carers. QED.

But the notion of gene selection improves on this tautology. We simply say that any gene for parental care has a 50 per cent chance of existing in the offspring, so it pays to look after offspring. In practice, this argument does not work unless the parents look after at least two offspring, but then, parents generally do produce two or more (if the lineage is to survive) even if they have them only one at a time.

But this 'selfish gene' explanation of parental care has much wider implications. To be sure, there is a 50 per cent chance that any particular gene in your genome is also contained in one of your children, and that is the 'biological' reason why you look after them. *But there is also a 50 per cent chance that any one of your genes is contained in your brother, or in your sister.* You devote an enormous amount of your time and effort to the upbringing of your own children. But it makes just as much biological sense to devote as much time to the well-being of your brothers and sisters. 'Am I my brother's keeper?' asked Cain (Genesis, 4:9). 'Yes', is the emphatic, sociobiological answer.

We can take this further. Your cousins also have genes in common with you. Your cousins do not have the same parents, but they have the same grandparents. The maths easily shows that each one of your genes has a 1 in 8 chance of being in your cousin as well, and a 1 in 32 chance of being in a second cousin. But by the time you get to third cousins, the relationship looks pretty thin. Each of your genes has only a 1 in 128 chance of being in a third cousin. Genetically, third cousins are not much closer than random members of the population.*

Fisher and Haldane first observed in a general way that it makes biological sense to invest in the welfare of siblings and at least for a cart-load of cousins. But Hamilton turned that observation into a rigorous and universal theory. He showed that there was advantage in looking after kin, and he quantified the advantage: such that two siblings = 8 cousins, or 8 great-grand-children. From this, he coined the expression that ever since has resounded through biology: *kin selection.* As the icing on the cake, startlingly and brilliantly, he showed that the care of parents for their children, which all of us perceive to be so 'natural' that it is hardly worth commenting upon, *is nothing more than a special case of kin selection.*

*A short diversion. I have a very good friend who is seeking to track down his own ancestors. There are people, indeed, who do genealogy for a living. The results are extremely interesting. My friend has some weird and wonderful ancestors including, it seems, at least one mediaeval king, and a miscellany of drunks. I am inclined to say, though, every now and again, 'Who doesn't?' Each of us has two parents, four grandparents, and so on. By the time we get back to the Middle Ages, our direct ancestors probably outnumber the population of the country in which we live, and by the time we get back to the time of Christ, it seems we have as many ancestors as there were people on Earth. In fact of course that is nonsense, because each of us is far more in-bred than we might care to acknowledge. Even so, we all have a very wide scattering of direct ancestors, and each of us – probably – can claim a poet or two, a drunkard, a marquis, a huge array of day-labourers and, almost beyond doubt, some socially miscegenous Royal. More to the point, however, in this context, is that each great-grand-parent, genetically speaking, is no more closely related to you than your first cousins; a 1 in 8 chance that any of your genes would be contained in him or her. By the time you have gone back a couple of hundred years or so, your direct ancestors are genetically no closer to you than fourth cousins, say; that is, than anyone you might sit next to on the bus. A depressing thought for those who like genealogy, but grandly encouraging for the rest of us who – within reason – can cheerfully envisage whatever ancestors we choose.

In fact, there can be advantage not simply in looking after kin, but in risking or even laying down one's life for kin. Thus, if I lay down my life to save the lives of twelve cousins, I have made a net genetic advantage. Such self-sacrifice to save the lives of others is generally termed *altruism*. 'Altruism', as employed by moral philosophers, generally has psychological implications: a person lays down his or her life in a spirit of self-sacrifice. 'Altruism' as coined by Hamilton has no such psychological overtones. It is employed again as a metaphor. The notion is simply that animals (or plants!) might be expected to act *as if* they were moral altruists if, by so doing, they benefit their own genes more than they would by staying alive.

'Kin selection' sounds very like 'group selection', and even some eminent biologists who should know better have at times confused the two. In fact, the two ideas have quite different origins, and quite different implications. Group selection implies that natural selection acts upon the club (the group) to which you belong, and that your actions are shaped by natural selection to protect that group. This once popular idea, as we have seen, is almost certainly *wrong*, and can be shown mathematically to be so. Kin selection by contrast derives entirely from the selfish gene idea (though in fact, the term 'kin selection' was coined before 'selfish gene').

Thus we have a very nice paradox: one of the pleasantest twists in all biology. We begin with the notion that genes are selfish, which seems an immensely depressing idea. But then we show, or rather Hamilton shows, that from this stark notion of genetic selfishness there emerges, as day follows night, the *fact* that creatures will risk their lives for each other, that they will behave *as if* they were altruists. We can envisage this most easily by imagining that there exists a gene for altruism. Such a gene has not been identified, but then, the science of genetics through most of this century has proceeded (and still proceeds!) without identifying particular stretches of DNA, so we need not let this worry us. This gene for altruism would spread through the population – provided only that the animals which carried that gene could recognise who was kin and who was not. We would predict, then, that such a gene *should* exist, because if it *did* exist, it would succeed, and in biology as in physics, whatever can exist – and would succeed if it did exist – often does exist. We would also predict that the altruism gene would be linked, in various ways, to the ability to distinguish kin from non-kin.

So we should look now for evidence, ask whether in fact the animal (or plant) world does contain evidence of altruism (apart from parental care, which we all take for granted), and, more specifically, seek out evidence for kin selection. But before we look we should add more refinements.

First, we might observe that if our altruistic gene was really clever (where 'clever' does not mean literally 'clever', but 'likely to succeed') then it would hedge its self-sacrifice with some fairly hefty caveats. For a start, it would observe that young individuals in general have longer to live than old ones, and that genes in post-reproductive individuals are doomed. So the altruistic gene would be more inclined to look after the young than the old. We would expect grandmothers to take more care of grandchildren than the reverse. Is this the case? Despite Little Red Riding Hood, yes it is. We would expect, too, that no individual would risk his or her life for another unless he or she was pretty sure that that other individual indeed carried a reasonable proportion of his or her genes. On this score, we might expect mothers to be more self-sacrificial than fathers, because they can be more certain that they are their children's parent. This is a powerful theme of literature: August Strindberg in particular was obsessed with fears of cuckoldry. We would expect, too, if our altruistic gene was clever, that it would weigh up the risks. On genetic grounds, an individual should risk his life to save two siblings, but not to save two cousins. But he might do *something* for two cousins. If there was a fire, for example, he might at least raise the alarm, provided it was not too inconvenient.

In practice, these apparently whimsical predictions have been borne out by observation a thousand times. Altruism and kin selection are part of the survival strategy of all animals. The entire art and science of natural history has been transformed by Hamilton's insight. Every naturalist, watching birds on the nest or lions snoozing in the noon-day sun, thinks in terms of who is doing what for whom, and what is the (selfish) genetic advantage thereof. Such questions are at once behavioural, ecological, evolutionary and, of course, genetic. Thus, as I said at the start of this chapter, have the great divisions of biology been brought together at a stroke. Many have contributed to this fusion, but

Hamilton's rigorous encapsulation of a difficult idea is at the heart.

Just a few examples; there are so many. The male lions in prides are generally related to each other, and the females are related to each other but not to the males. Brian Bertram worked out the relationships between the lions he observed in Africa and concluded that the degree of co-operation they showed each other was uncannily geared to their relationship. Herds of elephants are built around adult females, with the adult males dropping in only from time to time to mate. The females are related to each other, or at least are an extended family. Aunts look after the youngsters. Elephants are intelligent, and they may be *literal* altruists: it may be sensible to suppose that they are motivated by warm emotions. But it certainly is not necessary to suppose such a thing, for even the dumbest animals can behave altruistically (in the Hamilton sense). Aphids (which include greenfly) emit an 'alarm pheromone' if they are attacked, which warns nearby greenfly to scatter. Such pheromones form the basis of some modern (or at least experimental) pesticides, and potatoes are being produced by genetic engineering that produce such pheromones in their leaves. It takes metabolic energy to produce such a pheromone. Why do such a thing for the benefit of others? Well, aphids at least some of the time reproduce parthenogenetically. Aphids *en masse* are liable to be a clone: they have *all* their genes in common. 'Self-sacrifice', in genetic terms, means nothing in that context.

Then again, many young birds and mammals, from ground woodpeckers to wild dogs and monkeys, stay around with the parental group until they are almost adult themselves, and help to look after their siblings. Helping parents to look after the babies emphatically is not a human invention.

Recently, I.G. Jamieson and J.L. Craig pointed out that observations of young birds feeding their own younger siblings might, in fact, have nothing to do with 'helping'. They argue that young birds tend to hang about near their parents' nest because they cannot find territory of their own. Now suppose they have a genetic behavioural program – an instinct – which says, 'Feed any young bird in your territory.' Young birds could well have such a program: it would stand them in good stead for when they grew up and acquired territories of their own. But if they stayed in their parents' territory (because they could find no home of their own) their instincts would automatically lead them to feed their own siblings. They would indeed be helping, but it would be a general gene for feeding that was producing the effect, not a specific gene for helping siblings.

It is vital in science to put forward such counter-hypotheses. They provide a way of testing ideas that otherwise are accepted too complacently. But more recent studies of the white fronted bee-eater *Merops bullockoides* show that helpers almost always feed the offspring only of their close relatives, and if they are given a choice – 'feed either relatives or non-relatives' – they almost always prefer the former. This suggests that there is indeed a specific 'helper' gene and not just a 'feed babies' gene. (Quoted by Nicholas Davies in *The Tinbergen Legacy*, eds M.S. Dawkins, T.R. Halliday and R. Dawkins, Chapman & Hall, London, 1991, p. 25.)

Eusocial animals provide the extreme example and the best natural test of 'kin selection'. These include various bees, ants and wasps which belong to the insect order Hymenoptera; termites, which are insects of a quite unrelated group; and various creatures from other groups of animals including – among mammals – the naked mole rat, an extraordinary, hairless, subterranean rodent from arid Eastern and Southern Africa. 'Eusocial' means, in practice, 'extremely social': so social that only one of the females breeds (the queen) and the rest of the females are 'workers', and are either irreversibly sterile (as in the eusocial insects) or else are reproductively suppressed (naked mole rats). The males may be workers too (mole rats) or else, like male honey bees, idle their lives away, and are called upon just once, to impregnate the queen. Male bees are called 'drones', the name given by P.G. Wodehouse to his archetypal London club, the natural watering hole of such irredeemable wastrels as Bertie Wooster and Gussie Fink-Nottle.

The evolution of eusociality provides no problems for group selectionists, indeed it seems to demonstrate its validity. After all, the majority of *individuals* actually give up their right or ability to reproduce at all. Selection operating at the level of the individual could not possibly have produced such an end result. A bee that sacrifices its ability to breed cannot possibly be gaining a reproductive advantage.

To be sure, natural selection at the level of the individual does have a problem with eusocial animals. But natural selection operating at the level of the gene works very well indeed. The eusocial termite worker looks after her sisters. They share half her genes. They are as closely related to herself as her offspring would be. Colonial life is in general beneficial, so by dedicating herself to her larval siblings in a safe environment she does indeed do as much as she possibly could to further the cause of her own genes. To put the matter more precisely: the genes that bring such behaviour about maximise their own chances of replication. The eusocial colony as a whole is a supreme example of kin selection in action.

In short, a kin selectionist would *predict* that eusociality should have evolved, and indeed it has, not only in the extreme forms, as above, but also in a half-way form, as in wolves or common marmosets, in whom the reproduction of all but the dominant female is suppressed, though not irreversibly. As the kin selectionist might also have predicted, eusociality crops up here and there in all kinds of creatures. It is a possible survival strategy, and survival strategies that are theoretically possible are almost bound to turn up here and there in such a diversity of creatures.

In the Hymenoptera, however, eusociality is not just an occasional strategy. Many different hymenopterans are eusocial. We could simply suggest that the strategy evolved once in Hymenoptera, and then the eusocial types formed more species: after all, there are many different species of termite, and several different sub-species of wolf. But this is just not so. Eusociality has evolved separately in bees, ants, and wasps. Indeed, it seems to have evolved independently in Hymenoptera no fewer than eleven times. In short, Hymenoptera seem to have a particular predilection for eusociality. Why?

The reason (it transpires) lies in the way in which Hymenoptera determine sex. In many animals – mammals, birds, insects such as the cockroach – sex is determined by the possession or non-possession of particular genes, some of which are carried on specific sex chromosomes. In others, such as tortoises and crocodiles, sex is determined by the temperature at which the eggs are incubated. But in Hymenoptera, sex is determined by *ploidy*: the number of sets of chromosomes. Females (queens and workers) are diploid (with two sets of chromosomes), whereas males are haploid (only one set).

Consider, then, a colony of bees in full flow. The queen lays all the eggs. The drones in the colony are her sons, but they are far outnumbered at any one time by the female sterile workers, who do all the housework, look after the babies, gather food, stand on guard, and generally make the place tick. The sterile workers are each other's sisters. The queen is their mother. The drones are their brothers.

After a time, for whatever reason, the bees 'decide' that it is time to start a new colony. A larval female is encouraged to become fertile, by feeding her with royal jelly, and she becomes a new queen. When she is mature she sets off, accompanied by a gang of workers. These workers are her *sisters*. The queen mates just once, gathering enough sperm to propagate an entire new colony. The new colony is set up, and she starts laying. The eggs she lays are of course her own sons and daughters. The workers helping her at first are her sisters, so that the queen's new offspring are their nephews and nieces. Eventually, the founding workers die off, and the new queen's daughters take over, and thus re-establish the normal state.

Now consider the genetic relations between these nest inhabitants. The queen is diploid, and her worker daughters are also diploid, so each of the workers contains just half of the genes contained in their mother. The queen has mated only once, so all the workers also have the same father; they are full sisters, not half-sisters. But because the father is haploid each worker contains copies of *all* the genes contained in the father. In other words, each worker shares *three quarters* of her genes with each of her other, sister workers. Thus the worker sisters are *more* closely related to each other than they are to their mother. Hamilton's theory predicts that Hymenoptera should, therefore, have a particular tendency to evolve social systems in which sisters work co-operatively. And that is just how they have evolved, time and time again. Rarely has a theory of biology been so beautifully vindicated.

We can take the predictions further, and suggest, indeed, that because the worker sisters are more closely related to each other than to the mother, there could well be tension between the workers and the queen. The workers should not behave simply as slaves of the queen; she, rather, should be manipulated by them. Look closely (theory informing observation!) and this seems to be the case. After all, it is in the genetic interests of the queen to produce more queens: to produce more fertile offspring who can spread her own genes far and wide. But it is the workers who determine whether any of their larval sisters is or is not allowed to develop into a fertile queen; and in practice, the output of queens is strictly rationed. New queens are nurtured and new swarms take off much less often than seems theoretically possible. But then, the new swarm will have a new father. Its workers will be merely their nieces. No swarms can last for ever because queens do not live for ever, and it is better to have nieces than to have no young relatives at all. But in the short term, the selfish genes of the worker sisters do better by preventing the breakaway.

Finally, there is another, quite different form of altruism, which often will allow *non* relatives at least to take some risk on each other's behalf, though it should stop short of self-immolation. Animals may co-operate on a tit-for-tat basis. Thus, monkeys may groom each other; each spending time 'altruistically', on the understanding that the favour will be reciprocated. Thus, young male baboons have been seen to distract adult, dominant baboons so that other young whipper-snappers can sneak in and mate with one of the dominant male's harem. The detractor runs a risk, and he enables a potential rival to spread his genes. But he expects the same to be done for him. Such co-operation is clearly very close to the ethical principle much favoured in the early nineteenth century, of 'enlightened self-interest'. We will discuss this later.

First, we should move on to the second huge insight to emerge from the selfish gene view of animal behaviour: that of the 'evolutionarily stable strategy'.

EVOLUTIONARILY STABLE STRATEGY

This second contribution is due to John Maynard Smith. His general approach has been to apply to evolutionary problems the particular mathematics of von Neumann's game theory: to quantify the likely success of different survival strategies.

A simple but illuminating example involves two rival genes, one causing its possessor to behave very peaceably, like a 'dove'; the other producing a bullying individual, who sees what he wants and grabs it no-holds-barred, who for convenience is called a 'hawk'. (Note, in passing, that these terms are metaphors taken from the American military. Real hawks are no nastier than real doves. Being a meat-eater does not make you more 'aggressive' in any general sense, in the way that Robert Ardrey argued in another of his erroneous best-sellers.)

Doves waste no time fighting, and they run no risks. If two doves both need a piece of food or territory then, the theory is, they share the goods equitably. Game theory supposes that each action has some attendant advantages, which can be quantified, and some attendant disadvantages, which can also be quantified. Game theory calculations show that to be a dove in an all-dove society is the best possible way to be. Everyone gains, and the advantage to the society as a whole – measured as individual gain × number in the society – is maximal. Jesus Christ advocated universal dovishness in the Sermon on the Mount, but Peter later took him to task: ' "Lord, how oft shall my brother sin against me, and I forgive him? till seven times?" Jesus said unto him, "I say not unto thee, Until seven times; but, Until seventy times seven." ' (Matthew, 18: 21–22.) From this, Christ emerges as an excellent game theorist.

But we live in a world dominated by selfish genes. And even in the best-run gene pools, any gene is liable to mutate. Consider what happens, then, when one of the dovish genes mutates into a hawk gene. The resultant hawk, alone in an otherwise all-dove society, has a field-day. He struts about, shoving the peaceable doves out of the way, eating the best food, commandeering the best territory, and appropriating a wide selection of acquiescent wives. Game theory confirms, unsurprisingly, that the hawk strategy scores very highly indeed, but all the doves miss out. The total advantage to the society is reduced, but there is nothing to be done about it. Note that we immediately have an objection to the idea of group selection – because group selection theory predicts that the single hawk would mind his manners, to ensure that the group as a whole was not compromised.

However, group selection does not apply, and in practice our hawk produces a great many offspring, about half of whom harbour the hawk gene. We will assume that the gene is dominant, as advantageous genes tend to become, so possessors of the gene all behave as hawks. They too strut and grab. They too flourish and produce big broods of hawks.

Then things start to go wrong for the hawks. At the beginning, the doves greatly outnumbered the hawks. Indeed, we began with an all-dove society. The few hawks there were could strut with impunity. Whoever they bashed, they had no fear of reprisal. But when the hawks themselves become numerous, each strutting hawk starts encountering other hawks. He bashes, and is bashed back. This is very gloomy news indeed. One of the combating hawks is bound to lose, if not to die, and in either case his advantage rating plummets. A society that is all hawk is disastrous, both for the individuals, and in terms of total ratings.

So what happens? Frequency-dependent selection takes over. The hawks do well until there are so many hawks that hawkishness becomes bad news. The final result is a form of polymorphism, as we encountered in Chapter 1: two different forms of creature (in this case two that behave in different ways) living in balance in the same society.

Note, first, that if we apply game theory, and assume that hawkishness or dovishness are both feasible strategies, then this final polymorphic state – mostly doves, but with a proportion of hawks – is *bound* to arise. An all-dove society is not stable, because it is bound to be invaded, sooner or later, by a hawk. But an all-hawk society is self-destructive. The stable end-point is what Maynard Smith called the *evolutionarily stable strategy*, or ESS, and it is a *mixed* end point.

Note, too, that if you add up all the advantage points of all the hawks and doves in the evolutionarily stable society, the total is *less* than it would be in an all-dove society. Note indeed that neither the hawks nor the doves in the ESS society do as well as doves would do in an all-dove society. This again runs absolutely counter to the predictions of group selection. But then, the predictions of game theory, based on the selfish gene, are more robust. Perverse though it may seem, a society in which everyone loses to some extent, and the society as a whole loses quite a lot, is more stable than one in which the society as a whole maximises its advantage. Again we see that Christ was an excellent games theory theoretician.

In practice, of course, no animal (or human) acts as an out-and-out hawk, or an out-and-out dove. Most of us (like St Peter) tend in general to behave dovishly until somebody gives us reason not to, and then we behave like hawks. Even so, we can see in general terms how such game theory applies to human societies. In any society, it seems impossible to prevent the rise of a warrior, bandit, or samurai class – the professional hawks. In any society they will tend to dominate and be conspicuous, because the dove majority does nothing to prevent them. But any society in which the hawks become too numerous soon begins to break down. It is dangerous to apply such notions so glibly to human societies, but it is tempting none the less. There is a ring of truth, which will not be denied.

Again, the ramifications of game theory, the notion of ESS, and its combination with Hamilton's notion of altruism, have produced extraordinarily rich insights into the behaviour of animals – and, which is the final test of any theory, they have provided predictions. In fact, there is now in biology a satisfying dialogue of the kind that has long been apparent in physics: between quantified observation on the one hand, and prediction based upon mathematically inspired theory on the other. Let me give one last example, before we move on: part of the classic long-term study by Tim Clutton Brock and his colleagues of red deer, on the Scottish island of Rhum.

Red deer stags fight to build harems, that is common observation. It follows, therefore, that a successful male can produce more offspring than any female can. It also follows that males have to be very strong indeed if they are to have any offspring at all (apart from the odd sneak), but females can acquire a mate even if they are somewhat feeble. Close observation confirms that this is so. Game theory accordingly predicts that a female – especially a dominant female – should strive to produce sons, because sons can spread her genes further than daughters can do. She should also ensure that those sons are as strong as possible, because only the strongest are sure of mating at all. In practice, close observation again shows that the adult size and strength of deer is related to their plane of nourishment as babies. Close observation finally shows – exactly as game theory predicts – that hinds in practice allow their sons to suckle for about twice as long as they allow their daughters. (Quoted by Nicholas Davies in *The Tinbergen Legacy*, p. 22.)

IN THE BEGINNING WAS THE ANECDOTE

The Gospel of St John begins: 'In the beginning was the Word.' To discuss exhaustively all that this implies would take several volumes, but it seems at least to chime with the notion of René Descartes, 1600 years later: that words provide the indispensable medium of thought. The statement also seems to presage the twentieth-century conceit that the chief preoccupation of philosophy should be to define terms. Taken all in all, we have been brought up to believe that words are indeed the key to thought, and, more specifically, that narratives – stories – are ineluctably compounded from words, just as houses are compounded from bricks. No words, no thoughts, no stories. The cerebrations of animals, it seems, are a pretty hopeless mishmash.

All kinds of modern observations suggest that this rigorous, reductionist, Cartesian, apparently commonsensical view is simply untrue. Once biologists got past the constraints imposed by the behaviourist experimenters, they found it impossible to explain the variety and subtlety of animal behaviour without proposing that they 'think': indeed as Herb Terrace of Columbia University New York has commented, the task now is to explain how animals think *without* human language. But actually, we should also admit that we do not think exclusively in words, either. We set our brains problems, which for the most part the brain works out 'subconsciously'. That subconscious process may or may not be verbal. How could we tell? After all, the only part of the thinking process that we are conscious of is the part that *is* verbal.

Thus, thought is *not* dependent upon words – any more than houses are dependent upon bricks, because you can, after all, build houses out of wood or stone. If the brain did think exclusively in words, how could we possibly think as fast as we do? How do ideas occur 'in a flash', which subsequently may take hours to explain? The role of words is to provide precision; they enable us to set the subconscious thinking process of the brain precise tasks to address, and they enable us to some extent to access and monitor that essentially subconscious thinking process. These abilities are immensely valuable and give us a huge lead over all other animals. But we should not conclude, as

Descartes proposed, that our words give us a unique ability to think. They do, however, give us a unique ability to direct our thoughts.

Similarly, I believe we can argue that *stories* are not word-dependent, either. Why is it that the same themes recur, again and again, in fairy tales, novelettes, and great literature, from Greek myths to the modern tragedians? Carl Jung argued that these fundamental themes represented the 'collective unconscious', the thoughts of the people in a person's own culture, accumulated over centuries, and somehow tucked away in the brain. Those stories may or may not be stored in verbal form – who can tell? What is clear, however, is that to see the light of day, they have first to be translated into words, and it is part of Jung's thesis (and indeed of common observation) that the recurrent stories of any society become refined into verbal forms that become part of the tradition of the society.

I think we can go further. Earlier in this chapter we suggested that animal – including human! – brains contain programs, comparable with computer programs. This is a common heuristic device in biology: to compare things in nature which we do not understand with things in technology that we do understand (because we made those things in the first place). Thus in cruder times eyes were compared with telescopes, and then with cameras. Such comparisons are useful, a guide to understanding. But they are always imperfect. The only *true* model of life is life itself. The brain's programs are a little like computer programs, but only a little.

Once we move away from the notion that thoughts – and stories – need not be verbal, and indeed that words merely provide the means of reference, then we could reasonably argue that the programs in the animal's (or human's) brain are in fact equivalent to the narrative themes that run through all human societies, and are in turn equivalent to Jung's collective unconscious. We do not have verbal stories in our heads, like books on a library shelf. But we do have codified representations of those stories. Those codified representations, I suggest, are the brain's 'programs'. Translated into words – so that humans can deal with them – they become the themes that run through all of literature. But they exist also in the brains of animals, even though they are never actually translated into words.

In short, in every animal's and human's head are not so much programs, as *scenarios* which, when verbalised, become stories. In the beginning, in short, was not 'The Word'. In the beginning was the anecdote.

It is dangerous to extrapolate too athletically from animals to humans, or from humans to animals. But it is also tempting to suggest that the great themes of Shakespeare or of Chekhov – and even the details of literature, down to and including the sneak mating of French farce – are all encoded in the heads of our fellow creatures. I like the notion that an indecisive stag is a cervine Hamlet; that it harbours a gene or genes that encapsulate at least some of the sub-plots, which the all-discerning Shakespeare, sharing those genes, expressed in poetry. This is whimsy, but it is pleasant whimsy, and some day it may lead to something more solid.

There is more to say on sociobiology, in particular on the issue we hinted at at the beginning of this chapter, on its relationship to moral philosophy. I believe there is a relationship, which is important, but is also subtle, and I prefer to delay discussion to Chapter 10.

Meantime we should press on: further to discuss the crucial and resounding fact that within any one population genes from any one individual may be combined with genes from any other. The process is known as sex.

CHAPTER 5

Sex

Genetics would be a very boring subject without sex. Darwin could not have asked how it was that characters (such as blue eyes) disappear from human dynasties for generations and then turn up again, for this would not happen. The traditional methods of breeding livestock and improving crops – by crossing and selection – could not be carried out, because 'crossing' is *de facto* a sexual process. But then, without sex, creatures capable of thinking out a breeding programme or of writing a book could not have evolved in the first place. In short, a book on genetics must discuss sex.

Yet, for all its apparent advantages, sex is the most perverse of all the inventions of biology. As a means or at least as a component of reproduction it is inefficient and extremely dangerous, and although it does have some long-term advantages, it is difficult to see how those long-term advantages could have outweighed the obvious short-term drawbacks. Yet as we stressed in the last chapter, the natural selective forces that shape evolution deal only with what is immediately apparent: they have no power of foresight, and the short term is bound to pre-empt the long term. For all that, sex is by far the most usual means of reproduction among animals and plants; most fungi practise at least some sex, and even bacteria indulge in sex of a kind. So we should first ask two obvious questions. First, in what sense can we argue that sex is an unlikely way to reproduce? And second, if it is so awful, how has it caught on in the way that it has?

SEX IS INEFFICIENT, TIME CONSUMING, AND DANGEROUS

It is a characteristic of living things that they reproduce, and so replicate the DNA within them. One becomes many. The amoeba shows the way: it splits down the middle by the method which everyone learnt in the second form is called *binary fission*. The asexually reproducing stage of the malaria parasite goes one better: snug and well-fed within the blood cell of a human or a monkey it divides to form hundreds of offspring. Easy, logical, straightforward. When creatures reproduce sexually, by contrast, two different cells, usually from two different individuals, first fuse to become one. With an amoeba, reproducing *asexually*, one becomes two, which is the object of the reproductive exercise, whereas when rhinos or humans or orang-utans co-operate to fuse an egg with a sperm, two cells become one, and two individuals produce one. That is *anti* reproductive.

Furthermore, sex is self-evidently dangerous. At the very least, it takes time to find a mate, and time spent looking for mates is not spent feeding. Then again, animals looking for mates expose themselves to predators, and this is made worse because, as discussed below, animals adopt various forms of finery in order to attract a mate, which in theory may impair their ability to hide from or to fight off predators. In practice, too, any one potential mate is liable to be desired by more than one suitor, so the suitors may well have to fight in order to mate and reproduce at all. Elephant seals, red deer and lions are among the creatures in whom the fights are most spectacular and most obviously dangerous. Even plants take obvious risks in the interests of sex: the flowers that attract insect pollinators may also entice hungry herbivores.

Finally, when the suitor has found and identified a suitable partner, and evaded or beaten off all rivals, he (for it is usually 'he' in this context) may well find that the final gauntlet is the most daunting of all. For how does a fiercely predatory but not particularly intelligent creature like a female spider differentiate between a prospective mate (and male spiders are often smaller than the females, sometimes spectacularly so) and a self-sacrificial lunch? Not surprisingly, we find that male crab spiders (for example) semaphor frantically with their palps and legs as they approach their intended mates, while orb-web males vibrate the web of the female like a flamenco guitarist, and funnel spiders gingerly stroke the female's front legs, with eight weather eyes on the formidable fangs behind. Many male spiders have evolved elaborate techniques to de-fuse the possible attack: many bear gifts (as indeed do the males of many other kinds of animal, from insects to birds of prey), and one resourceful species of spider commandeers an immature female and mates with her as soon as she completes her final moult – which is a way of excluding other males, but also of catching her before she has gathered her full strength. Nevertheless, male spiders (and scorpions, and insects such as mantids) are sometimes killed and eaten after mating. Other male animals, such as the squid, simply die after mating. Given that sex exists, and is a fact, it makes sense for surplus males to depart: their job done, they become protein, or at least cease to compete for food, and their demise thus benefits their offspring, which is the 'object' of the whole exercise. All the same, it seems to add to the perversity. Why not just bud new offspring?

As if all that were not enough, sex often involves close physical contact between individuals, and a wet and sticky contact at that, and such contact is a perfect medium for the spread of infection. In the last chapter we discussed 'the games that animals play', but, of course, bacteria and viruses play exactly the same games, to essentially the same rules. The world belongs to them as much as to us, and the contact made from time to time between two hosts obliges the parasites perfectly. Most infectious diseases are more likely to cause trouble if the population is high, or dense. Thus, the measles virus runs through a community quickly and burns its own bridges as it goes, because the victims either die or become immune – and in either case are rendered unsuitable as future hosts – so unless a population is enormous, measles cannot persist within it. But the HIV virus of AIDS is lovingly passed from individual to individual: not necessarily today, or tomorrow, but some time, and then it persists for years before it kills its host, and in that quiescent time is passed to others. A disease transmitted in such a way, and operating at a leisurely pace, can survive even in very small populations. Here are two of several reasons why HIV will be very difficult indeed to eliminate. But the transmission and hence the existence of this fearful disease *depends* upon the sexual habits of its hosts. Here is yet another reason, one might reasonably think, to reproduce asexually.

Yet all the above argument assumes that each sexually reproducing animal is able to find a suitable partner. Animals that are sedentary, or are rare, may have extreme difficulty in doing so. Thus, as Darwin first observed (much to his own amazement) some barnacles – which are sedentary, distant relatives of crabs and shrimps (that is, are crustacea) – have parasitic males: nothing more than bags of sperm, which attach themselves permanently to the female, thus compensating for the fact that neither adult barnacle is able to move to the other. Parasitic males occur, too, in some animals that are bound to be thin on the ground, because the food supply is so scarce. Thus, some deep-sea fish such as the angler-fish have parasitic males. Others solve the partner problem by being hermaphrodite, which is generally the case with snails, although not all hermaphrodite snails are self-fertilising. When animals become rare because of environmental pressures, their demise might be accelerated simply because they cannot find partners. The last seven dusky seaside sparrows on this Earth (in Florida) were all males, and obviously could find no partners. Many a Sumatran rhinoceros wandering around fragmented forest is doubtless unable to find a mate. To be able to reproduce without a sexual partner would, in such cases, clearly be an advantage.

We could go on. With a little lateral thinking it would easily be possible to fill an entire book with the disadvantages of sex. In fact, much of the literature of all nations focuses upon precisely this. After all, the finest love affairs of literature tend to end with death (as in *Anna Karenina* or *Romeo and Juliet*), or with betrayal (as in *Troilus and Cressida*), and it would be only a little mischievous, and not entirely foolish, to argue (*vide* the mantis) that the biological roots of such human disaster run deep. So what on Earth is the advantage of sex? How could such a monstrous innovation have become ubiquitous?

WHY IS SEX POPULAR?

The astute reader would not argue that sex is popular because animals like it. Such an observation would raise the obvious objection 'What about plants?', for they presumably are somewhat less sensuous than animals. More to the point, such an observation would be extremely unbiological. Animals would not like sex unless natural selection had favoured them to like sex. Natural selection would not have favoured them to like sex unless there was some biological advantage in liking it. The predilection for sex is definitely *post hoc*. We might rather put the matter the other way around: the disadvantages of sex are so great, the risks so enormous, that unless animals were very highly motivated indeed, they would not undertake it at all. Sexual desire and all that goes with it is nature's way of ensuring that animals that cannot reproduce asexually do, in fact, reproduce. Without such desire, men would doubtless puff away an amiable existence with tweed jacket and vintage car, but very definitely without issue, while women – well, that's for them to decide.

To find the reasons for sex, then, we must dig deeper. In essence, two outstanding explanations have been put forward. One is true (insofar as we can say that anything in science is true), overwhelmingly important, extremely convincing, and yet *cannot* (paradoxically) serve as an explanation for the origins of sex. The other seems both humdrum and unlikely and yet the more people look at it the more convincing it becomes.

The first, satisfying but in the end unsatisfactory explanation for the evolution of sex is – evolution. As was discussed in earlier chapters, the modern, neodarwinian view of evolution is that creatures accumulate variation by means of genetic mutation, and then natural selection eliminates those variants that are not well adapted to the prevailing conditions, and so by default favours the ones that are better adapted. We also saw, however, that mutation is a rare event; that most mutations are harmful; and, we might add, that the potential benefits of mutations can in general be realised only when those mutations are in proper combination with others. For example, a mutation that produced a super-efficient enzyme would not realise its full potential except in the company of comparably efficient enzymes within the same metabolic pathway.

However, if an organism reproduces *only* by asexual means, then its chances of accumulating favourable combinations of mutation are limited. Creature 1 might, say, undergo mutation A. Its offspring would then contain that mutation; and if by some miracle A was a good thing, then those offspring would be favoured, and produce an entire lineage with the A mutation. Some time later one of the As might undergo mutation B, and B might (by remote chance) be good, and might (by even remoter chance) work well with A, to produce a super-race of ABs. If, however, one of the As instead underwent mutation C, then the resulting ACs might prove to be failures. On the other hand, C might work well with AB; so if the ABs underwent mutation C, the resulting ABCs would be very good. However, given that any mutation is rare, and most are deleterious, the chances (a) of accumulating beneficial mutations and (b) of accumulating them in an acceptable order, are very low indeed. Organisms that reproduce *only* asexually, then, are doomed to evolve very slowly indeed. At least, they may (and do) make *ad hoc* changes from one generation to the next, but the cumulative addition of characters which over time turned bacteria-like creatures into rhinoceroses or oak trees would hardly be possible. It would be possible by asexual means to produce an oak tree from a bacterium, but not, probably, within the total lifetime of the Universe; just as it is theoretically possible for a monkey to write *Hamlet* on a typewriter but not, so statistical analysis suggests, before the Universe finally implodes.

Suppose, by contrast, that our hypothetical organisms are reproducing sexually; that is, that they are swapping genetic material in any one generation. Then we might imagine that while one individual is undergoing mutation A, another might be undergoing mutation B. If As mate with Bs, then the very next generation will contain the highly favoured ABs. As do not have to wait for a second mutation to become ABs. They simply acquire B from someone else. B on its own might not be particularly advantageous, but never mind, within one generation it has found a partner, A, to bring the best out in it. Meanwhile, others in the population might be undergoing mutations C, D, E, F and so on. Most mutations will be poor. Most will be extremely poor on their own, but quite helpful when permutated with some other mutations that are current in the population. Sexual reproduction provides the means whereby the different combinations can be brought into being, and hence subject to natural selection. Mutation is indeed rare. But in a sexually reproducing population – particularly one that is large, and in which 'outbreeding' (the opposite of 'inbreeding') is the norm – the combined effects of the different rare events can be made common, and the variation that Darwin perceived as the pre-requisite of natural selection, is constantly on line.

Finally, and importantly, we should note that sexually reproducing organisms spend at least part of their life-cycle in a diploid state. There are two copies – or two versions – of each allele. Most mutations are deleterious, but in a diploid organism, deleterious mutations can be stored in recessive form. Recessive alleles that are out-and-out bad cause genetic disorder if brought together in a homozygote, and that seems to be a disadvantage. But most mutations are not disastrously bad. Besides, badness is very often a matter of context; for example, an enzyme that does badly in one metabolic pathway may come into its own if other enzymes in the pathway are also changed by mutation. Thus, the phenomenon of recessiveness enables heterozygous individuals to store mutations, and the gene pool as a whole in outbreeding, heterozygous individuals will thus be stacked with covert genetic variation, waiting to be called upon: waiting for recombination, or some new mutation, or both, to bring out the best in the recessive alleles. Thus do diploid outbreeding populations hoard the raw material for evolutionary change.

On the face of things, then, this is the perfect explanation for sex. Asexually reproducing organisms, such as amoebae, clearly have their niche in life, and are clearly successful. They have been around for many millions of years. But there are other niches to explore as well: to be a daphnia, and eat amoebae; to be a fish, and eat daphnia; to be a heron, and eat fish. But these alternative niches could not be explored by creatures that exclusively reproduce asexually. Without sexual reproduction – a means of sharing genetic information within any one generation – these super-beasts could not have evolved.

For many years this kind of explanation satisfied biologists. Some kinds of biological advantage cannot be achieved without sex; natural selection favours creatures that are at a biological advantage; QED, natural selection favours sex.

However, John Maynard Smith and others have pointed out that such an explanation, satisfying though it is, cannot explain the evolution of sex. Certainly the ability to evolve rapidly and efficiently may benefit sexually reproducing organisms once they have evolved. But it cannot have moulded their sexuality in the first place.

The reason that it cannot comes back to the innate inefficiency of sex. For example, codfish produce about two million eggs. But as it takes two codfish to produce any young codfish at all, we should say that, for each codfish that exists, only one million offspring are produced.

But each individual animal can expend only a limited amount of energy on reproduction. If they try to expend too much, they die in the attempt. If they expend too little, then some of their peers, more adventurous and self-sacrificial, will produce more or bigger offspring, and those offspring will out-compete those of the more niggardly parent.

It follows from all this that if a codfish were to arise that had mastered the art of *parthenogenesis* – that is, of producing offspring from unfertilised eggs* – in the short term it would be bound to out-gun the codfish that stuck to straightforward sex. One parthenogenetic codfish would produce as many offspring as two sexual codfish. More to the point, the offspring of the parthenogenetic codfish would each produce two million offspring, whereas the offspring of the sexual codfish would produce only a million offspring each. Clearly, within a very short time, the parthenogenetic lineage would swamp the sexual lineage.

The overall point of the above arguments is that the theoretical, long-term, evolutionary advantage of sexuality would not in practice have a chance to become manifest. To be sure, codfish that reproduce sexually have within them the potential to evolve into super-codfish. By contrast, the distant descendants of the parthenogenetic codfish would be virtually the same in a million years' time, as they are now. But the sexually reproducing codfish would never have the opportunity to realise their evolutionary potential, if once faced with a parthenogenetic codfish. Within a few generations, the sexual codfish would be wiped out. To put the matter in the language of Chapter 4: a gene that enables codfish to reproduce parthenogenetically would soon sweep through the population. If the parthenogenetic codfish were themselves wiped out a million years later because they could not adapt to changing circumstance – so what? Selfish genes do not look ahead. Natural selection does not work to a grand plan. It merely selects from what is available whatever happens to flourish at the time.

We know that many groups of animals do, in fact, include some parthenogenetic species. Some fish and lizards reproduce parthenogenetically and cannot reproduce any other way. Among birds, turkeys are known to be able to reproduce parthenogenetically. Many other animals reproduce parthenogenetically

*Note that parthenogenesis is a form of asexual reproduction, even though it is adapted from the apparatus – eggs – of sexual reproduction.

some of the time, such as bees and their relatives (ants, wasps and sawflies), and aphids. Some plants are parthenogenetic (such as the dandelion), and many other plants employ other asexual means of reproduction, as well as (and in some cases instead of) sexual means.

So parthenogenesis does arise, and having arisen, the above argument says that it ought to succeed, and swamp the more ponderously reproducing sexual animals, even though the latter are in the long term 'better'.

The fact that this does not happen – the fact that parthenogenesis or other forms of asexuality have not swamped sexuality – implies that sex *must carry a short-term, as well as a long-term, advantage.* The question is, what is it? To this question, there have been two kinds of answer, often known as the 'Tangled Bank' hypothesis, and the 'Red Queen' hypothesis.

SEX IN THE SHORT TERM

'Tangled Bank' takes its name from the last paragraph of Darwin's *Origin of Species*:

> It is interesting to contemplate an entangled bank, clothed with many plants of many kinds, with birds singing on the bushes, with various insects flitting about, and with worms crawling through the damp earth, and to reflect that these elaborately constructed forms, so different from each other, and dependent on each other in so complex a manner, have all been produced by laws acting around us. (Penguin, London, 1982, p. 459.)

In its modern form, the hypothesis owes most to the American biologist George Williams. He points out that the immediate environment of any one creature is extremely variable – like an entangled bank. One seed of a tree may land in a swamp. Another may land on well-drained heath. To maximise its chances of continuance, it pays the tree to produce variable offspring. For the reasons discussed in Chapter 1, the offspring of sexually reproducing parents are bound to be variable – provided only that the parents are themselves heterozygous and/or unrelated. Hence, to go back to our earlier example, the million (*per capita*) offspring of the sexual codfish may in practice out-compete the two million *per capita* codfish of the asexual codfish because, in reality, only a few of either kind of litter will survive, and the variable offspring, produced sexually, have a greater chance of including some survivors, than have the uniform, asexually-produced offspring, even though the latter are more numerous to begin with.

Thus, the tangled bank hypothesis supposes that the environment is variable from place to place, at any one time. The Red Queen hypothesis supposes, rather, that the environment varies from time to time: it takes its name from the Red Queen in Lewis Carroll's *Through the Looking-Glass*, who had to run very fast just to stay in the same place.

The Red Queen hypothesis not only supposes change through time. It supposes that conditions vary so rapidly that it pays the individual of any one generation not simply to be different from each other, but to be different from the parent generation. What on Earth could produce this short-term variation – and by so doing, exert such enormous natural selective pressure as to justify the danger and absurdity of sex?

The answer has been formally provided yet again by William Hamilton, and is irreducibly mundane. Parasites are the answer: parasites defined broadly to include everything from a virus (or indeed a sub-viral particle) to a metazoan worm or mite. To be sure, parasites such as viruses and bacteria cannot evolve (or, at least, obviously have not evolved) into oak trees or hippopotamuses. But they do vary enormously none the less, within any one generation, and from generation to generation. Some of them mutate rapidly; they almost all reproduce extremely quickly, often with generation times of only a few minutes; and many do exchange genetic information (that is, practise a form of sex). I suspect, too, that they do not accumulate genetic information (and hence remain fairly simple) merely because they are small and cannot physically accumulate much DNA, but that selection favours smallness because, for example, big viruses could not easily invade the nucleus of plants and animal cells which is their usual *modus vivendi*. Be that as it may, the fact is that the viruses even within one epidemic vary significantly (and HIV, like influenza, is immensely variable); and that pathogens (disease-causing parasites) change significantly as an epidemic unfolds, so that the myxoma viruses found a few weeks after the start of a myxomatosis outbreak in rabbits are significantly different from those that began the outbreak.

Clearly, different individuals within a plant or animal population vary in their ability to cope with a particular parasite. Many people died from the plague bacillus of the Black Death, but many did not; and the difference was in part due to genetic differences between the victims. Much effort in modern plant breeding (and to a lesser extent in livestock breeding) is intended to produce strains resistant to particular infections. We can see the relationship between parasites and hosts as a race. The parasite 'strives' to adapt to the host: that is, to break down its defences. The host 'strives' in its turn to develop new defences, to defeat the parasites. The parasites seem in general to have the advantage, because they mutate quickly, produce huge numbers of offspring, and have a rapid generation time. The hosts in general are far bigger than the parasites, and work on a much longer time-scale. But sex provides some kind of an answer. By shuffling the genes from generation to generation, the hosts at least assure that there is some variability, to confuse the parasites. The parasite that is supremely adapted to one host will not be so well adapted to life within that host's offspring, or within its siblings.

The constant pressure of parasites, then, ensures that sex confers short-term as well as long-term advantage, enough to out-weigh the short-term disadvantage of sexual reproduction's inefficiency. The idea seems unlikely, even distasteful, for sexuality at its best is a noble thing, demanding finery and self-sacrifice, and it is deflating to suggest that microbes and worms have promoted it. I believe, too, that the tangled bank hypothesis also applies: that variability has tangled bank as well as Red Queen advantages. But there is enormous and growing evidence that Hamilton's parasite hypothesis is very strong indeed, evidence that we will explore later.

All that we have established so far, however, is that it pays organisms to produce variable offspring, which they can do by linking their own genes, in a somewhat jumbled manner, with those of another individual. Yet that still does not explain why we have men and women, cocks and hens, bulls and cows.

SEX YES, BUT WHY GENDER?

If it pays individuals to link their genes to other individuals, surely they should give themselves the widest possible choice? Sex is demonstrably useful, but why *gender*? Any individual surely would do best if it could – if it chose – link its genes directly with any other individual in the population. But in practice it cannot – or not directly at least. Males can link their genes only with females, and *vice versa*. Two beautifully compatible males, or incomparably matched females, must make do with each other's sons or daughters, whose genomes have already been jumbled and diluted. It all seems most perverse.

In practice, there appear to be two main kinds of explanation. First, if it pays to produce new combinations, it obviously pays most to combine two genomes that are dissimilar. As we have already seen, heterozygosity is usually better than homozygosity, and inbreeding is usually a bad thing. We can envisage from a very early stage, then, the evolution of mechanisms to provide at least some reproductive barriers between individuals in a population that was too similar. Gender to some extent can be seen as a special kind of reproductive barrier.

Second, and on a quite different tack: DNA does not exist as a naked molecule. In all modern organisms it is cocooned in elaborate membranes. To bring two DNA molecules together the organisms that contain them have to meet. Relatively simple maths shows that if two organisms are looking for each other, whether parents and the children they have lost at the zoo, or two gametes in the sea, they are most likely to meet if one stays still and the other wanders. In the case of gametes, the one that stays still, since it is not going anywhere anyway, can evolve mechanisms that further promote the welfare of the offspring – such as yolk. The wanderer should be as mobile as possible. So static gametes become eggs, and mobile ones become sperm. Vague intermediaries vaguely wandering would be much less efficient.

By similar arguments, it is easy to see how natural selection must act differently upon individuals who deliver sperm and individuals who carry eggs. Sperm bearers will evolve as efficient deliverers of sperm, and also evolve ways of competing with rivals; egg bearers will become more efficient receivers, and protectors of the offspring they bear, and each type would evolve the appropriate accoutrements. Hermaphroditism – two sexes in one – is of course possible, and occurs in most plants and some animals, such as garden snails and earthworms. But we might suppose that hermaphrodite mammals, say, would lose out to specialist (female) child bearers working in partnership with specialist (male) competitors.

I have presented the arguments in brief, general terms, but the maths shows that they work. So, we have two sexes.

THE CONSEQUENCES OF HAVING TWO SEXES

We mentioned above the cautiousness with which the male crab spider must broach the issue of coitus when approaching the female. Wrong signals mean death. If, by mistake, he approached a female funnel-web spider with the same repertoire, he would certainly bite the dust. Contrariwise, a curmudgeonly female crab spider who ate *every* semaphoring suitor would die without offspring, and the genes that promoted such standoffishness would soon die out. Contrariwise, too, an indiscriminate female funnel-web spider who welcomed a short-sighted male crab spider with open arms would also die without issue. If she did produce offspring the resulting hybrids would be neither one thing nor the other: neither crab spiders able to ambush their prey by lying camouflaged in flowers, nor funnel-webs able to build a functional underground web. (In practice, crab spiders and funnel-web spiders are far too different to produce viable hybrid offspring. But many hybrids of other kinds of animals show all kinds of physiological and behavioural shortcomings.)

Three consequences follow from these simple observations, and from each of those consequences flows a vast wealth of biological lore. First, it is obvious that *each* sex of a sexually reproducing species *must* evolve ways of recognising the other sex of the same species, and must evolve ways of being recognised. This is true even if the two partners never come into direct contact, as is the case with many marine animals that shed their sperm, or eggs and sperm, into the sea, or is the case with flowering plants. The receiving organs must recognise suitable gametes from elsewhere.

Second (which is a corollary of the first point), each sex must recognise that suitors of alien species are, in general, *not* suitable. Inter-species promiscuity, leading to danger, time waste, and (at best) to offspring that are not able to compete, is clearly bad.

Third, organisms cannot evolve the means to recognise and be recognised, or to shun other species, in isolation. It takes two to recognise and be recognised; it takes two to shun or be shunned. Hence, while the male crab spider (say) must evolve the apparatus and behaviour for semaphor, the female of the same species must simultaneously *co-evolve* the mental apparatus which tells her that such signalling should be taken notice of.

In practice, the German-American biologist Ernst Mayr suggested many years ago that all creatures had to evolve 'isolating mechanisms' which would make it unlikely (or impossible) for them to mate successfully with creatures of different species, and indeed would make it extremely unlikely that they would even try to mate. (Although some pairs of animals that would not normally mate do sometimes do so if forced by lack of normal contacts. For example, wolves *in extremis* will mate with domestic dogs although in general they prefer other wolves; more strikingly, there have been occasional zoo hybrids between, say, lions and tigers). By contrast, Hugh Patterson of the University of the Transvaal in South Africa has emphasised that animals must merely evolve mechanisms for recognising their own kind: he points out that animals do indeed evolve such mechanisms even on islands, say, where there may be no other species with which there could possibly be confusion. Some say that these two arguments are merely different sides of the same coin, and others suggest that each kind of argument has its own set of philosophical and biological ramifications. My own feeling is that recognition mechanisms must have both kinds of connotation. It is vital that individuals of the same species do recognise each other, and each other's intentions; it is also vital that they are behaviourally isolated from others. In either case, the generalisation is worth dwelling upon: that the behaviour of an individual lineage – even a solitary individual – cannot evolve in isolation. In any sexually reproducing population there has to be co-evolution, of different anatomical features and behaviours in different individuals, because complementarity is all.

As I have already mentioned, however – and as everybody

knows – in the real world, animals must compete for mates, and in the plant world, too, although in animals the conflicts are more obvious. In short, it is not always or usually good enough to be recognised by your potential partner. You must come up to scratch. You must meet some preconception of what a partner ought to be like, and you must out-compete your rivals.

In general, though by no means always, the most overt competition for mates is between males for females. The roars and the crashing antlers of the rutting stag are the archetype. Because this is so, and perhaps because books of science are more often written by men than by women, the impression is gained that males merely have to beat off rivals, and thereby demonstrate that they are the best, whereupon the females come running. However, although the males are often (though by no means always) bigger than the females, they do not commonly find that they thereby have a *right* to the females. Very rarely is the female simply taken by force, and to my knowledge it is only in some human societies that females are regarded essentially as chattels, and as property, and given to males who thereby have a *right* to them. In nature, male animals may fight and strut. But in the end, the females in general *choose*.

Most females, in making their choice, are in the end pragmatists. They choose males which, by whatever process, they decide will provide the genes that make the best complement to their own genes: that is, will produce offspring in which their own genes have the greatest chance of survival. For example, male orang-utans are solitary and territorial. Those adult males who succeed in staking out a territory can be pretty sure that females will acknowledge their obvious superiority and seek them out. All they really need to do is issue the 'long call': a sonorous roar that echoes through the forest, amplified by the mighty throat pouch which is a secondary sex character of the male, and proclaiming mastery. In fact, such dominant male orangs are so confident that if any virgin female puts in an appearance, he sends her away. Orangs consort with each other, sometimes for some weeks, before they mate, and territorial males do not waste time with females who have not already proved their ability to produce offspring. The territorial male demands proof of fertility, and far prefers to consort with a female who already has a child in tow. You may ask, then, how virgin females ever break their duck; to which the answer is that they have to make do with young males, not yet big and clever enough to command territories of their own. However, although the virgin females are virgins, they are also females, and have no interest in males who have no territories – it seems like deadlock, Catch 22. So the young males take the virgin females by force. This is one of the few acknowledged instances of 'rape' in nature (which is indeed applied in this context as a technical term). The female's compensation is (a) that she thereby acquires the means to attract a territorial male at a later stage, and (b) at least knows that the father of her first offspring is strong enough to force his attentions upon her.

Primates that habitually live in groups, such as baboons and many other monkeys, are more subtle. It looks at first sight as if the big males simply decide among themselves who is dominant, and that the dominant ones simply take the females they fancy. Closer observation reveals that the males in practice spend much of their time trying to gain the approval of the females, which male baboons do, *inter alia*, by seeking to establish eye contact, fluttering their coloured eyelids. Dominance among males is indeed in large part decided by the females – for the dominant ones to a large extent are the males they approve of. Furthermore (as observed for example among the woolly monkeys of Cornwall's Monkey Sanctuary) the approved dominant males are not necessarily the biggest and most aggressive, but tend to be the most sensible and confident: the ones, for example, that tolerate the company of infant monkeys, and take care of them without letting them get out of hand. Female woollies do not merely favour brute strength.

Many classes of animal practise several different kinds of mating system, depending on species. Thus, among birds, some males are polygynous and mate with several females in a season; some females are polyandrous, with more than one husband; many mate and remain faithful for one season; and some mate and remain faithful for life. In all cases it is reasonably easy to find evolutionary explanations. For example, female hedgesparrows are among many species of bird who are now known to produce eggs from several different fathers in any one season – even though they *appear* to have one faithful mate. Thus they get the best of all possible worlds: a partner to help them raise the babies, but also a choice of complementary genes with which to combine their own genes. The fact that the eggs in any one nest may be from different fathers is revealed by genetic fingerprinting, discussed in Chapter 8. Bryan Nelson, in *Living with Seabirds* (Edinburgh University Press, Edinburgh, 1986), has explored why it is that gannets (which are big sea-birds that nest on rocks in the middle of the sea) are faithful to their mates throughout their lives, whereas boobies (which are big sea-birds that nest on islands in the middle of the sea) choose a new mate every year. He concludes that it is largely a matter of latitude. Gannets, in northern temperate latitudes, can have no doubt about what season it is, and if they do not mate at the right season their young have no chance of survival. Furthermore, food is abundant in northern seas, though again the supply is seasonal. It follows that gannets must mate at a particular time of year, and are able to do so because the changing day-length tells them what time of year it is. As they also have a good sense of direction, the male and female find it easy to turn up at the same place and same time every year, and it is much more convenient to stay with last year's mate than to go through the traumas of finding a new mate, and a new nesting place. Boobies, by contrast, are tropical. It does not matter much when they breed – the tropical seas provide only a meagre food supply all year round. Besides, the unchanging day-length gives little clue as to the time of year. Any one bird is liable to turn up at the same place to breed each year (for they have no time to search round for new places), but as no bird knows what time of year it is, its visit is unlikely to coincide with that of its mate from the previous year. So it is obliged to find a new partner.

Gannets and boobies, like many birds, nest in colonies, which means they mate in colonies too. To some extent this is forced upon them; there is only a limited number of suitable rocks and islands. But it is also advantageous. If gannets or boobies are seeking a mate, they know where to look. The ocean would be a very large hunting ground if there were no special places within it. Humans, of course, also tend to find mates in places of congregation. The disadvantage, of course, is that the various suitors do not have the field to themselves. If they seek mates when in company, they cannot avoid competition.

These two facts – that it is advantageous to recognise specific mating places, and that competition is thereby inevitable – are formalised by many birds (and some other animals) in the form of a *lek*. In these birds – peafowl, great bustards and ruffs are among them – the males congregate and collectively display their sexual finery. In the case of the peacock, of course, this consists mainly of a huge and shimmering tail (in reality incorporating the feathers of the back). The peacocks stand a little apart from each other, often hidden from each other's view, and the females strut up and down, very obviously choosing the best. In general (as is suggested not least by studies at Whipsnade Wild Animal Park), the most dominant female selects the most glamorous male, and having mated, she may then drive other females away from him. Presumably by so doing she ensures that her own offspring do not have to endure competition from other birds who also carry his extra-special genes.

However, in the case of orang-utans, it is not too difficult to see why the females choose the males they do. They choose the ones with something positive to offer: a territory and, more abstractly but none the less impressively, the physical wherewithal and general prowess to drive off others who might contest that territory. Dominant orangs, in short, are impressive beasts. But peahens choose their mates by their feathers, feathers which (so common sense suggests) would actually *reduce* the overall 'fitness' of their possessor, because a long train must reduce the ability to fly, and hence make him more susceptible to predators. Charles Darwin was the first formally to address the question – how did the peacock evolve his tail? This is, of course, a general question – how do characters evolve that apparently aid mating, and yet equally may seem to detract from overall fitness?

HOW DID THE PEACOCK GET HIS TAIL?

In practice, the answer to this question has generated four main kinds of answer, and the truth seems to contain elements of all of them. The arguments are made beautifully clear in Helena Cronin's *The Ant and the Peacock* (Cambridge University Press, Cambridge, 1992).

Darwin himself, sometimes presented by critics as the arch rationalist, in fact offered a somewhat romantic answer. He concluded that it was necessary in practice to differentiate between 'natural selection' (which was the main shaping force in evolution) and 'sexual selection' (which shaped the characters that attract mates). He concluded, after much cogitation, that peahens favour 'beauty for beauty's sake', which seems like a terrible duck-out – although, as we will see, there is more to it than meets the eye.

Darwin's friend and collaborator, Alfred Russel Wallace, was an obvious romantic, and yet in some ways was more coolly pragmatical than Darwin. He argued, effectively, that nature had no time for frivolity. If a peahen selects the showiest peacock, then this is because he has some quality that is observable to her, but may not be observable to us. The showy feathers must, in fact, be an advertisement, and the point about advertisements is that they must advertise *something*, for you cannot fool all of the people all of the time. So showy peacocks must have some extra quality.

Light on the problem was thrown, as was so often the case, by Sir Ronald Fisher. He proposed what has become known as 'Fisher's Runaway'. We may start simply with the observations that we have made earlier: that each sex must evolve features which make it recognisable to the other, and that each must evolve a means of recognising the other. As the generations pass, natural selection will tend inevitably to reinforce the strength of the signal, and the depth of the response. Suppose, now, too (it is not strictly speaking necessary to add this, but it does not offend the facts) that the female favours the male which is the more easy to recognise. Then the reinforcement will be even stronger.

The way this works is as follows. Suppose that a female bird of paradise favours the male with the longest tail (which research shows is in fact the case). Her own sons will inherit their father's long tail, and her own daughters will inherit her predilection for long tails. The sons now find themselves competing with rivals who all have long tails, the daughters with rivals who all have a predilection for long tails. As time passes, then, tails inevitably get longer, and predilection grows deeper, a process that must continue until the males are simply too long-tailed to escape predators, or to find enough food to spare enough protein to grow them.

It seems to me that Fisher's Runaway must operate; once it is explained, it appears self-evident. It does indeed explain why an initial preference for a slightly longer tail should, some generations later, result in tails of inordinate length. It does, however, leave open the question, did the first females to show a predilection for longer tails do so because they favoured 'beauty for beauty's sake', or because they 'thought' that a long-tailed male must have some other, more obviously useful qualities? That is, was their first choice Darwinian or Wallacian? We should observe, however, that Fisher's Runaway can alone vindicate Darwin's idea. Once the process is in train, the females are locked into the selection of 'beauty for beauty's sake' – and the males are doomed to go on becoming more and more beautiful.

There is a final idea, the Handicap Principle, first developed by A. Zahavi in the 1970s (e.g. 'Mate selection – a selection for a handicap', *Journal of Theoretical Biology* 53, pp. 205–14, 1975). At first (like so many good ideas in biology) it seems somewhat ridiculous. The proposal is that males with long tails are 'deliberately' handicapping themselves. They are in effect saying, 'even with this ridiculous tail weighing me down, I am better than you'. It is important to the theory that the tail does actually *cost* its possessor energy to produce and maintain. It is important, too, that the possessor of the tail really does have strength and energy, for if the advertisement were a fake, then natural selection would punish the offspring of the females who fell for it, and the whole co-evolution of signalling and signal reception would break down. Unlikely though Zahavi's idea may seem, later work supports it, not least the mathematical models of Alan Grafen at Oxford University. Indeed, Grafen has suggested that a similar principle can be seen in operation in pubs, where men may jostle to buy the first drink. They are saying, 'even though I throw away all my cash, I am still richer than you'. This may explain why the people with the least money often finish up buying the drinks, while millionaires confine themselves to drinking, at least in my experience. In any case, and surprisingly (to me), mathematical models suggest that the Handicap Principle works, and its operation could indeed produce long tails in peacocks. (Zahafi is quoted by John Krebs in *The Tinbergen Legacy*, pp. 65 *et seq.*)

But we should return to Wallace. If showy feathers do indeed denote some 'hidden strength', what may that strength be? William Hamilton, who first proposed that sex was driven to evolve by parasites, evokes parasites again. Birds with showy feathers thereby reveal that they are in control of their parasites, he argues. Drab-feathered individuals are clearly being given a bad time. He has gone on to argue that, among water fowl, those which have bright-coloured males should belong to the species that have the longest lists of known parasites. Strange though it may seem, this prediction seems to be borne out by the facts.

So the answer to the question 'How did the peacock get its tail?' might run as follows. In general, peacocks that are free of parasites are brighter than those that are beleaguered by them. Years ago, natural selection favoured females who mated with the bright ones, because this tended to produce offspring with more powerful immune systems. They could not, however, conduct an immunological examination. They were obliged to go for external appearance, and in fact, the brightest birds are the most parasite-free. Fisher's Runaway then took over. The showiest males were selected, and so, too, were the females who showed the most preference for show. Fisher's Runaway was reinforced further by the Handicap Principle: the 'statement' by the showiest males that they are able to survive despite their feathers (which in turn indicate their parasite-free status). That at least is an overall synthesis of what otherwise seem to be rival ideas. I do not know how the different notions could ultimately be teased apart. But at least we need not feel that the peacock's tail is entirely implausible. Reasons of a biological nature can be found for it.

There is one other aspect of mate choice which we ought to discuss, but has nothing very directly to do with extravagant tails. Several experiments (including those on quails by Professor Pat Bateson at Cambridge) suggest that when given a choice, animals tend to prefer mates that are similar to themselves, but not too similar. Many studies in many kinds of animal have shown that animals show no sexual interest in other individuals that they perceive to be siblings. Thus female cheetahs want no truck with males whom they recognise from old, and small children brought up communally in *kibbutzim* do not subsequently marry each other. On the other hand, when given a choice of mate, animals tend to eschew other individuals who are too different. Thus (for example) wolves tend to ignore domestic dogs as mates, even though dogs are direct descendants of *Canis lupus*. Professor Bateson's quails, given a choice, tended to select their first cousins in preference to siblings or completely unrelated individuals. Simple cogitation provides a simple genetic reason for this. Mating between siblings tends to produce inbreeding depression; mating between very different types, tending to break up successful gene combinations, can lead to outbreeding depression. It pays to steer a middle course.

One final observation. Competition between males for mates does not end with mating. Inside the female tract, the battle continues: between spermatozoa. Different sperm from the same organism may actually co-operate to bring about a successful fertilisation: another example of kin selection. It now seems, for example, that some of the deformed spermatozoa that are common in many animals, such as cheetahs and lions, may not simply reflect loss of fertility, as is generally assumed. The deformed sperm may in various ways help to prepare the way for their more mobile kin. But *sperm competition* between different males, all mating with the same female, is commonplace.

Males seek to overcome this in various ways. One is to ensure that the female remains faithful during her period of heat, which is why male orangs 'consort' their females at least until they are pregnant (and then abandon them). Male chimpanzees also consort, but since chimps are more sociable than orangs and female chimps are more promiscuous, they do not always succeed in ensuring exclusivity. They overcome this by producing a great deal of sperm, and have correspondingly large testicles. Gorillas are social too but, like male orangs, the big, dominant, 'silverback' males do generally seem to ensure that no inferior male gets to the females. Gorillas do not need to produce vast amounts of sperm therefore, and their testicles are correspondingly tiny. Men and women in a vague sort of way seem to feel that large testicles, like large penises, are a sign of virility. However, if a man could see off rivals as effectively as a male gorilla does, he wouldn't need large testicles. There is more than one way to be macho.

Overall, then, we have established that sex is a good thing, even though it does lead to problems. We might then address a very basic question. What makes a male a male, and a female a female?

WHAT MAKES MALES MALE, AND FEMALES FEMALE?

As is already clear, natural selection has provided more than one mechanism for differentiating the sexes. Male and female mammals have different sex chromosomes: XY in males, XX in females. Female is the 'default' sex: individuals who are born with a chromosome anomaly, and have only one X gene with no partner (and are called 'XO'), are female in general form, although they are not fertile. XXY individuals, on the other hand, are basically male, even though Xs outnumber Ys. Birds, too, have special sex chromosomes, known as Z and W, but in birds the males are the homozygotes (ZZ) and the default sex is male. In bees, as we have seen, females are diploid, and males are haploid.

In many reptiles, astonishingly, sex is determined not by differences in chromosomes*, but by the temperature of incubation. Thus, both crocodile and tortoise eggs need to be incubated at about 38°C – roughly the temperature of mammalian blood. But if crocodile eggs are incubated a little above this temperature, they turn into males; if below, they become females. With tortoises it is the other way round: the warmer eggs become females, the cooler become the males.

Evolutionary biologists have produced an 'arm-waving' explanation for these differences. In practice, croc eggs or tortoise eggs do not need to be maintained throughout incubation at high temperatures to become male or female: a surge of warmth (though not too warm!) at a critical stage of incubation does the trick. This surge apparently sets a physiological pattern which lasts throughout the animals' life and affects other characters besides sex. Reptiles are poikolotherms: that is, their temperature is generally close to that of their surroundings. Reptiles that stay warm metabolise quicker than those that stay cool, which means they also need more food, but they grow quicker, and finish up with a larger body size. Crocs or tortoises that are warmed as eggs develop a *preference* for warm places after they are hatched, though those kept cool as eggs remain happy with cooler temperatures.

*It is not strictly true to say that sex in any animals is 'determined' by chromosomes. It is however determined at least in part by particular genes carried on particular sex chromosomes. So biologists do speak of 'chromosome-determined sex' as a shorthand.

Thus male crocs grow faster and become bigger than female crocs, while in tortoises the females are bigger. Arm-waving theory says that male crocs *need* to be big to increase their chances of finding a mate, but in tortoises the female has the greater need to be the larger because the number and size of her eggs is limited by the size of her shell. Whatever the explanation, temperature does the trick in each case.

Finally, we may note that some fish, such as wrasse, actually change sex as they get older. Commonly they begin as males, and become females when they are big enough to bear plenty of eggs.

At first sight, it seems that chromosome-determined sex and temperature-determined sex are completely different mechanisms. Closer investigation suggests, however, that the two are really much the same. In both cases, sex is determined not by single genes doing simple things, but by whole cascades of genes which are triggered by particular events at particular stages of development. Many of the genes in these cascades could be similar in the different vertebrate classes – but the triggers are different. In mammals, a particular group of genes known as SRY (sex-region, Y chromosome) chime in early in the cascade; in crocs and tortoises, some other gene, somewhere else in the cascade, is heat-sensitive, and affects the behaviour of the rest.

The fact that in crocs and tortoises the two sexes may have exactly the same genes, and the fact that some fish can change sex, suggests that the genetic difference between the two sexes is not as profound as might be imagined. There is a twist to this story, however. In mammals at least, the same gene may behave differently, depending on whether it is in a male or a female.

WHY MEN ARE A GOOD THING

One rag week when I was at University female undergraduates from the faculty of physiology marched the length of the town with a banner that read: 'Men are biologically superfluous!'

Well, it just ain't true. For one thing, sociobiological/game theory arguments suggest that selfish DNA does best under the present arrangement, with only half the population actually giving birth while the other half competes for the right to involve its own

DNA in this process. The argument that *all* the individuals in a society ought to give birth (which presumably would be true in an all-female society) would be true only if selfish DNA (or indeed the population or species as a whole, for those who still prefer group selection arguments!) was being held back by lack of fecundity. But it is not. The human species, like every successful organism, can already give rise to far more individuals than can possibly be supported, and to increase the potential birth-rate by cutting out the non-child-bearers is not in itself a good survival strategy.

But there is a second reason why men are necessary, which is much more intriguing. Research since the mid-1980s has shown that in mammals, DNA from a male and DNA from a female do not behave in the same way. This has two connotations.

First, we may note that although parthenogenesis is possible in many invertebrates (such as bees and aphids) and in some vertebrates (some fish, some lizards, and even the turkey), it has never been observed in mammals. Even among some animals that do not normally practise parthenogenesis it is possible to stimulate the eggs to begin division by simulating fertilisation; a frog's egg, for example, will divide if poked with a needle. In general, however, mammalian eggs do not oblige.

Perhaps it is not so surprising, however, that a haploid unfertilised nucleus is reluctant to divide. So let us try something more subtle. What happens if we simulate fertilisation (by suitably treating the egg surface) but then microinject a pronucleus* not from a sperm but from another egg?

This, in fact, is the kind of experiment undertaken in mice by Dr Azim Surani and his colleagues at the Institute of Animal Physiology and Genetics Research at Cambridge. The answer is that the two female pronuclei will fuse, and embryo development does begin. But it does not get very far. The embryo itself develops, but the essential subsidiary tissues of the *trophoblast*, which include the placenta, do not develop. Contrariwise, an embryo created by the fusion of two male pronuclei forms magnificent trophoblast tissue, but only a feeble embryo. More broadly, it seems that maternally derived genes are needed to lay down the broad tissue types in the embryo – muscle, gut, etc – but paternally derived

*'Pronucleus' is the term given to the haploid nucleus of a gamete.

genes are needed for these tissues to proliferate and grow.

What's going on? After all, any genes present in the female can be present in the male, except for those few concerned with sex determination. In inbred animals, such as laboratory mice, the genomes of male and female can be virtually the same. Surely, then, the two genomes ought to be equivalent?

Although the male and female may indeed share most of the same genes, it is now clear that the *expression* of many of those genes is profoundly influenced by the sex of the animal that contains them: that is, by the presence or absence of other genes that are concerned directly with sex. This phenomenon – sex influencing gene expression – is called *genomic imprinting*.

Genomic imprinting has many profound, theoretical implications: throwing new light on gene expression in general and tissue differentiation in particular, as discussed in Chapter 3. It has immediate practical implications, too. Clearly it is of interest to breeders of livestock: for example, we know that some characters manifest differently depending on whether they are inherited through the female or the male line. Mitochondrial genes, which are carried only by females, may provide the explanation. But genomic imprinting must be involved as well. 'Genetic engineers' seeking to transfer genes directly between organisms (which is explored further in Chapter 7) should presumably take genomic imprinting into account, because the same gene may behave differently in a new host, depending on the sex of the donor – or indeed on the sex of the host. The influence of genomic imprinting in medicine is already becoming apparent. For example, if the single-gene disorder known as Huntington's chorea is inherited through the male line, it appears earlier in life with each generation. If it is passed through the female line, it appears later in the succeeding generation.

Finally, we may whimsically observe that the virgin birth of Christ was even more miraculous than it has been presented, for it is very difficult to see how an all-female gamete could have developed into a viable fetus. This is a frivolous point, however. After all, many theologians argue that the virgin birth is not meant to be taken literally, while others might reasonably suggest that miracles occur when God chooses to suspend the natural laws; and if once the laws are suspended, then anything can happen.

It is tempting to write about the ramifications of sex indefinitely.

For biologists, as for the world at large, it is the ever-interesting topic. In general, though, it is time to give this book a practical twist. In the next two chapters we will discuss attempts to manipulate the genomes of animals and plants: to create new livestock, new crops, and a whole lot more besides.

CHAPTER 6

Breeding

Living things survive on this planet either by taking life as it comes, or by altering their surroundings to suit themselves. The animals and plants that form the plankton of the sea are accepters: either they are adapted to what their environment offers, or they die. Human beings are the most accomplished of all alterers. If we are ever so slightly dissatisfied with the things around us, we change them. Over the past 30,000 years – the era of people who not only looked like us but could obviously think like us – the rate, range and accuracy of alteration has accelerated.

All creatures are liable to influence the evolution – which means the genetic content – of the creatures around them. The creatures of the plankton that seem merely to serve as food for others have prompted the evolution of a host of filter feeders, from jellyfish to whales. Animals that hunt provoke a chain of evolution in their prey which leads to evasiveness. Without such provocation, potential prey animals remain complaisant – like many an innocent island creature, popped into the pot by passing sailors.

Most animals live by hunting other animals or by browsing wild plants. Many have taken these exercises one step further, and look after or even cultivate the creatures they exploit. Some coral fish stand guard over the patches of algae that they graze upon – and so influence the form, growth habits, and degrees of tolerance of those algae. Leaf-cutter ants cultivate fungi on composted fragments of leaf; termites have a comparable strategy. Many species of ant 'milk' the honeydew from aphids which serve them virtually as 'cows'. Aphid evolution has in turn been influenced by the ants' attentions.

Human beings, versatile creatures that we are, can effectively emulate all the tricks of all other animals. We are the deadliest of hunters, the most astute of gatherers – and the most various farmers. We have influenced the evolution of wild animals and plants by our hunting and gathering, as all other creatures have done. But the changes brought about by farming – arable, pastoral, horticultural – have been profound, in many instances startling, and in many cases far more rapid than any other evolutionary change in the history of the planet.

I have not begun this chapter in this quasi-philosophical vein purely out of whimsy. To be sure, ours is the age of biotechnology, and of the particular technologies of genetic engineering, and everything that came before may seem simply to be quaint. But the modern genetic engineers are not starting from scratch; as Isaac Newton said of himself, they are standing on the shoulders of giants. It is important to understand what those giants have achieved – and to appreciate that they are still active: the traditional ways of altering other living things still represent the mainstream of crop and livestock 'improvement', and they will always be essential. In short, if we want to understand what is really happening in the world, and what is liable to happen in the future, it is well to begin at the beginning.

HOW THE WORLD'S FIRST HUMAN FARMERS ALTERED OTHER ANIMALS AND PLANTS

Archaeologists have generally maintained that agriculture began in the Middle East about 10,000 years ago (ya). Later, and independently, it arose in China, India, North and South America, and several times in the Pacific.

The more modern view – promulgated for example by John Yellen of the National Science Foundation, Washington – is that agriculture (broadly defined!) first began about 30,000 ya, probably in Africa, and very possibly in the forests. To be sure, there are no obvious archaeological signs of the earliest endeavours, but then, plants rot, especially in tropical forests. Farming is revealed in the archaeological record only after it was very well established – because *all* archaeological records are patchy, and we would be unlikely to unearth an endeavour when it was still new, and therefore uncommon. In general, too, archaeologists find relicts only in places that favour preservation. So the oldest known agricultural remains – those of the Middle East of 10,000 ya – do not represent the first agriculture, and it would be very unlikely if they did. They represent agriculture already well established in a place that happens to favour preservation.

It is a mistake to assume, after all, that agriculture began all of a piece. Modern agriculture involves many different processes – but that does not mean that all agriculture has to involve all of those processes. The modern farmer begins with crops (or animals) that have been deliberately bred for farming. In the case of crops, he first breaks (cultivates) the ground, then he plants, then he waters and fertilises, then he protects – by excluding weeds, pathogens, pests and predators – then he harvests and starts again. The modern livestock farmer feeds his carefully bred animals on carefully bred grass and clover, or proprietary feed made from carefully bred crops.

But there are infinite shades between the farmers who perform all these tasks, and the hunter-gatherer. The traditional pastoralist, as still seen on plains and mountains from Africa to Greece and from India to Switzerland, does not try to improve the pasture. He simply lets his animals graze on what is there. Modern 'organic' farmers may give their crops little or no specific protection against pests. But they are all 'farmers' (defined broadly).

Indeed, it is obvious that farming did not *begin* with all its modern components. Rather it evolved step by step. We can easily envisage at least two plausible origins, both dating from tens of thousands of years before the traditionally assumed beginning in the Middle East.

First – the John Yellen scenario – the first farmers may simply have thrust twigs from favoured food plants into the ground, whereupon, in a moist tropical environment, they would sprout. The first farmers may indeed have planted seeds rather than twigs; after all, the seeds of many tropical forest trees germinate before they leave the tree, so the gatherer can *see* that a seed turns into a plant. These first farmers – who in this scenario are more likely to have been women than men, since women are the traditional gatherers – would then have protected the young crop, as a fish may protect a patch of coral. The next conceptual step would be to clear away rival plants to make more space – which effectively is weeding, and this clearance could well have been achieved by fire, which human beings have been using for about a million years. If the ash of the cleared plants is left *in situ* (as it almost certainly would have been) then we have the beginnings of fertilisation. If the ground is poked a little to make it easier to plant, then we have the beginnings of cultivation. And so on.

Another possible route to farming is demonstrated by the

modern Aborigines of Australia. They burn away the dried bush – 'rank vegetation' – to enable new green plants to poke through, which is not so 'unnatural' as it may seem because it is now clear that many natural landscapes, including many kinds of forest as well as grassland, depend on regular (natural) fires. So astute are the Aboriginals in their use of fire, that Dr Rhys Jones of the Australian National University, Canberra, calls them 'firestick farmers', and they have clearly been 'firestick farmers' for at least 40,000 years.

But if Africans began farming (of a kind) 30,000 years ago, and Australian Aborigines began 40,000 years ago, why did they not emerge clearly as the world's first farmers? The Aborigines, after all, are among the few modern peoples who on the face of things seem *not* to have taken to agriculture. There are many possible answers, but the main *general* reply is that farming (in the modern sense) simply is not necessarily the most appropriate way to exploit all environments. Australia is subject to long periods of drought, and huge numbers of modern farmers have come to grief in its capricious landscapes, trying naively to apply the methods that work so well in temperate Europe. In the attempt, they have in two centuries made a hideous wreck of land that Aborigines have worked successfully for forty millennia. Similarly, John Yellen has found that the Bushmen of the modern Kalahari 'toy' with agriculture. They may keep a few goats for a few years, but then decide (probably for sound ecological reasons) that it would be better to revert to hunting for a decade or so. Farming, in short, should not be seen simply as an 'improvement' on hunting-gathering. It achieves some ends that hunting-gathering cannot – notably, a huge output of food fit for humans, per unit area. But it should properly be seen as a tactic, suitable in some circumstances and not in others. It should *not* be seen as an inevitable stage in some inexorable human development.

The point, however, is not to preach an anthropological sermon but to look at the genetic consequences of human action. Many are immediately obvious. If the first farmers – let us say, forest women in Africa – protect their crops from rivals and from predators, then they remove the natural selective pressure which otherwise would have encouraged those plants to be toxic or prickly. It is costly to a plant – in energy terms – to produce toxins or prickles (which are significant physical structures), so a plant that does not produce them can grow more quickly and efficiently than one that does – if, that is, some obliging human being is providing protection. Then again, the world's first farmers would not have grasped the principles of planting unless they realised (as any bright creature might do!) that like begets like. So they would naturally have re-planted the plants that were *least* toxic and prickly, and were *most* succulent and tasty. In genetic terms, they would thus have generated a population of plants which lacked individuals that contained the genes that produce toxins or prickles – or at least, that produce them in large amounts. The first farmers would, in short, have created a population with a gene pool that was measurably different from the wild one (if there had been anyone about to do the measuring).

Thus, the world's first farmers would have set in train the processes that have led to modern crop breeding. They would have replaced natural selection with *artificial* selection: that is, with selection based upon conscious human decision. Such artificial selection would have produced changes that were quite different from those favoured by natural selection: a *lack* of toxins and prickles, as opposed to an abundance. And, since artificial selection can be imposed so rigorously and constantly, it would have brought about changes more *rapidly* than would normally be achieved through natural selection. Yet the world's first farmers, in their African woodland, would have achieved all this without making any huge or implausible conceptual leaps.

The 'first' farmers of the Middle East – or those traditionally taken to be the first – had already inherited 20,000 years of endeavour and ingenuity. This, at least, is the John Yellen scenario, and I for one find it far more plausible than the 'traditional' archaeologists' account. The seeds of wheat and barley which the Middle Easterners left behind were acknowledged to be cultivated because they clearly differed from the wild types – otherwise for all we could tell they might simply have been gathered, as a squirrel may gather nuts. They differed from the wild precisely because they had been subjected to generations of selection.

We can envisage an exactly parallel series of events leading to livestock farming, albeit with different timings. Thus, our hunting ancestors may quickly have learned that wild herds could be caught more easily if first directed into particular swamps or creeks; from this evolved the corral; from this, a closer and closer involvement with the animal until all aspects of its life is controlled. Again, artificial but essentially inadvertent selection would quickly have favoured strains of animal that were easier to handle than their wild counterparts (more 'tame') and able to make do on a more limited diet than they would expect in the wild (thus, the first domestic animals are known to be domestic because they are smaller than their wild relatives, though modern domestics are often far bigger). Neoteny has played a key part in domestication, as we noted in Chapter 3. Domestic animals, especially pet animals such as dogs, retain much of their juvenile or infantile gentleness and dependency.

In *The Covenant of the Wild* (William Morrow, New York, 1992) Stephen Budiansky suggests that (some) animals actually co-operated in this domestication process. After all, he argues, those that did co-operate got to breed more prolifically than those that were left out in the cold, even if individuals did, subsequently, meet a premature death. Looked at from the point of view of the individual, this seems slightly distasteful, if not implausible: how *could* animals have sold out in this way? Didn't they realise that human beings were offering them slavery in return for security? However, once we invoke the notion of the selfish gene, Budiansky's argument makes perfect sense. Genes that encouraged their possessors to behave subserviently would indeed have spread more than those that encouraged proud freedom and standoffishness. Domestic cattle and sheep are now the commonest of all hoofed animals, while chickens – which originated as esoteric fowl on the forest floors of India – have become most widespread of all land vertebrates, apart from rats and human beings.

These, then, were the initial processes: beginning, in the case of livestock at least, with co-adaptation, but leading quickly to one of artificial selection, as farmers kept the animals and plants that suited them, and rejected those that did not. And once farmers had begun to bring their charges – plant and animal – more fully under their control, they could begin a further series of processes which still form the basis of modern crop and livestock 'improvement'.

BREEDING BEFORE BREEDERS

Modern breeders – increasingly abetted by modern science – alter crops and animals dramatically in ways we will explore later. But before I discuss what they do, we should look at the kinds of changes made by farmers, in the natural course of cultivation.

I have already mentioned selection, a process that effectively began the very first time a gatherer decided to plant a twig or a sprouting seed. The effect of selection is to *reduce* the gene pool. Desirable characters are not added by selection, they are simply given a chance to become manifest, as undesirable characters (and the genes that underpin those characters) are removed.

The gene pool of the domestics – crops or livestock – may be further reduced by genetic drift. This implies, in the main, either that individuals with rare alleles die before they reproduce, taking their alleles with them; or that, by chance, they fail to pass on particular alleles (those that do happen to be shuffled into the gametes) to their offspring. As was shown in Chapter I, genetic drift occurs more rapidly in small populations than in large, so there would be very little drift in crops growing *en masse*. But drift must have played a large part in the initial shaping of crops and livestock when the domestic populations were still small.

There are also two processes that add to the gene pool – even before the professional breeders get to work. The first is mutation. Most mutations, as has been shown, are deleterious, and that is true in domestic creatures as well as wild. However, some mutations that would be deleterious (or neutral) in the wild are seized upon by farmers. For example, the seed-stalk – 'rachis' – of wild wheat shatters easily; that is how the wild plant spreads its seed. Early in the history of domestication a mutant arose with a strong rachis that does not shatter. In the wild, such a mutant would be at a grave disadvantage, for its seed would not be spread. In a domestic crop it was a godsend, for the farmer could then gather the ripe wheat without losing the seeds – and then harvest the seed tidily by threshing. Similarly, some time in its prehistory, a strain of maize appeared in which the seeds were firmly encased within a cob. Again, in the wild this would be fatal – precluding seed distribution. Again, in a domestic crop it was a boon. The farmers, not wild nature, were in charge, and they favoured the mutants. The farmers also took care of seed distribution. The history of horticulture provides thousands of such examples. Indeed, breeders of flowers positively scour their fields and greenhouses to find some 'sport' that can form the basis of a novel cultivar.

The same applies to domestic livestock, whether bred for serious agriculture, or for frivolous purposes. For example, whereas hill farmers need agile animals that can escape from bad weather and seek out the scarce grazing, lowland farmers prefer flocks of lounge-lizards that do not stray too far and can be confined within small fences, which are cheaper than high fences. So when, in New England in 1791, a ram was born with short, bent legs, the farmer, Seth Wright, did not knock it on the head, as surely would have been the fate of a wild sheep with such a deformity. Instead, he used it to found a new short-legged breed – the Ancon – which offered enormous savings on capital costs.

Early domestic crops (and even recent ones!) also gain genes by one other major route: *introgression*. The point here is that many domestic crops are prone to cross-breed with related plants living in the wild. This is particularly true of plants that are natural 'out-breeders' and 'prefer' to mate with individuals that are not identical with themselves. Often such introgression has been a menace – as recalled by William Cobbett, one of the first of the 'greens' who extolled the virtues of the countryside, in 1822:

> This very year, I have some Swedish turnips, so called, about 7,000 in number, and should, if the seed had been true, have had about twenty tons weight, instead of which I have about three. Indeed, they are not Swedish turnips, but a sort of mixture between that plant and rape. I am sure that the seedsman did not wilfully deceive me. He was deceived himself. The truth is that seedsmen are compelled to buy their seeds of this plant. Farmers save it; and they but too often pay very little attention to the manner of doing it. The best way is to get a dozen of fine turnip plants, perfect in all respects, and plant them in a situation where the smell of blossoms of nothing of the cabbage or rape or even of the charlock kind, can reach them.★

★*Cottage Economy*, Oxford University Press, 1979, p. 89.

For 'smell of blossoms' read 'pollen', and you have it. Cabbage, rape and turnips are all brassicas – indeed, all belong to the same species, *Brassica oleracea* – while charlock is closely related within the family Cruciferae. Brassicas are committed outbreeders, and some are self-incompatible, so they are 'anxious' to accept pollen from elsewhere. Modern seedsmen take care, as Cobbett advises, to ensure that the seeds they sell on to the farmers are not thus contaminated.

But the addition of genes by introgression has not been all bad. Often it may have led to the creation of quite new crop lines. Most notably, modern domestic bread wheat is hexaploid and contains the entire original genomes from no fewer than three (diploid) grass ancestors. The first stage in such an amalgamation may well have been introgression: an early domestic (diploid) wheat hybridising with some grass growing as a weed in the same field, on some early farm. A parallel in animals might be that of the husky dog, which at times has been deliberately outbred with wild wolves, to toughen it up.

There is one final process that occurs naturally among plants, and has produced many new crops – some of them in the fields of farmers – without the intervention of professional breeders. This is polyploidy.

INSTANT NEW SPECIES: POLYPLOIDY

Sometimes (in plants, though rarely, it seems, in animals) the chromosomes double by mitosis, but the cells then fail to divide. The result is a cell with four sets of chromosomes – tetraploid – rather than two (a diploid). In plants, any cell in the growing tip can in principle give rise to an entire shoot, and cells within that shoot can differentiate into ova or pollen. Hence you sometimes find tetraploid branches in diploid plants, and these tetraploid branches sometimes give rise to entire tetraploid lineages. The tetraploid offspring are not necessarily able to hybridise with the diploid parents that gave rise to them – or if they do, the hybrids are not immediately fertile – and so they constitute a quite new species.

However, polyploidy is of most interest when it turns an infertile hybrid into a fertile one. Consider the situation in which two related but different species of plant hybridise, either naturally or in a farmer's field. Because their chromosomes are different, the resulting diploid hybrids are sterile: the two different chromosome sets cannot come together to enact meiosis. Suppose now, however, that polyploidy takes place. Now we have *two* sets of the maternal chromosomes plus *two* sets of the paternal chromosomes. Each chromosome does now have a compatible partner with which to combine in meiosis. Hence the hybrid with tetraploid cells may well be fertile, even though it contains genomes from two different parents. Cells that result simply from a doubling of chromosomes in a pure-bred plant are called *autotetraploid*, and those that result from doubling of a hybrid diploid cell are *allotetraploid*. Allotetraploids are an important source of new plant species both in nature and in farming.

Polyploidy has clearly played a great part in the evolution of plants. Consider, for example, the aloes: those succulent house-plants that sprout like pineapple-tops on many a suburban window-sill, and grow wild in Africa. Scientists from the Royal Botanic Gardens, Kew, have shown that some wild aloe species have only a few chromosomes; some are clearly autopolyploids of those 'basic' types; some are allopolyploids of the basic types, and some seem to be polyploid hybrids of species that were already autopolyploids.

This process has also given rise to many extremely important domestic crops. The most primitive kind of domestic wheat is *Triticum monococcum*, otherwise known as 'einkorn', which is diploid, with 14 (2 × 7) chromosomes. Its probable ancestor is the diploid wild grass *Aegilops speltoides*. The hybrid of einkorn and *A. speltoides* is sterile, but such hybrids underwent polyploidy some time in the distant past, and their allotetraploid descendants became *Triticum turgidum*. *T. turgidum* is known as 'emmer', and was much favoured in the western world; its modern subspecies is *durum*, which is used to make pasta. *T. turgidum* of course has 28 chromosomes (7 × 2 × 2). Bread wheat, *T. aestivum*, is hexaploid: it is evidently a hybrid of *T. turgidum* with the wild goat grass *Aegilops squarrosa*, which is another diploid with 14 chromosomes. So bread wheat has 42 chromosomes (7 × 2 × 3).

There are many more examples among domestic crops, and in some groups, such as the extraordinarily complicated series of the genus *Rubus* – raspberries, blackberries, boysenberries, loganberries, etc – the interrelationships are quite mind-blowing (as described by D.L. Jennings in *Evolution of Crop Plants*, edited by N.W. Simmonds, Longman, London, 1976). The cultivated banana is an oddity: it is triploid (three sets of chromosomes) and is therefore sterile, but produces fleshy (but seedless) fruits anyway, and is of course reproduced vegetatively (that is, by cuttings). Bananas, in short, are monocotyledous triploid parthenocarps; be sure to ask your greengrocer for some.

The oddest of all is modern sugar cane. The original cultivated cane was *Saccharum officinarum*, which has 80 chromosomes, and may well be octoploid (eight sets of chromosomes, presumably accumulated through hybridisation and polyploidy of various wild ancestors). In the early decades of this century breeders crossed *S. officinarum* with the wild *S. spontaneum* to produce offspring with chromosome numbers ranging from 40 to 128 – the point being that when each individual chromosome exists in many different versions, some can get lost without killing the plant. These hybrid offspring were then crossed again ('back-crossed') with the original *S. officinarum* octoploids to produce an extraordinary mixture of plants, some of which have 100–125 chromosomes, of which 5–10 per cent are derived directly from the wild *S. spontaneum*. Many of these back-crossed hybrids have an odd number of chromosomes, and are sterile, but they grow extremely well and are reproduced asexually. Organisms with peculiar numbers of chromosomes are said to be *aneuploid*. Most aneuploids are severely disadvantaged; Down's syndrome in humans, for example, is caused by the presence of an extra chromosome. But aneuploid sugar canes are a success, and indeed are the only commercially valuable aneuploid food crops.

As will be shown later, modern breeders provoke polyploidy, and manipulate it. But my point here is that polyploidy is one of the processes that occurs naturally in farmers' fields, just as it does in the wild, and if the farmers are alert, as farmers tend to be, then over the years they continue to select the plants that serve them best (some mutants, some hybridised with local wild plants, some polyploid) and so produce varieties that are quite different from the wild ancestors and which are also well adapted to the fields in which they have been developed. Such 'local' varieties, produced informally by alert farmers as they cash in on the processes of nature, are called *landraces*.

Since the early days, when crops were developed only by the people who grew them, there have been three significant conceptual advances. The first has been the advent of the professional breeder: formally undertaking the selection and cross-breeding that traditional farmers carried out as they went along. The second has been the advent of Mendelian genetics, which in various ways has enormously refined and accelerated the traditional breeding processes. The third has been the rise of a whole mass of auxiliary sciences and technologies which have enabled the Mendelian breeder to rise to even greater heights, even more rapidly. One of these 'auxiliary technologies' – genetic engineering – is so special, it deserves a chapter of its own. The rest will be covered in the next few pages.

THE ADVENT OF THE BREEDERS

When wild type plants and animals are converted into domestic crops and livestock that suit human purpose, it is termed 'improvement'. It may not be improvement in the eyes of God, and many crops and livestock, taken all in all, are much feebler than their wild ancestors, but 'improvement' is the term, so we might as well lie back and enjoy it.

From all that I have said above, it is clear that a great deal of 'improvement' can occur without the farmer doing very much thinking. If he or she simply stays alert, and re-plants cuttings or seeds from the individual plants that are most pleasing, the crops (or animals) will be changed over time, and very significantly – in general getting closer and closer to what is required. That this is so is extremely important. The first farmers could not have *anticipated* that their attentions would alter their crops and animals in desirable ways. That, however, was the effect, and it was an enormous bonus.

It is safe to assume, on statistical grounds, that some of the earliest farmers were geniuses, and must have realised very quickly that 'improvements' could be planned and worked towards; it was not necessary simply to wait for them to happen. I have no idea when farmers first realised that two individual plants which had complementary qualities (one succulent, say, and one big) could be deliberately mated – *crossed* – to produce offspring that might combine the good qualities of both. Certainly, horse-breeders were working to this principle in classical times, and knew the dangers of inbreeding and the rest. Crossing of individual crop plants is in essence similar to introgression, and results in a new combination of genes, except that crossing is deliberate, and aimed in a prescribed direction, while introgression is a matter of time and chance, and for every vital input from some agreeable, wild neighbour there are 100 irritating incursions akin to Cobbett's 'smell . . . of the charlock kind'. The deliberate crossing of individuals with complementary qualities, combined with deliberate selection of their most desirable offspring, is the essence of all breeding, and its origins lie deeper than written history.

In detail, however, there is a great deal more to the improvement of crops or livestock than a simple crossing and selecting, or, as modern breeders say, of 'crossing the best with the best and hoping for the best'. Modern plant or livestock breeding should be seen as an extrapolation and refinement of the phenomena that can occur in peasant fields, but it is important none the less to appreciate how great that extrapolation has been.

HOW MODERN BREEDING HAS ADVANCED UPON PEASANT FARMING

Selecting and crossing, crossing and selecting, these indeed are the twin components of all breeding. In genetic terms, selection is a deliberate extirpation of genes that produce undesirable characters, and crossing is intended to create new combinations of desirable genes.

Peasant farmers may do both of these essential things, but they have severe limitations. First, the gene pool they begin with, from or within which to make their selection and crosses, is liable to be extremely limited: the population of plants or animals that happens to live in their particular valley. Professional breeders can cast their nets more widely. Thus British pig breeders in the eighteenth century introduced Chinese strains because they were fatter – and the fashion then was for fat. Modern British pig breeders are now turning to China again to acquire the enormous prolificacy of some Chinese breeds, which have up to thirty piglets per year. This has serious welfare implications, which will be discussed later. More acceptably, breeders from the International

Crops Research Institute for the Semi-Arid Tropics (ICRISAT) in the Sahel recently created a mildew-resistant sorghum for the Sahel which contained genes from a sorghum strain that grows in India. Even to know that such a strain exists requires a worldwide network of information.

Second, the peasant farmer cannot easily measure the improvements he makes. Of course, he can see that an offspring plant is bigger or juicier than its parents, but even this can be deceptive, and there are far more subtle qualities of equal value that are far more difficult to quantify. For example, 'ordinary' fields on 'ordinary' farms are not uniform: some places have better soil, or better drainage, or more shelter or sun than others. Some individuals may grow better than others simply because they are in a better spot. Only the professional breeder, raising plants under strictly controlled, uniform conditions, can really tell that perceived improvements truly have a genetic basis. Qualities such as disease resistance are extremely difficult to quantify unless the controls are very refined indeed, and the trials are carried out on a very large scale. After all, any crops growing in a natural field may be subject to scores of different parasites and pathogens, many of which (for example many bacteria) occur in many different strains. Virtually all parasites tend to come and go in waves, or epidemics – there are 'good years' and 'bad years'. Thus a peasant farmer who assumed that he had produced a mildew-free sorghum simply because his crop did well in a particular year, might be in for a very nasty shock the following year. In modern breeding stations such as ICRISAT, disease resistance is quantified by exposing groups of plants to known doses of carefully identified pathogens – and to nothing else! – under controlled conditions. When ICRISAT scientists say they have produced a cereal resistant to such-and-such a pathogen, the farmer can be sure that this is the case. It is easy to appreciate, however, that the *back-up* required to carry out such trials – with input from pathologists and a wide range of technologists as well as from geneticists – is prodigious.

A few words more about pest and pathogen resistance. Its importance in crop breeding can hardly be overstated. Farmers in the beleaguered Sahel commonly lose one half of their sorghum *from mildew alone*, but mildew is only one pathogen among many. A half or more of any crop that is stored in a tropical Third World country is liable to be damaged by fungi or insects. Such losses are obviously serious, but the knock-on effects are devastating. After all, if a crop is halved, then the farmer must cultivate twice as much to achieve the same end. One of the overwhelming problems of the Sahel (and many other semi-arid areas worldwide) is over-cultivation: land is dug that would be best left alone, and then blows away on the wind. Bigger crops on the most favoured areas would go a long way to solve this. Worldwide, there are many thousands of breeding programmes in progress, aiming to refine many thousands of different, desirable characters. But in all of those programmes, parasite resistance is a very high priority.

In general, resistance to parasites is conferred in three main ways. Some plants are simply unsuitable hosts because they are chemically all wrong, or have the wrong shaped cells to invade, or whatever. Thus the wasps that make galls in oak trees do not attack, say, onions. Second, plants may contain *specific* mechanisms of resistance both to pests and diseases, conferred by specific genes. These genes operate in many different ways. For example, scientists at AFRC's Institute of Arable Crops Research at Long Ashton have isolated and characterised a protein from the tropical millet-like cereal known as 'Job's tears' which inhibits a vital digestive enzyme (alpha-amylase) in the guts of locusts, and also can digest the chitin (outer protective protein) of fungi. (See *AFRC Annual Report, 1989–90*, Swindon.) In general, it is the genes that provide such *specific* resistance which are the subject of breeding programmes. Resistant 'strains' are individuals that contain specific resistance genes. In general, too, a constant evolutionary arms-race is waged between plants (or plant breeders) endeavouring to refine specific resistance mechanisms and pathogens mutating to overcome them.

Third, groups of plants – domestic or wild – may acquire a general resistance simply by being variable (a point first made in Chapter 5). Any one pathogen, whatever the host – maize, oak tree, human being – will by chance find it much easier to invade some individuals than others. Crops in general are more vulnerable to disease than wild plants because so many individuals of the same general kind grow together, but if the crop is highly variable, it will not so easily be wiped out by any one pathogen in any one year, as it would be if it were entirely uniform.

Landraces, produced over aeons by peasant farmers, tend to

have considerable resistance to local parasites. There are two reasons for this. First, the peasant farmer does not so much 'breed' plants as modify the processes of natural selection. A crop could not be developed at all *unless* it had some resistance to local parasites. Second, and importantly, landraces of the kind that are reproduced by seed (as opposed to tubers or cuttings) tend to be fairly variable; any one pathogen, in any one year, takes out only a proportion of the whole.

The qualities of landraces should not be underestimated. Many a landrace worldwide has been replaced by some modern, 'improved', high-yielding variety which has then succumbed to local disease that the landrace could resist. However, we should not over-react to this fact and conclude – as people tended to conclude in the 'back-to-nature' 70s – that landraces are always better than modern varieties. True, a landrace is unlikely to be wiped out entirely by any one epidemic in any one year, but the losses they do suffer may none the less be enormous (including the 50 per cent loss of Sahelian sorghum to mildew). The ideal, then, is to apply modern breeding, but to apply it with humility. Thus a landrace that retained the best of its original qualities *but also was given a battery of genes that conferred specific resistance to specific local parasites* would be better than the original landrace. Such a plant would be a prize indeed.

In genetic terms, an ideal crop is one that is as variable as possible in characters that do not matter (the variety conferring general disease resistance and survivability), but is not variable in the genes that confer the most desirable qualities (such as flavour or protein content), and also contains a *variety* of specific resistance genes that confer resistance to a variety of pathogens, in each case by a variety of different mechanisms. That is a very tall order, but it is the kind of tall order to which the most sophisticated breeders who appreciate the problems of poor farmers (such as the ICRISAT breeders) are working to fill. Note that to meet this tall order requires very high science indeed, everything mentioned in this chapter and the next, taken to the *n*th degree. Note also that the main beneficiaries of such very high science are liable to be the poorest farmers, working in the most difficult conditions. One of the key economic conundrums of the modern world is to apply the highest (and therefore the most expensive) science to the true needs of some of the world's poorest people who, in practice, could be its main beneficiaries. I will return to this issue in Chapter 11.

Thus modern breeders differ from peasant farmers in their application of selection and of technology: a deep and wide knowledge of the reproductive caprices of different creatures; an ever-growing knowledge of the genetics underlying particular characters, and an ability to induce, accelerate, greatly extend, and to some extent control the various mechanisms that the peasant farmer must hope occur spontaneously – including an ability to widen the gene pool by inducing mutation, and to effect crosses that the peasant farmer could not ever achieve. It is time to leave peasant farmers behind, brilliant though they are, and look at the moderns in more detail.

INBREEDERS AND OUTBREEDERS

As has already been noted, some plants 'prefer' to mate only with others which, though of course of the same general type, are none the less genetically different from themselves. Such plants are called *outbreeders*. Cabbage, runner beans and millet are examples. Some – like many cherries and plums – are such emphatic outbreeders that they are actually *self-incompatible*. They will not accept pollen from other individuals of identical, or too-similar, genotype.

Other plants, by contrast, can be fertilised by pollen from individuals genetically identical to themselves. These are the natural *inbreeders*. Wheat, barley and sorghum are examples. Some are such committed inbreeders that ova are fertilised by pollen from the same flower, as in Mendel's garden peas. Some plants, through the history of their cultivation, have shifted from outbreeding to inbreeding. Wild tomatoes are generally outbred, while cultivated ones can be inbred.

Outbreeding and inbreeding can be regarded as survival strategies, shaped by natural selection to exploit different advantages and to cope with different restraints. We have seen that in general – in the wild – it pays animals or plants to be fairly heterozygous; that way they are more likely to avoid inheriting a double dose of any 'bad genes' (deleterious alleles) that may be in the gene pool. It also pays wild populations to be variable, not least to make life difficult for parasites. In general, then, it pays *not* to mate with your own twins or siblings. Many human societies evolve taboos to avoid incest, many animals show no sexual interest in their own siblings – provided they know they are siblings. Plants cannot make such 'intelligent' choices, but many instead avoid incest through various chemical mechanisms.

Yet over-insistence on outbreeding can raise problems. Chapter 5 showed how difficult it can be to find a mate. Sometimes it can be so difficult that it is safer to forget the niceties of outbreeding and rely on whoever is to hand – or, indeed, to employ pollen from the very same flower. Thus we may speculate that some of the grassy ancestors of wheat had a variable tolerance of outbreeding or inbreeding, but that in the uncertain circumstances of the arid Middle Eastern hills the unfussy inbreeders left more offspring than the more pernickety outbreeders, simply because they had a better chance of being fertilised. Scientists at Australia's CSIRO Tropical Forest Research Centre have shown experimentally that some trees in the forest of Queensland clearly prefer to be outbred, and their young flowers are structured to avoid self-pollination, but if pollination is not achieved, by bats, birds, or insects, then the tree shrugs (metaphorically speaking) and allows self-pollination to take place.

Obviously, from the principles described in Chapter 1, inbreeders tend quickly to become highly homozygous, whereas outbreeders tend to remain heterozygous. The inbreeders could not survive their homozygosity unless the gene pools of which they partake are free, or almost free, of deleterious alleles. If there are no such 'bad genes' in the pool, they have no chance of inheriting any, in double dose or single. That is a tremendous advantage. On the other hand, 'hybrid vigour' does not apparently result simply from an absence of deleterious alleles. Sometimes complementary alleles boost each other (for example in the phenomenon of 'superdominance') which can produce an extra *positive* quality, not just an absence of a negative one. Homozygous animals and plants clearly miss out on this particular boon.

Breeders of plants or animals seek two general ends. They seek to produce individuals that have a whole collection – a *suite* – of desirable characters. Equally important, they seek to produce distinct *varieties* (in the case of plants) or *breeds* (in the case of animals) in which the individuals are reasonably uniform.

After all, the whole point of growing a named variety (or breed), is that the farmer (or consumer) knows what to expect.

Clearly it is much easier to produce uniform varieties of natural inbreeders, than of natural outbreeders. Natural inbreeders are highly homozygous. Their offspring can therefore inherit only one version of each particular gene. So their offspring closely resemble their parents, and each other. So do the offspring's offspring. Thus, once a breeder has produced a good example of an inbreeding plant, such as a wheat, he has only to multiply it up, and there he has it: a population of plants with desirable qualities that is also uniform: a variety. By the time the breeder has carried out the initial cross, then selected and re-selected several generations of the most promising offspring, it takes him about twelve growing seasons to produce a new wheat variety, even if he begins with two varieties that are already very promising. A lot can happen in twelve years. In Europe, the entire market can change. In Africa, entire populations can be devastated by famine. There is an urgency about plant breeding, but plant breeding is not easily rushed.

Outbred plants, on the other hand, are highly heterozygous. Thus, their offspring could inherit different versions of any one gene, and the gene combinations within the offspring as a whole are legion. You may feel, then, that it is impossible to achieve uniformity within a population of natural outbreeders, and that it is therefore impossible to produce a true 'variety'. Well, it is impossible to produce outbreeder varieties that are as uniform as inbreeder varieties, but it is possible to refine the outbreeder pool to the point where subpopulations within it are uniform enough to serve as varieties. So there are many varieties of cabbage, though most cabbages are committed outbreeders. The point is that although each individual in the cabbage variety is reasonably heterozygous, none the less the *total* gene pool of which that variety partakes is not as broad as the total gene pool of cabbages as a whole, and the gene pool that is manifest in one variety of cabbage is different from that of another cabbage variety. In general, too, there would be fewer deleterious alleles in a cabbage variety gene pool than in the entire gene pool, so that the cultivated plant would be more tolerant of homozygosity than its wild counterpart; and, indeed, some genes within the cabbage variety gene pool would be present in only one version (that is, would be *fixed*) even though others were present in two or more versions. If the 'fixed' genes in the cabbage variety were of the kind that influence appearance, then the variety could *look* reasonably uniform even though, in details of metabolism, it was variable.

Overall, though, production of varieties of outbreeders is something of a compromise. Various strategies are adopted, but one used at ICRISAT in the early stages of millet production is simply to grow millet plants in a group, then select the best and let them go on breeding with each other, and with selected plants from other groups, and so on and so on. It is remarkable what improvements can be wrought in a few generations by such 'mass selection'.

BACK-CROSSING

Suppose a breeder has produced a plant that grows beautifully in his own country but lacks some vital quality, and suppose that quality is to be found in some other variety growing in another country (to which it is suited) or within a wild plant. The breeder wants to acquire the gene that produces the desirable character in the foreign plant but he certainly does not want all its other genes, which undoubtedly would lead to no good.

He achieves this by first crossing his own, generally outstanding, variety (which from now on we will call the *recurrent* variety) with the foreigner from which he wants the single gene (henceforward called the *donor*). This first cross produces F1 offspring with 50 per cent of the recurrent's genes, and 50 per cent of the donor's.

He then selects F1 hybrids that possess the character he wants, and crosses these again with pure-bred recurrent plants. The offspring of this cross contain 75 per cent of the recurrent genes and 25 per cent of the donor's. Again he selects the individuals that possess the desirable quality, and again he crosses them with pure-bred recurrents. After about five such back-crosses, very few of the donor's genes remain, but provided the breeder selects the appropriate cross-bred plants in each generation, the few donor genes that do remain will include the one that confers the desirable quality. Clearly, however, it takes about five seasons at least to complete the whole cycle. Plant breeding is time-consuming.

Back-crossing plays an enormous part in crop improvement. Sugar cane, mentioned earlier, provides one of thousands of examples. British wheat breeders have employed back-crossing to acquire genes from American wheat varieties that produce high-quality protein of the kind suitable for bread-making, but without also acquiring the various growth characteristics (including a different response to day-length and fertiliser) that would make the American varieties unsuitable as they are.

F1 HYBRIDS

Breeders, however, always seek refinements. Indeed, whatever logical refinement you might care to think of, they have probably acted upon already. For example, the ideal surely would be to produce a crop (or a population of livestock) that on the one hand was highly heterozygous but on the other was uniform. Can this be done?

Yes it can. Such a crop is called an *F1 hybrid* (where F1 means 'first generation'). The breeder starts off by producing two strains of plant (or animal) that are uniform and also fairly (or highly) homozygous. If the parent plants are natural inbreeders they will not 'mind' being highly homozygous, but if they are natural outbreeders they may look somewhat enfeebled.

The breeder then crosses males from one population with females from the other population. The offspring are then all highly heterozygous, having inherited quite different alleles from each parent. But they are also uniform, because each of their parents was highly homozygous. Although each homozygous parent contains two copies of each gene (being diploid), those two copies are the same allele. Thus all the different offspring can inherit only one version of each gene from each parent. Thus, high heterozygosity in each individual, yet perfect uniformity between individuals. Of course, if the grower of F1 hybrids decides to save the seed from them, he is in for a nasty surprise, because the F2 offspring of the highly heterozygous F1s will be very variable indeed. For one generation, however, the F1 hybrids do very well. The grower therefore benefits, but so does the breeder, because the grower can obtain more supplies of the F1 seed only from the breeder, who alone holds the parent stocks that are parents to those F1 hybrids. Professional and amateur growers alike generally find that F1 seed is extremely expensive. This is partly because the breeder has the grower over a barrel (since the grower cannot produce the seed himself) but partly because the breeder has to go to a great deal of trouble to produce the F1 seed. So all is fair. F1 hybrids have in practice become pernicious only when introduced into fragile Third World economies, in which some farmers (usually the majority) are extremely poor. They find they cannot afford the F1 seed, or save it if they do buy it, and are quickly driven to the wall by richer farmers who can afford it. The point again is that economies must adjust to technologies, or technologies to economies. Mismatch between the two is disastrous. This will be discussed again in Chapter 11.

Livestock farmers make use of F1 hybrids, too, although they do not always use the term. Many modern commercial pigs are F1 hybrids, with their genes monitored as carefully as those of any wheat. Traditional cattle types such as Herefords × Friesians are also F1 hybrids: a 'beef-type' bull (the Hereford) is crossed with a dairy cow – albeit a big one, the Friesian – to produce a calf that is good for beef, from a cow whose prime function is not to produce calves at all but to produce milk. Note, however – in animals and in plants – that not all marriages between different lines produce desirable offspring. Some hybrids are decidedly feeble, and this is not surprising, as we can envisage that some combinations of genes work well and some do not. Herefords and Friesians happen to go together very well. As will be seen later, too, many 'rare' breeds of livestock could have a valuable future not necessarily as pure-breds but as fathers of F1 hybrids when crossed with particular common breeds.

F1 hybrids therefore solve many problems. They also raise a few, however, not the least of which is how to produce them tidily in the first place. How, after all, does the breeder of plants ensure that the males from one parent population pollinate only the females from the other parent population, and not other individuals of the same population? How, contrariwise, does he ensure that the females in one population are pollinated only by males in the other population?

By several means, is the answer. Maize, the third most important cereal crop in the world (after rice and wheat) and one of the two great staples contributed by the Americas (the other being the potato), is one of nature's outbreeders. Some modern lines of maize have been largely purged of their deleterious alleles by repeated inbreeding combined with rigorous selection, but sixty years ago maize that was not highly heterozygous tended to suffer badly. By the end of the nineteenth century breeders were effecting improvements by making judicious crosses to gain 'hybrid vigour', but the great leap forward came after World War I when breeders started crossing two different hybrids to produce 'double hybrids'. These dominated American maize growing until the 1950s, after which the gene pool was sufficiently purged of 'bad genes' to enable ordinary F1 hybrids, or even pure-bred types, to take over. In principle, at least, the peculiar anatomy of the maize plant enabled the F1 to be produced easily and tidily. For maize carries its stamens in a tassel, high up in the plant, and its female flowers in inflorescences lower down, which later mature into the cobs. To produce F1 hybrids, the breeders had merely to grow the plants that were to grow the cobs next to the ones that were to provide the pollen – and remove all the tassels from the female recipients as they grew. Conceptually this is irreducibly simple, but as maize is a major arable crop, 'de-tasselling' at one point employed many tens of thousands of people.

Producers of F1 hybrids can also exploit the phenomenon of 'male sterility'. I was brought up as a zoologist myself, and continue to find plants amazing. I am amazed, for example, that a great many plants contain mitochondrial genes that render their possessors male-sterile. Mitochondrial genes are, of course, inherited only through the female line, otherwise it is difficult to see how such genes could be passed on at all. Almost as amazing as the plants are the generations of assiduous botanists who identified such phenomena, and the plant breeders who make use of them. Male-sterility genes have been introduced into various crops, and these provide lines which are admirable recipients for pollen from other lines, and hence make fine parents for F1 hybrids.

A third approach to the production of F1 hybrids is chemical. With appropriate treatment, pollen can be aborted, while the female part of the plant remains intact. By this means, breeders are now producing F1 hybrid wheats. It may seem surprising that a plant such as wheat which is so well adjusted to homozygosity could benefit from the heterozygosity conferred by hybridisation. But it can, not least, presumably, through the expression of overdominance.

In short, F1 hybrids play a major and probably increasing role in crop production, and also in livestock production. The only question is whether the advantages always outweigh the extra bother, and hence cost, of producing them, to which the answer is sometimes yes, and sometimes no.

It has been noted already that peasant farmers can make use only of the gene pool that is immediately to hand – apart from the odd pollen that blows over the hill and is liable to do more harm than good. The modern breeder, by contrast, can take active steps to expand his genetic resource, and his main methods should be explored.

BROADENING THE GENE POOL

Conceptually, the easiest way to widen the genetic options is to accelerate mutation; mutations, after all, provide the new variety which Darwin perceived to be the necessary raw material for natural selection. In the 1920s H.J. Muller showed that X-rays would induce mutation. So, too, it later transpired, would high-energy (short-wave) radiation of any kind, up to ultra-violet. So, too, would a wide range of chemicals, known as alkylating agents. In practice, to create useful variation by inducing mutation is not quite as straightforward as it has sometimes seemed. Most mutations are harmful, and so are most of the agents that induce them: the margin between induction of mutation and death is narrow. Nevertheless, the list of useful, artificially induced mutants continues to grow. Several varieties of barley, for example, now benefit from artificially mutated genes which are raising yield, reducing sensitivity to daylength, and increasing resistance to mildew.

Of more general interest are the deliberate attempts to bring in genes from elsewhere – essentially a form of controlled introgression. The first stage, as has been noted, is simply to identify other plants that do contain useful genes. The next stage is to cross those plants with the crop in hand. If the plants with the beneficial genes are of the same species as the target crop, then there should be no problem.

But breeders increasingly seek to bring in genes from other plants, of other species, and this obviously becomes harder and harder, depending on the degree of difference. Thus in 1971 Drs Jack Harlan and Jan de Wet of the University of Illinois proposed that each plant should be considered to belong to three concentric gene pools. (*Taxon*, 1971, vol. 20, pp. 509–17). Other plants in its immediate, or *primary gene pool* – GP1 – were members of the same species: plants with whom it could mate by normal sexual means to produce fully fertile offspring. Plants in the *secondary gene pool*, GP2, were of related species. Crossing with them was possible, but the offspring generally would not be fully fertile, though some would be fertile to some extent. Then there was an outer ring of plants, in the *tertiary gene pool* or GP3, with which some degree of crossing is possible, but only by extraordinary means, and the offspring are infertile, or indeed difficult to raise to maturity. Harlan and de Wet produced their paper in the year before the term 'genetic engineering' was coined. Genetic engineering, as will be shown in the next chapter, enables genes to be brought in from any other organism, at least in theory. So perhaps we should now regard all organisms, from barley to jellyfish, as belonging to the same quaternary gene pool, or GP4. I will return to this in the next chapter.

The domestic tomato, *Lycopersicon esculentum* – Graeco-Latin for 'edible wolf-apple' – illustrates the point. Several other *Lycopersicon* species (including the still-extant ancestor of *L. esculentum*) belong to its primary gene pool. It freely interbreeds with them: they are its GP1. The domestic tomato will also interbreed somewhat less freely with three more species of *Lycopersicon*, plus one member of the potato genus, *Solanum pennelli*, and they belong to its GP2. Since the domestic tomato cross-breeds with at least one species of potato more easily than it does with some species of tomato, we might suppose that the botanists had classified them wrongly. Do not be hard on the taxonomists, however. They should be looking for broad genetic similarities and differences, but as was shown in Chapter 1 with reference to *Partula* snails, very small genetic differences between populations can sometimes render them unable to mate successfully – which by definition places them in different species – whereas it sometimes happens that relatively large genetic differences do not, by chance, happen to affect the critical mechanisms of mating. Anyway, the two remaining *Lycopersicon* species, plus two more potatoes (including the domestic potato, *S. tuberosum*) complete the GP3 pool of the domestic tomato.

Beyond doubt, peasant farmers have many times gathered genes for their crops from beyond GP1, but they had to wait for chance crosses to occur (most of which would have come to nothing). Modern breeders can arrange such crosses systematically, to see what turns up, and have various means both to overcome the species barriers that otherwise would prevent successful mating, or at best provide infertile offspring. The chief of these techniques is to induce *polyploidy*: that is, to stimulate in the laboratory the process of chromosome replication that occurs naturally, but only spasmodically, in the wild.

NEW CROPS BY POLYPLOIDY

The main way to induce polyploidy is to make use of an extraordinary chemical – another piece of biological serendipity – that is produced by the autumn crocus, probably as a pest-repellent. The 'Latin' name of the autumn crocus is *Colchicum*, and the chemical, an alkaloid, accordingly is called *colchicine*. The specific effect of colchicine is to interrupt cell division. When a cell is treated with colchicine, mitosis stops half way: the chromosomes divide, but the subsequent division into two cells does not occur. Hence, colchicine induces polyploidy effectively to order. Colchicine was isolated by a Frenchman, Pierre Givaudon, in 1937. It has been a boon ever since, for experimental biologists in general and especially for plant breeders.

In practice, the value of polyploids in plant breeding is threefold. First, some autotetraploids are more vigorous than their parent diploids. This is not always the case, for if it were, the world would already be overrun with tetraploids, and indeed with octoploids, and in fact it is not. But some autotetraploids – like the domestic potato – are good value. Second, polyploidy as we have seen can convert a sterile hybrid into a fertile hybrid that effectively is a new species, and may well be more vigorous than either parent. Many domestic crops, such as the swede and various crosses of raspberry and blackberry such as loganberry and boysenberry, are allotetraploids.

Third, polyploid plants contain several versions of the same chromosome. Autotetraploids contain four copies of each chromosome, and the copies are said to be *homologous* to each other. Related plant species may not have identical chromosomes but they may well have *equivalent* chromosomes. Each cell in an allotetraploid plant possesses two sets of chromosomes: one set from each parent. The *equivalent* chromosome pairs from each parent are said to be *homoeologous.* In diploid plants (as in animals) loss of a chromosome is typically fatal, while the addition of a chromosome causes severe disruption; in humans, for example, an extra copy of chromosome 21 results in Down's syndrome. But polyploid plants begin with four or so copies of each chromosome, and if one is lost or broken up there are three others – either homologous or homoeologous – to fill the gap. Hence it is possible to manipulate the chromosomes of polyploid plants without killing them. Such chromosome manipulations also play an enormous part in plant breeding, and indeed can properly be seen as an essential precursor of genetic engineering (together with industrial microbiology and plant and animal tissue culture).

Sugar cane, triticale and wheat – all grasses, as it happens – are outstanding examples of crops that have been created (to some extent naturally and to some extent artificially) by polyploidy, and have also been greatly influenced and improved by additional manipulation of chromosomes. Sugar cane was explored earlier in this chapter. Triticale (pronounced tritty-kailly) is a brand new cereal crop, now making an impact in both Americas and in Eastern Europe and China. In fact it is a man-made genus: a marriage of wheat (*Triticum*) and rye (*Secale*). On the face of things, such a marriage is indeed attractive, for wheat is the most versatile of cereals and, with rice, is usually considered to be the finest, while rye has an extra hardiness that has made it one of the great standbys in harsh climates, and accounts for the plethora of beautiful rye breads from Eastern Europe.

The Scottish botanist Stephen Wilson first tried to marry the two great cereals in 1875, but the offspring were not exciting, and were frequently sterile. Sterility does not matter in a crop like sugar cane, in which the crop is the stem, and which can be reproduced asexually. But it is crucial in a cereal, where the crop is the seed. Wilson was ahead of his time. He could not have known that rye has only 14 chromosomes – a very standard diploid grass – whereas modern wheats are either allotetraploids with 28 chromosomes, or allohexaploids with 42. Tetraploid wheats, broadly speaking, are the durums, and the hexaploids are the bread wheats. So a hybrid of rye and durum has 21 chromosomes (7 + 14) and a hybrid of rye and bread wheat has 28 (7 + 21). Rye chromosomes are homoeologous with wheat chromosomes (and the haploid number in each case is 7) but none the less, in simple hybrids, the rye chromosomes have no partners to line up with in meiosis and the resulting plant is therefore sterile.

Enter colchicine. In the 1950s, F.G. O'Mara of Iowa State University applied it to a hybrid of rye and durum, and so for the first time produced a fertile hybrid – an allohexaploid (given that durum is tetraploid already) with 42 chromosomes (2 × 21). Now, rye and bread wheat octoploids have also been produced, with 56 chromosomes (2 × 28).

These first fertile hybrids did not solve all the problems, however. They do in fact illustrate part of the reason why polyploids in general have not taken over the world. Meiosis can raise problems in polyploids, particularly allopolyploids. The chromosomes can find themselves with an embarrassment of choice, and may line up alongside homoeologous chromosomes instead of their proper homologues. Chromosomes of different species may also undergo meiosis at different rates, so timing in allopolyploids can be a problem.

Yet, with further breeding (not least, a matter of selection), most of the problems have been solved. Today's triticale varieties do behave as perfectly good, fertile plants, and may indeed produce large, seedy heads that in the particular conditions where they are grown out-do both wheat and rye for yield and quality.

To revert to an earlier point: wheat belongs to rye's GP2, and rye belongs to the GP2 of wheat. Peasant farmers may occasionally see natural hybrids, though these in general would have been poor. But to modern breeders, GP2s are fair game. Then again, once the modern breeder has a polyploid (whether or not he created it himself), he can manipulate whole or pieces of chromosome, and transfer them between organisms.

PLAYING WITH CHROMOSOMES

The transfer of chromosomes or pieces thereof between organisms – that is, the bodily transfer of groups of genes – plays a large part in the breeding of some crops that are polyploid. These techniques play little or no part that I know of in animal breeding, because polyploidy is not a trick of animals, and unless organisms have chromosomes to spare they resent interference with those they have. Plants and animals really are very different: as a botanist once commented wrily to me, 'Plants are not green animals.'

Bread wheat has so many sets of chromosomes that it can readily tolerate varying degrees of aneuploidy, and this has provided the route whereby breeders, since well before the days of genetic engineering, have been able to draw in desirable genes from the farthest reaches of wheat's GP3. *Inter alia*, grass of the genus *Aegilops* has been persuaded to contribute genes that confer resistance to eyespot, and *Agropyron* has contributed resistance genes to stem-rust.

The *modus operandi* goes roughly as follows. The grass donating the desired gene is diploid, with 14 chromosomes. The wheat-grass hybrid thus has 28 chromosomes – 7 from the grass and 21 from the wheat. Colchicine treatment then produces a fertile tetraploid version, with 56 chromosomes.

This tetraploid hybrid is then back-crossed with wheat. The hybrid supplies 28 chromosomes – 7 from its grass parent and 21 from its wheat parent – and the wheat provides 21 chromosomes. So this back-crossed hybrid contains 49 chromosomes of which 42 are wheat and 7 are grass.

This 49-chromosome hybrid is partially fertile. With so many chromosomes milling around, most of which do have compatible partners, some 'spare' chromosomes can simply be carried over into the gametes. Thus these aneuploid hybrids produce a mixed bag of gametes with variable numbers of chromosomes. They can therefore be back-crossed with wheat yet again, to produce offspring that also have a variable number of chromosomes. Some of these offspring have 42 chromosomes, and are in fact pure wheat; some have 49, with a complete haploid of grass chromosomes. But some have 43 chromosomes: 42 wheat and 1 grass. These are called *addition lines*. Indeed it is possible to produce a complete set of seven addition lines – each containing a different grass chromosome.

Wheat breeders also maintain lines (for breeding as opposed to commercial purposes) known as *monosomics*: individuals that lack just one chromosome. This would almost certainly be lethal in a diploid plant (as in a diploid animal) but not necessarily in a hexaploid, where every chromosome is present in several copies (both homologues and homoeologues). Again, they maintain complete sets of monosomics, each one lacking a different chromosome.

If an appropriate addition line is crossed with an appropriate monosomic the result is a 42-chromosome plant that contains 41 wheat chromosomes and just one grass chromosome. The grass chromosome does not of course have a homologue, but it does have a wheat homoeologue. Homoeologous chromosomes do not normally form partners during meiosis, but in some cases (those lacking a specific gene that normally prevents such alliances) they will do so.

Hence we have a fertile lineage, founded by a plant that contains just one grass chromosome. That grass chromosome contains the desirable gene, because the breeder has been careful to select the plants that have the required quality. The grass chromosome becomes broken up by crossing over as the next generation is produced, allowing the breeder to select out lines that contain the required gene, but as few of the other grass genes as possible. He can then cross these newly made lines with commercial wheat strains, and then, by back-crossing, finally produce commercial strains that contain the required grass gene, and very little else.

This process may sound incredibly cumbersome but it works. It depends, however, on fairly high technology, and, perhaps above all, on the ability to do things under extremely rigorous conditions and on a very large scale, because only a small proportion of the plants produced in each generation actually have the required qualities, so the breeder needs to produce a lot, and to be able to monitor each one both for its chromosomes and for its physical attributes.

There are, however, some simplicities in plant breeding, notably the exploitation of asexual reproduction.

ASEXUAL MULTIPLICATION

As has been shown, outbreeders of all kinds, and F1 hybrids, raise problems of uniformity. Outbreeders are always a compromise between too much heterozygosity (which tends to reduce uniformity) and too much homozygosity (which reduces viability). F1 hybrids cannot be re-multiplied without losing uniformity altogether.

All these problems are solved, however, if the plant in question can be reproduced asexually. An excellent individual plant (in some contexts called an *elite* plant) is simply multiplied by cuttings, bulbs, tubers, or whatever. All the individuals thus produced are genetically identical, and all the identical daughter plants, produced asexually, collectively form a *clone*. All the roses of any one variety are a 'clone', multiplied by cuttings, so are all the Cox's orange pippin apples, produced from a particularly felicitous cross in the nineteenth century; so indeed are most familiar fruits, and all potatoes, multiplied *en masse* by tubers. The main snag with clones is their uniform susceptibility to disease, and their propensity to collapse entirely in epidemic – just as the potatoes of Ireland and Western Scotland collapsed with the fungus blight, *Phythophtera infestans*, in the 1840s.

The other snag is quite opposite: some plants that it would be highly desirable to produce asexually do not supply suitable cuttings or other 'propagules', and can be cultivated only by seed. Coconuts and oil palms are among the crops that it would be highly desirable to produce by cuttings. In fact, asexual reproduction of coconuts is only now becoming possible, not by traditional means but by tissue culture. Tissue culture is a vital technology – another essential precursor of genetic engineering.

A GREAT LEAP FORWARD: PLANT TISSUE CULTURE

As was briefly touched upon in Chapter 2, plants differ from most animals in ways more profound than is often acknowledged. One of the deepest differences lies in the degree of flexibility that the cells retain. All multicellular plants contain groups of cells that are *totipotent*: able when suitably encouraged to generate an entire plant, with all its various tissues. The *meristem* cells immediately behind the growing tip of many flowering plants obviously have this capacity: they do indeed generate entire new shoots with all the attendant leaves and flowers, and those shoots, if cut and planted, may generate new roots. Now it seems that almost any plant cell that has a functioning nucleus is at least vicariously totipotent, for plants when wounded produce undifferentiated *callus* tissue which can be cultured to produce whole plants.

By contrast, few cells in animals are totipotent, past the early embryo stage, although totipotency is found for example in creatures such as corals, which multiply asexually (in a very plant-like way). In most adult animals, even the most versatile cells are only *pluripotent*, which means that although they can give rise to more than one kind of cell, they can give rise only to a few kinds. For example, the 'erythropoietic' cells in the bone marrow give rise to the red and various white cells of the blood. But most mammalian cells either do not divide at all after a certain age (such as nerve cells) or else can divide only to produce replicas of themselves – so that liver cells divide to form more liver cells, for example. The notion of cloning human beings from fragments of body ('somatic') cells remains a science fantasy, albeit a much-bruited one, even though every nucleated cell of the body contains a copy of the entire genome.

Because plant cells are totipotent, each of them – by definition – can in principle be turned into a whole new plant. It seems only to be a question of finding the right conditions. To a large extent, finding the right conditions is a matter of empiricism: finding a 'recipe' that works. For many plants, however, the appropriate recipes have now been found, and the search for 'recipes' is increasingly underpinned by theory-based guiding principles.

Plant cells are cultured by offering them the kinds of conditions within a glass vessel that they would normally experience as part of a growing plant: a nutrient medium with sugar (usually sucrose) as normally provided by the plant's photosynthetic activities; amino acids, and suitable growth factors, or 'hormones'. In practice, then, attempts to culture plant cells generally failed until the 1930s when plant hormones were first isolated (the first being the auxin IAA, or indole acetic acid, in 1934). The first true success came in 1939, when isolated callus tissue of carrot and tobacco was kept ticking along in culture.

The crucial discovery came in the 1950s: that the ratio of hormones (auxin to cytokinin) was important. In 1958, a mass of cultured carrot cells was induced, for the first time, to generate a complete plant. In France two years later Georges Morel produced a clone of orchids of the genus *Cymbidium* from single cells, and thus gave rise to an industry that has now brought orchids to everyone and saved at least some of those in tropical forests from the depredations of commercial gatherers. Two years on again came the first published 'recipe' for the culture of tobacco cells: not a universal formula (for there can be no *universal* formula) but certainly a good starting point. Complete tobacco plants were generated from single cells in 1965.

The ability to generate entire plants from single cells has many applications. First, as Georges Morel anticipated, single plant cells that are free from virus will give rise to entire virus-free plants and hence to virus-free lines, which with many plants is very difficult to arrange by conventional means. Among the first important crops to benefit was cassava, a commercial crop in California (as a basis for fructose and fuel alcohol) but an important staple in Africa. In India, cultures of meristems (rather than single cells) provide important disease-free clones of potatoes.

Second, breeders can generate entire *clones* from different cells taken from the same tissue (clones being groups of genetically identical individuals). To be sure, some crop plants are easily cloned, simply by cuttings: apples and roses are an example. But other extremely valuable crops that would with advantage be cloned do not produce shoots, or any other kind of 'propagule', from which cuttings can be taken. For such plants, lacking any natural means of asexual reproduction, cloning from tissue culture is a godsend. Orchids provide one such example, but of far more practical value are palms.

Perhaps the most important of all the palms – if not necessarily in straight cash terms, then certainly in social significance – is the coconut; source of 'meat', fibre, and oil. But coconuts are devilish difficult to 'improve'. They are natural outbreeders – itself a difficulty for the breeders – and have a long generation time, taking five to six years to bear fruit, and ten to show whether they are of good quality, with a total useful life of sixty to seventy. They obviously have the genetic potential to improve, however, as the best trees in any one grove (for example in Kerala, southern

India, or in the Philippines) may yield more than 400 nuts per season, while the average (in Kerala) is only around 35. But, unlike Cox's Orange Pippin and many another tree, coconuts yield no suckers or other natural propagules from which to multiply the 'elite', and hence simply to by-pass the need to breed.

So scientists both at Wye College of the University of London, and at Hindustan Lever in Bombay, have for many years been seeking to multiply coconut trees from tissue culture. They both begin with leaf tissue. Culture has not proved easy – palms are not in general as amenable to such treatment as the Umbelliferae (carrot family) or Solanaceae (tobacco-tomato-potato family). But the research has now succeeded. Cloned, elite coconut palms are now growing commercially. Different clones are grown side by side. It is unwise to plant huge numbers of a single type together, for fear of epidemic.

The cloning of plants from single cells does reveal an oddity, however, one that is on the one hand an embarrassment, but on the other provides opportunist biologists with yet another bonus. For the individuals in the cultured clones are not invariably identical – even though they are all generated from cells that are presumed to be genetically identical. This is the phenomenon of *somaclonal variation*.

SOMACLONAL VARIATION

Somaclonal variation probably has more than one cause. Cells in normal growing tissue mutate from time to time. Normally the small differences thus brought about remain unnoticed, until the cell is cultured to produce an entire plant. Sometimes, however, this effect can be seen in a growing plant, as is the case when a mutated cell in a meristem gives rise to an unusual branch, of the kind that every now and again can be seen on some wayside tree. Perhaps in long-lived plants such as trees such meristem mutations have survival value, as the novel branches may be less susceptible to prevailing pathogens than the rest of the tree (and every branch is able to bear fruit and set seed on its own account). It is also possible that culturing *induces* mutation. Or perhaps the method of culture affects not the structure but the expression of particular genes.

In practice, most mutations are harmful, and most somaclonal variants come to no good. But some do, and these provide a steady source of new material for the breeder to work with. For example, scientists at what used to be the Plant Breeding Institute at Cambridge have produced somaclonal variants of the popular potato Maris Piper that show increased resistance to common scab and to some viruses. In Hawaii and Fiji, sugar cane cultures have yielded somaclonal variants with increased resistance to the fungus disease red rot which can wipe out entire crops. No resistant strains have been found in the wild. Scientists at Hindustan Lever are seeking similar variants, and also seek to produce somaclonal sugar-cane lines resistant to salinity. Hindustan Lever's Dr S. Bhaskaran has indeed suggested that somaclonal variation may be one of the most valuable sources of new crops.

Cell culture offers yet another source of variety. For some years biologists have explored the possibility of *fusing* cells from different species to produce hybrids that could not be generated by normal, sexual means.

SOMATIC HYBRIDS

A crucial stage in the serious manipulation of plant cells is to remove the cellulose integument that encases them: the cell walls. This can be done with enzymes, to produce a naked *protoplast*.

The protoplast is, however, surrounded by a membrane (as in an animal cell), and adjacent protoplasts can be fused, by breaking down that membrane (temporarily) with chemicals (including calcium) or even electric pulses. Fused protoplasts are bags of cytoplasm containing two nuclei.

Events can become interesting when normal cell division (mitosis) takes place. The two sets of chromosomes can begin to act as one, finally to produce a tetraploid cell. Obviously, such a marriage occurs most easily when the two naked protoplasts emanate from the same plant, or kind of plant. Much more interesting, however, is that complete fusion can sometimes occur between cells from different plants, to produce true hybrids.

When this technique was first developed, in the 1950s, hopes ran high that any kind of hybridisation might be possible: wheat with barley, even barley with beans. This in general proved unrealistic for many reasons, not least the fact (as has been observed before) that genomes from different species may in detail behave differently at mitosis. Interest then waned somewhat because, after all, plants that are closely related can generally be fused by normal sexual means.

Yet the technique of somatic hybridisation remains in the plant breeder's armamentarium for various reasons. One (as has been observed in Chapter 5) is that the ability of plants or animals to cross sexually with other plants or animals is somewhat capricious. Sometimes species that seem very closely related cannot mate successfully, and sometimes more distant species hybridise very well. It all seems to depend on the presence or absence of just a few particular genes. Sometimes, then, breeders find that they cannot make certain sexual crosses that they would like to make, and then somatic hybridisation can come to the rescue. Thus in the 1980s scientists at Rothamsted Experimental Station in the UK sought to fuse the domestic potato *Solanum tuberosum* with the wild *S. brevidens*, which contains resistance genes to potato leaf roll virus; two plants that seem very close but cannot mate sexually. The procedure was complicated, not least because *S. brevidens* is diploid, like many wild potatoes, while *S. tuberosum* is tetraploid. But it worked.

Dr S. Bhaskaran and his colleagues at Hindustan Lever have explored a different kind of somatic hybrid, in coconuts. They seek to develop strains resistant to potato root wilt. However, normal breeding in coconuts is slow, as has been shown, and besides, the particular resistance genes to potato root wilt are carried in the DNA of the cell mitochondria, within the cytoplasm. The solution is to take cell nuclei from cells of elite individuals (the kind that produce 400 coconuts) and put them into cytoplasm containing the appropriate mitochondria.

Dr Bhaskaran's research illustrates several crucial points: very high technology can be *extremely* appropriate to the needs of poor countries, and indeed may bring far more immediate benefit than in rich countries. It shows, too, that wherever possible the research should be done by the beneficiaries themselves, in their own countries. India is short of funds for research, but its scientific talent is unbounded, as testified both by the sheer number of science graduates, and by the lengthening string of Indian Nobel Prizewinners. It illustrates, furthermore, that people in poor countries are best able to judge what is really worth doing and what is not. This will be discussed further later. For now, we should move on to animals.

LIVESTOCK

To a large extent domestic animals, like crops, select themselves. Some species do better in captivity than others, and some individuals within those species reproduce better than others: they are the ones that tend to prevail. But in the main, from a very early age, farmers shaped the livestock they kept. They selected them largely by eye – what they looked like – and at first they certainly selected them according to behaviour. Farmers certainly prefer individuals that are easy to handle. Bulls, horses, pigs, dogs, even sheep, can be lethal if they cannot be tamed.

For many hundreds of years, with livestock as with crops, farmers and breeders judged the quality of their charges largely by their appearance. To a large extent of course this is eminently reasonable. For example, wild bulls are big in the shoulder and lean in the rear, a secondary sex characteristic that stands them in good stead in battles for mates. But the meat of the rump is leaner and juicier, so breeders of beef cattle have always quite reasonably preferred bulls with a 'square' profile, at least as big in the hindquarters as in the shoulder. To some extent, however, reliance on eye seems whimsical, as in the ankole cattle of Africa, which put so much of their energy into the growth of extraordinary horns. But then, to African herdsmen, cattle are a source of wealth more than of food, and who are outsiders to judge?

Sometimes, to be sure, whimsicality seems to be taken to ridiculous extremes. Britain, for example, has a wonderful old-fashioned breed of pig called the Gloucester Old Spot. No pig without spots can reasonably be admitted to the breed. Yet the solemn averment by one leading breeder that Gloucester Old Spots should have precisely *two* spots, not more nor less, does seem to border on the obsessive. Neither is it clear why the Portland rams of Dorset were supposed to need a 'challenging' expression. In all western farming nations such preferences were reinforced from the eighteenth century until well into the twentieth by the show-ring, at which judges pronounced on the value of this and the worthiness of that almost entirely according to such preferences. But although these judgements could be taken too far, they were hardly foolish. Berkshire pigs, for example, are predominantly black, but are required also to have four white feet – surely a feature of no value whatever. Yet as pointed out by Stephen Hall of the physiological laboratory at Cambridge University, if you see a black pig with four white feet you can be fairly sure that it is, indeed, a Berkshire, even if you do not know the breeder who is selling it to you. If it is a true Berkshire, then it will have a whole number of other qualities – good mothering, ability to thrive outdoors – that certainly are of value. In short, white feet serve as *markers* of other, more cryptic qualities. Markers – features that have no intrinsic significance but which indicate the presence of good or bad characters that are significant – play a very large part in all genetic pursuits.

If the domestic animals are required just for decoration, selection by eye seems good enough, except, of course, that animals selected *purely* by eye often finish up with built-in genetic defects that can cause them a great deal of distress. Thus, dogs bred with short noses, like pekes and pugs, tend inevitably to suffer respiratory problems, while the quaint oriental eyes of chows are brought about by in-turned eyelashes that permanently grate against the cornea. In Britain at least, some vets are now urging dog-breeders to give proper attention to health and well-being, and judges of dog-shows are also keener to avoid what has been frank cruelty.

Breeders of farm livestock have moved away from judgement by eye – and hence from the show-ring – for more hard-headed reasons. Now they measure the specific qualities they require, from growth rate in the first days of life, to the protein content of milk, good mothering, and, in pigs, the ability of the flesh to stay intact in the stressful time before slaughter (for the flesh of some breeds degenerates into 'pale, watery pork' if the animal is panicked in its last few hours).

Livestock breeding and production for the past few centuries at least has been dominated by the use of just a few 'stud' males, elite beasts which, aided by artificial insemination, may nowadays sire many thousands of offspring. What counts in these studs is not their own appearance, or even their own performance, but the performance of their offspring. They may possess fine qualities, but if those qualities happen in that particular animal to have a peculiar and complex genetic basis, then they will not be inherited reliably. So modern stud animals are 'progeny tested'. The performance of their first few, trial offspring is assiduously monitored, and only if they perform well is their sire's semen spread through the herd at large.

Obsessive measurement of specified qualities, progeny-testing, AI, and modern knowledge of genetics and hence of patterns of intelligence – these have moved livestock breeding away from show-ring judgements since the 1950s and wrought a transformation.

Livestock breeders are currently seeking to 'improve' hundreds of qualities, some of the more intriguing of which will be examined below. But one notion that hangs over all is 'efficiency'.

'EFFICIENCY'

'Efficiency' means in livestock farming precisely what it means in engineering: work (or energy) got out, divided by work (or energy) put in. What the farmer puts in can be measured in food energy, area of land, farm labour, or straight cash. What he gets out is measured in kilos of flesh, number of eggs, or litres of milk of appropriate quality.

Suppose, for example, a farmer is raising young pigs, lambs, or calves for beef, which we will henceforth refer to collectively as 'offspring'. 'Energy in' equals the amount of feed given to these offspring by the time they reach slaughter, *plus* the amount given to each animal's mother during the whole period of her pregnancy and lactation, plus a proportion of the amount given to her when she herself was a youngster, and in the intervals between pregnancies. After all, the only cash the farmer receives is from sale of the offspring, so the cost of keeping the mother has to be ascribed to those offspring.

We can increase efficiency by raising 'energy out', and/or reducing 'energy in'. 'Energy out' in this context can in practice be gauged as 'weight of the offspring at slaughter'; we can clearly increase this by (a) raising the weight of individual offspring at slaughter and (b) increasing the number of offspring.

We can reduce 'energy in' in three ways: (a) by ensuring that the offspring reach their slaughter weight in the minimum possible time; (b) by increasing the number of offspring produced in a given time, and (c) by reducing the amount of feed given to the mother. Pertinent to this last requirement is that small mothers eat less than big mothers.

It follows, then, that the ideal mother – whatever the species – would be tiny, and eat very little, but she would give birth to huge litters of offspring that grow enormously fast. A cow the size of a collie dog that gave birth to half a dozen calves that grew to half a tonne in about three months would be ideal – from the farmer's point of view.

In practice, such a cow with such an output is probably a biological impossibility. But breeders and farmers, working together, none the less strive to move as closely as possible to the ideals of efficiency. Thus lambs in Britain are traditionally born on the uplands, where their mothers live permanently, and they are fattened for slaughter on the lush grass of the lowlands. The mothers are small, agile mountain sheep, such as Welsh Mountain ewes, while their fathers are big meaty rams, such as Border Leicesters or Suffolks. (In fact, traditional sheep breeding is far more complicated than this, involving cross-bred ewes and pure-bred rams. But this is the essence.) The lambs then grow faster than would a pure-bred Welsh Mountain lamb. If the ewe can be persuaded to have twins, then this is two for the price of one, a definite bonus. One development in sheep breeding is, therefore, to cross ewes such as Welsh Mountain with highly prolific breeds such as Finns, which commonly produce quins, or to cross them with breeds such as Dorset Horn, which commonly breed twice a year, autumn as well as spring.

Cattle breeders in modern times are in a slightly different dilemma. In Europe at least, most beef animals are the offspring of dairy cows, for it is not economic in most of Europe to raise cows *purely* as a source of calves. Milche cows tend to be very bony creatures, and often rather small (as with Channel Island cattle); their main job, after all, is to 'convert' feed into milk, and not into flesh. In practice the commonest cow in Europe is the Friesian, which is a big animal (effectively 'dual purpose' – meat and beef) but pure-bred Friesians are bonier than required. So most dairy cows most of the time are crossed with a bull from a meaty beef breed. Friesians in Britain are very commonly crossed with Herefords. Nowadays, too, big European breeds such as

Charolais and Simmental – originally bred as draught cattle – may provide the beef input.

Cattle naturally produce only one calf per year, sheep only one or a few lambs. But wild pigs will produce four to six piglets per litter and domestic pigs have a dozen or more. Wild pigs give birth once per year but domestic pigs, with a pregnancy of three months and early weaning, are expected to produce two litters per year and then embark on a third pregnancy – and they are now expected to fit in 2.4 litters per year. So domestic pigs commonly average twenty or more piglets per year. In this case, the farmer does not have to worry whether the sow is big or small relative to the individual offspring. *Numbers* matter far more than relative size. Modern pigs are now being crossed with extremely prolific breeds, such as the Meishan from China, to make them even more prolific. (*AFRC Annual Report, 1990–91*, Swindon, p. 23.) Thirty offspring per sow per year is the present intensive commercial goal.

There are many drawbacks. If a Welsh Mountain ewe has twins, that is good news, because she will usually suckle two lambs quite happily. But if a hill sheep has triplets – as a cross-bred Welsh Mountain/Finn could well do – the farmer has to raise the third one by hand, which greatly increases labour costs. In general, though, sheep farmers increase efficiency by crossing big rams with small ewes and hoping for twins.

More serious are the humanitarian aspects. If a small dairy cow such as a Jersey is crossed with a very big bull such as a Charolais (which is easily achieved by AI), the calf can be too big for her to bear comfortably. Increasing in popularity is the Belgian Blue, in which a genetic defect known as 'double muscling' has been put to commercial use: in such animals certain muscles of the hindquarters are duplicated to give enormous quantities of high quality beef. Belgian Blue calves born to small dairy cows often have to be delivered by Caesarian section, and in Belgium, where the Blue is common, some dairy cows have had up to ten Caesarians. Pigs that give birth to thirty offspring per year are producing six or seven times as many as their wild ancestors; equivalent, perhaps, to a woman giving birth to triplets every year. It is a horrendous strain, with a correspondingly high morbidity.

Equally serious, and potentially inhumane, are the attempts

to increase efficiency by raising the growth rate of the offspring. The growth of a mammal or a bird should be an orderly process. Nature's intent is to ensure that the young animal is always able to function well, even though it is growing fast. Thus, in young animals, the skeleton at first grows faster than the muscles. This is why puppies and toddlers have such big knees, and why foals are 'all legs'. Only when the frame is well established does muscle growth begin to catch up, and adolescents, particularly adolescent males, are lean and muscular. Finally, as the animal matures, it begins to accumulate fat. A layer of fat is a secondary sex characteristic, obviously in females but also in males: a signal to the female that the male is reasonably successful. Fat also has obvious survival value as a food store and as an insulator. Human beings in affluent societies and many domestic animals illustrate, however, that nature's careful plans can be abused.

Farmers, however, sell only meat; meat with a covering of fat, as dictated by the market fashion of the time. In the eighteenth century fat was at a premium, and British pigs were bred (again *via* crosses with Chinese breeds) to resemble sacks of lard on legs, as depicted in many an agricultural print. Nowadays we like pigs leaner, though still much fattier, in general, than a wild animal. Modern beef may not look fat (compared to that of forty years ago) but it is none the less 'marbled', with plenty of fat between the muscle fibres. This is not seen in the wild animal.

Thus, farmers interested in efficiency cannot simply let an animal grow at its own pace. A beef animal that grew as nature intended – first skeleton, then muscle, then fat, with periods of no-growth in winter – would be four years old by the time it was ready for the butcher. Modern beeves are destined for slaughter at eighteen months if fed on grass, and as little as ten months if fed on cereals ('barley beef'). The modern breeder has therefore contrived to concertina the animals' normal growth cycle, producing creatures which – if appropriately fed by the farmer – lay down muscle before they have grown a good strong skeleton, and accumulate fat before they have the muscle to carry it. In genetic terms (though inadvertently) the breeders have selected genes that alter the *timing* of events within the genomes. This effectively is the *opposite* of neotony: the imposition of adult features into a young, growing animal.

The animals pay the price of these endeavours. We have all

seen modern beef bulls lumbering around the show-ring, barely able to pick up their feet; very different from the agile creatures of the wild, which are much closer in shape and demeanour to the animals of the Spanish bull-ring. More seriously, some modern pigs can hardly stand. The worst sufferers in this are probably birds, whose skeletons grow differently from those of mammals, and whose young bones are tipped with fast-growing cartilage that is later infiltrated by bone. University studies in Britain suggest that about one fifth of the 500 million broiler chickens raised in Britain suffer to some degree because of leg weakness brought about by over-rapid growth. A total of about 25 million broilers are severely crippled. Turkeys, which in the modern form have extremely heavy bodies (though their wild ancestors can fly perfectly well!) suffer a comparable deformity of the hip. Bristol University scientists found that up to 70 per cent of the biggest and heaviest turkeys were affected. This is a monstrous toll.

In the next chapter I will discuss genetic engineering, and how that, too, can be applied to animals. I will not argue that genetic engineering is intrinsically 'worse' than traditional breeding, or that it raises qualitatively new issues. But it will enable breeders to make bigger changes more quickly, which seems inevitably to increase the chances that the changes could be inappropriate. We can legitimately take liberties with plants, I believe. There is no reason to suppose they are truly sentient. But animals are our fellow creatures. All the evidence, when looked at through unprejudiced eyes, proclaims that they are both sentient and intelligent. Perhaps we are obliged to farm them for food, and perhaps that is no less humane than hunting. But we do not have a right to take liberties. Breeders and farmers must be guided by humanitarian principles, as well as by the grail of 'efficiency'.

Fortunately, 'efficiency' is not the preoccupation of the modern world, even though it has often been the predominant one. People also seek to create a world that is more agreeable: more humane and more pleasing aesthetically. With these motives, more and more farmers are interested in 'traditional' breeds, many of which are now rare, and they are discovering that 'rare breeds' have a great deal to offer besides aesthetic pleasure.

RARE BREEDS

In the years that followed World War II people sought to establish a brand new world: more socially just and more 'tidy' than what had gone before. In the cities, 'planners' replaced the higgledy-piggledy houses with geometrically simple and sanitised blocks of apartments. In the countryside, agriculturalists asked why there were scores of breeds of sheep (in Britain at least), plus at least half a dozen breeds each of cattle and pigs. Surely it was better, they said, to decide precisely what was needed from a sheep, a cow or a pig, and simply develop one all-purpose breed of each.

Anyone can see obvious snags in such a draconian approach: that with only one breed it would never again be possible to gain the advantages of 'hybrid vigour'; that little ewes ought to give birth to fast-growing lambs; that different regions require different breeds for local conditions. Yet the new-wave agriculturalists had a point. Modern breeds out-grew and out-yielded some 'traditional' types by 100 per cent or more – modern cows commonly yield 5,000–10,000 litres of milk per lactation where old types may have given only 2,000 – and a world short of protein and energy could not waste time on sentiment. So the 'rationalisation' began. In Britain the black-and-white Friesian (essentially the same as the European and North American 'Holstein') largely replaced the Ayrshire and the Dairy Shorthorn, and the Large White began to dominate the piggeries.

But a few visionaries saw drawbacks, commercial and practical ones as well as aesthetic. Markets and fashions could change, as they always have, since commercial livestock farming began. Conditions could change: climate, disease, available food. For all such reasons it was vital, some thought, to conserve the genetic variety contained in the traditional breeds. A pioneer in Britain was the zoologist Sir Solly Zuckerman, later Lord Zuckerman, who persuaded Whipsnade Zoo to offer sanctuary to Britain's last remaining flock of Norfolk Horns, which in Domesday times were the principal sheep of East Anglia. Finally, in 1973, a group of farmers, scientists and general enthusiasts established the Rare Breeds Survival Trust, which is now probably the principal focus of rare breeds expertise in the world.

The generalisations – that genetic variation is a good thing and that rare breeds are agriculture's repositories of such variation – are undeniable. Many of the breeds themselves, however, really do fall far short of the moderns in performance and in 'efficiency'. The reasons for this are clear. A breeder cannot maximise performance and efficiency unless he breeds for those qualities specifically. But performance and efficiency are genetically complex, are often extremely variable from individual to individual even among litter mates, and can be extremely difficult and time-consuming to measure. For example, as has been shown, the potential of bulls must be assessed by progeny-testing. In short, performance and efficiency can be maximised only by professional geneticists operating effectively on a worldwide basis, with large herds to play with under controlled conditions. Traditional breeds, by contrast, were for the most part developed effectively as land-races, beginning with small gene pools though with occasional *ad hoc* imports (e.g. Chinese pigs to Britain in the eighteenth century, or Arab horses), and were to a large extent selected not by rigorous performance-testing but by eye, with the show-bench judge deciding what was best. Whimsy played a large part in the development: long horns in British Longhorns, white feet in Berkshires, curly coats in Leicester Longwool sheep. To be sure, these characters served largely as 'markers' – proofs of pedigree – and each of those breeds has commercially desirable characters as well. Even so, show-bench is not progeny-testing.

However, there is a lot more to many rare breeds than meets the eye. Many perform extremely well, and have fallen out of fashion largely through historical accident. British White cattle, for example, are excellent both for milk and beef. There is a vicious circle, here, too, for once a breed becomes rare, there are too few individuals to choose between, and it is impossible to undertake the rigorous selection needed to effect further improvement. In addition, various rare breeds produce excellent offspring – showing hybrid vigour – when crossed with common breeds. British White and Longhorns both cross excellently with Friesians.

More generally, rare breeds are the repositories of physiological quirks and behavioural niceties that might, one day, prove invaluable. Ronaldsay sheep from the Orkney Islands are adapted to live largely on seaweed. Other sheep become copper-deficient when fed on seaweed, because the cell walls of seaweed are rich in alginates which inhibit copper absorption. But Ronaldsays absorb copper with extraordinary efficiency. Mineral deficiencies of various kinds are a problem worldwide and such abilities as this could well be pressed into service. Many traditional breeds of pig – Saddleback, Large Black, Gloucester Old Spot – do very well on a diet high in grass. They are also very good mothers. So all three are pre-adapted to free-range husbandry of the kind which, for humanitarian reasons, could soon replace the present-day 'factory farms'. On a more hard-headed note, many people are now eating *less* meat, for reasons largely of health. Eating less, they can afford to pay more for higher quality. Many rare breeds are especially fine to eat. Gloucester Old Spot pork is wonderful, and Ronaldsay lamb is excellent when smoked. Finally, as societies grow in wealth and become more leisured, people may begin to demand that their livestock contribute more than protein. Whimsy could well creep back. Belted Galloway cattle with a white belt against a black background; Herdwick sheep, which are born black and then turn a whole variety of colours; or the delightful, delicate, pale red Castelmilk Moorit sheep could well be bred for their enhancement of countryside, as they were in large part in the past.

However, those who raise rare breeds face a difficult dilemma. Many of the breeds cannot realise their full commercial potential unless they are further 'improved', as conventional breeds have been. But improvement implies selection, and selection implies a narrowing of gene pools. Thus, if the rare breeds are indeed 'improved', then their value diminishes as repositories of rare genes. Rare breeds should, indeed, be bred as if they were wild animals (see Chapter 8) – to *maximise* genetic diversity, which implies the opposite of selection. But keepers of rare breeds are often commercial farmers who cannot afford to raise 'inferior' animals simply for the recondite genes they contain; and besides, as farmers they *like* to 'improve' their animals. In the words of Joe Henson, founder-chairman of the Rare Breeds Survival Trust and co-director of the Cotswold Farm Park: 'It would depress me beyond measure to breed from a bad male, when I really want to breed from animals which take my eye.' ('Variety in Vogue', *New Scientist*, 18 March 1989, pp. 50–3.) But Lawrence Alderson, who contrives to organise the Trust's breeding programmes, just as zoos organise wild animal breeding (see again Chapter 8), is not too worried. Rare-breed farmers might indeed break the Trust's carefully devised rules: 'But not all breeders are trying to do the same thing. If they have different ideas, the breed maintains a lot of variability.' This does not quite meet the standards of modern zoos in breeding wild animals, but it probably answers the problem as far as is practical.

Note, finally, how the word 'breeding' is used to describe two quite different exercises. Breeders of endangered wild animals seek to maximise genetic diversity within the population available, and do this by the methods described in Chapter 8: equalising family sizes, outbreeding, and so on. Breeders of domestic livestock seek 'improvement' and *narrow* the genetic pool as they home in towards a preconceived ideal. In the past, breeders of wild animals often mistakenly employed the methods of farmers, and Przewalski's horses, for example, were at first selected largely for their willingness to breed in captivity. Inbreeding resulted, as indeed it put paid to many a zoo-breeding endeavour in less enlightened days. Breeders of rare domestic livestock in practice fall between the two stools.

However, the most important breeding exercises in the world are not those that seek efficiency at all costs, or that seek to revivify the glories of yesteryear. The people who could benefit most from 'improved' livestock are those of the Third World. Unfortunately, efforts there – like early efforts in crop improvement – have often done more good than harm. To end this chapter, then, we should ask what 'improvement' really means.

WHAT IS 'IMPROVEMENT'?

It is very easy to list the biological qualities of plants or animals that theoretically could be enhanced. Resistance to pests and diseases must be at or near the top of every list. Resistance to other physical stresses must also be high: drought, heat, frost or – for plants in particular – lack of nitrogen or phosphorus, or the presence of toxins (loosely termed 'salinity'). Then comes yield, nutritional quality, flavour, 'shelf-life' and appearance. Those who breed for the supermarket have tended to put shelf-life and appearance first. Flavour has often taken second place, so that some crops in some areas (such as the US tomato) are reduced to decoration. But flavour in general is supremely important. People in poor countries can die of malnourishment if they perceive the available food to be unpalatable.

It is a huge mistake (and one that is all too often made) to assume that *all* enhancement of biological qualities is worthwhile. This has been emphasised many times in recent decades, but all too often falls on the deaf ears of multinational companies, international agencies, and governments, who are working to their own agenda. Dr E.R. 'Bob' Orskov has summarised the issues excellently in *Reality in Rural Development Aid* (Rowett Research Institute, Aberdeen, 1993). The first task, as he points out, is to identify the *constraints*, and this is all too rarely done properly. What actually is preventing agricultural success? Does the fault lie with the plants or animals at all, or is the farmer simply held back by a lack of good banking, or of a reliable market? Note what lengths we go to in the affluent countries of the west to ensure that farmers get a good deal: they may be subsidised (as often in Britain), or their crops may be bought at guaranteed prices (as in the EEC), they may have *carte blanche* to meet specified quotas (EEC again) and, throughout, they have the best and most reliable back-up from banks that any business could ask. Without this support, they would be lost; indeed, if any component of this support is threatened, they take to the streets, combine harvesters and all. The Third World farmer for the most part has *none* of these supports. In place of the bank is the money-lender. If his crops fail, too bad. If there is a glut (which is just as bad economically), too bad again. Yet the people of the cosseted West have often dared to suggest that poor farmers, operating under these all-but impossible conditions, fail because they are 'backward'. At times, our arrogance and insouciance are truly staggering.

Where the constraint does lie with technology – or is theoretically open to improvement by technology – those who seek to help must again avoid glibness. Thus, in the west we breed cereal crops with a high 'harvest index': the total biomass of the crop may not be great (and may even be less than in old-fashioned crops) but the *ratio* of edible parts (grain) to non-edible parts (straw) is high. This seems to make perfect sense, yet we cannot assume that any particular Third World farmer would benefit from such a crop. For a small arable farmer of, say, India (of whom there are millions), straw is an essential commodity. It is the principal feed for his cows or oxen, without which there is no transport and no plough, and hence no crops at all. The ideal crop could be one with an abundance of soft straw, and perhaps with a (relatively) high protein content, although in a West European context, protein that remained in straw after the grain was ripe would be a terrible waste.

Indeed, livestock play several vital roles in poor, tropical countries. Plants, rather than animals, are the chief source of calories and proteins (as they would be in rich countries, too, if we had any sense), but meat and milk in poor countries do and should act as guarantors of quality. Plant protein tends to be deficient in certain amino acids which small amounts of meat can make good. Meat provides some essential (unsaturated) fatty acids, as well as non-essential (saturated) ones, and is also the major source of some minerals, such as zinc (and calcium in the case of milk), and of vitamins such as B12 which are difficult to obtain from other sources. The traditional South-east Asian diet – primarily rice, with vegetables, meat and fish as garnish – is, in essence, the unimprovable human diet. A vegan diet is not.

Of course, livestock raised on grass or cereals clearly does not provide as much energy and protein as is contained in the fodder itself. But animals – especially those bred for the Third World – also make use of provender that human beings cannot eat at all, and live on land that cannot in practice be cultivated. Thus do the cattle of India subsist on rubbish (literally), or the straw of rice or the woody stalks of pigeon-peas, while the ducks in the traditional paddy fields of China live on the algae and snails that grow in the ditches, while the carp live on the excrement of the ducks. Cattle, goats, sheep, asses and camels in Africa graze in mixed herds, and between them utilise the tussocks and acacias of semi-deserts that could never bear a conventional crop. Third World animals, especially cattle, also work for a living. The bullock cart and plough have been with us for 3,000 years and could well be around in 3,000 years' time. Cows originally became sacred in India because they were the mothers of those bullocks, and now, increasingly, the cows themselves are being pressed into transport duties.

So the 'ideal' Third World animal should, above all, be tough.

It must withstand tropical heat, make do on poor fodder, and often make do with very little water. Typically, however, western entrepreneurs generally try to 'improve' Third World livestock by introducing cattle and poultry that have done well in the soft pastures of the North – only to see them dried to a crisp at the first tropical drought. Yet this does not mean that western breeding *techniques* have no place, only that they must be applied with care. Traditional breeds and methods of husbandry must not be thrown lightly aside, but that does not mean they are beyond improvement. That this is so is proved by the variation in performance. The best cows and bullocks of the Third World out-perform the worst, in strength and general toughness, several times over.

Fortunately, there are enlightened attempts to improve tropical livestock. Bob Orskov carries out research to produce cattle for the Third World that will grow well even though they are fed only on rice straw and pigeon-pea stalks: for example, animals which have extremely big stomachs (as tropical cattle generally do) to hold poor fodder in sufficient quantity. Scientists of Britain's Agricultural and Food Research Council co-operate with scientists in Addis Ababa to produce all-purpose Third World cattle: *cows* (as opposed simply to bullocks) that are able to pull a cart or a plough, and yet can produce a calf each year and also supply enough surplus milk to feed a family. This is much more appropriate than a 5,000-litre Friesian.

In general, however, the improvement of both crops and livestock must be a compromise. If the protein content of a plant is maximised, yield will go down because, weight for weight, it takes more energy to produce protein than starch. A cow that does well on low-grade fodder may not yield particularly heavily even when fed on high-grade concentrates. It's horses for courses: improvement can be defined only by context.

But whatever improvements we seek, in crops or livestock, will in future be achieved more easily and more radically by the emerging techniques to transfer genes, with great precision, from organism to organism. This is 'genetic engineering': the subject of the next chapter.

CHAPTER 7

The Age of Genetic Engineering

This is a chapter of wonder and of promise, for genetic engineering is truly wonderful, and the promise is of transformation – of living things, of human life, and the prospects for survival on this planet, of us and all our fellow creatures. The problems, the caveats, the huge implications and, above all (for we must not despair!) the *challenge*, I will discuss in later chapters. For the time being we can wallow in achievement.

The wonder is several-fold. First, most obviously and most extraordinarily, genetic engineering enables us in principle to arrange the marriage and combine the qualities of any two creatures – or indeed of any number of creatures – into one, irrespective of species or indeed of kingdom. Genes may be transferred from beans into cabbages, or from beans into cows or bacteria, or *vice versa*, or all three, or any other combination you may care to think of. In the last chapter I described concentric gene pools, culminating in G3: distantly related organisms between which genes could be transferred by heroic exercises in tissue culture. Genetic engineering dwarfs all of that (although by no means does it supersede the techniques). Every living thing has become the gene pool of every other thing. For you and me and everything else, monkeys, worms, oak trees and microbes are – if we choose – our G4.

Furthermore, the new techniques – in principle, and with caveats which must again be discussed later – enable us to be *precise*. Conventional breeders generally combine entire genomes, or at best transfer large pieces of chromosome from organism to organism (if such cytological manoeuvres can be called 'conventional'). But lengthy programmes are then required to eliminate the unwanted genes that come with the few desirable ones. Genetic engineers can transfer just the gene they require, and no other.

But the genetic engineer is not confined to the genes provided by Nature, ingenious and bountiful though Nature is. Any given gene can be modified. It may be mutated (not simply by crude assaults with X-rays and other mutagens, but precisely; changing specific sequences of nucleotides), or (I give an example later) two or more genes from different organisms may be hybridised, before introduction into a third host. Thus the engineer can try out the effects of known alterations, which is highly instructive, and can also in principle improve upon what Nature has to provide, just as a pharmacologist may modify a natural agent from a plant to create a truly precise and efficacious drug.

In principle, too, engineers can create quite new genes, and when we understand better the relationship between gene structure and gene function, and between gene function and the effect on the whole organism, they will be able to instal not a gene so much as an *ability*, guaranteed to be provided by the gene in question.

In parallel with all this, genetic engineers are learning not simply to add new genes, but precisely to modify the behaviour of the organism's existing genes, and if the genes that are modified are themselves controllers, overseeing the rest of the genome, then the effects of such intervention can be profound indeed. As again discussed later, *antisense genes* are now among the leading means to achieve this.

Such power is becoming manifest in all aspects of life, and in the next few decades will surely become predominant. Genetic engineering promises agriculture that is unprecedentedly productive and yet is benign, in environments where crop production hitherto has at best been fitful. In medicine we are already seeing a new generation of diagnosis and therapy, including several new generations of drugs. Overall, we see what can properly be called a new 'paradigm' of medical science. Industry is being transformed, with profligate and polluting factories that now thrive only at huge temperatures and pressures replaced by processes that operate with truly dazzling efficiency in the mild environments that nurture living things. Old-style pollution can be avoided and, to a large extent, the mess that is there already cleaned up. All this and more is the promise – the real prize that is there almost for the taking, if only (as discussed later) we can conduct our affairs appropriately.

What then is 'genetic engineering'? We should begin at the beginning.

THE BASIC TOOLS OF GENETIC ENGINEERING

Genetic engineering at its most basic implies that a specified piece of DNA – in general corresponding to a desired gene – is snipped out of one organism and snipped into another. The very fact of it, let alone the implications, is awesome. In practice the scientists who carry out these techniques tend indeed to be dextrous; the cack-handed need not apply. But there is also an underlying logic and simplicity which can make the process seem, in essence, like advanced cookery.

Genetic engineering depends, as does so much in applied biology, upon the opportunistic exploitation of agents and mechanisms that occur in nature. The agents in question are two main sets of enzymes: those that occur in all cells, and whose job it is to help to assemble and repair DNA; and, crucially, enzymes found in bacteria whose job it is to destroy DNA.

Bacteria, some of which are the agents of a number of the world's most deadly diseases, are themselves attacked by viruses known as phages. Any host involved in prolonged conflict with a pathogen tends in time to evolve specific defences. With a virus pathogen there is very little (if anything) for the host to attack, apart from the virus's DNA. So bacteria have evolved enzymes that attack and sever the invader's DNA. These enzymes thus restrict the virus attack and are known accordingly as *restriction enzymes.**

That such mechanisms existed first became evident from observations in the 1950s, through the work of Salvador Luria. Then, in 1970, Hamilton Smith at Johns Hopkins University, Baltimore, isolated the first actual enzyme. It came from the bacterium *Haemophilus influenzae*, and as it happened this enzyme attacked the DNA not of a phage, but of the bacterium *Escherichia coli.* However, the enzyme did not attack the DNA of the *H. influenzae* that produced it. As is known now (though it was not known then), the vulnerable portions of the DNA of a bacterium that actually produces a restriction enzyme are chemically protected (by additional methyl groups).

*I still find the expression 'restriction enzyme' confusing, but the term refers to the general effect of the enzyme upon the virus attack; it does not allude to its specific effect upon DNA.

Crucially, the *H. influenzae* restriction enzyme attacked the *E. coli* DNA only at specific points: only, in fact, when it confronted the sequence: G, T, any pyrimidine, then any purine, A, C. The enzyme simply regarded such a series of six as its appropriate target. Other sequences, of other lengths, it simply did not recognise. The net effect of its depredations is to cut the invading DNA into a series of lengths, each cut being at a specific, targeted spot. Each such length is called a *restriction fragment.*

Now, several hundred different kinds of bacteria are known to produce comparable restriction enzymes, but they do not all act in the same way. There are others that attack specific sequences of six nucleotides, but different enzymes home in on different sequences. Others again attack specific sequences of four nucleotides – again, different ones attacking different sets of four. Yet others attack at specific sequences of five, and others of eight. Since a given sequence of four is more likely to occur in any particular length of DNA than is a given sequence of six, restriction enzymes that attack sequences of four cut the DNA at more places, and hence into shorter lengths. Restriction enzymes that attack only at eight-nucleotide sequences produce very long lengths of DNA; useful in analysis of, for example, the human genome, as discussed in Chapter 9.

Then again, while many restriction enzymes cut the double-stranded DNA straight across, leaving blunt ends, others make staggered cuts. Thus each severed end is left with two nucleotides from one of the strands protruding beyond the other. These protrusions are known as *sticky ends* because the two nucleotides poking out from one piece of DNA can be joined up to a complementary pair poking out from another piece, *even if the two pieces originally came from quite different sources.* This mechanism is thus of crucial importance to the genetic engineer.

Nowadays, restriction enzymes from many different sources and with many modes of behaviour are produced routinely by specialist companies for the genetic engineer. They are the standard engineering equipment.

Because different restriction enzymes will cut any given sample of DNA at different known points and into different lengths, they provide between them an obvious and superb mechanism for analysing DNA sequences. Simply produce a series of different fragments from the same DNA with different enzymes; identify the nucleotides within each fragment; and then work out how the different fragments overlap. Thousands of different DNA sequences have now been published.

However, the chopping of DNA into bits is only a preliminary. The second requirement is to stitch together the DNA ('genetic tailoring' really would be a better expression than 'genetic engineering'!) from different individuals or organisms. This process is called – obviously enough – *recombination*. DNA formed by joining sections from different organisms is called *recombinant DNA*. The expression 'recombinant DNA' has echoed this past decade through chambers of commerce and through parliaments. I sometimes wonder how many of those who bandy the term and are so exercised by it actually know what it means.

The first recombinant DNA to be created in orderly fashion was produced at Stanford University, south of San Francisco, in 1972, by Janet Mertz and Ron Davis. They first employed restriction enzymes to produce staggered cuts in DNA. Then they mixed different samples of DNA, each with staggered cuts made at similar places by similar enzymes. The sticky ends of the different fragments cleaved together, or *annealed*. Fragments with different origins were just as likely to anneal as those with the same origin. So long as the different samples had been cut with the same restriction enzyme, they would be bound to have similar sticky ends; and complementary sticky ends link together, irrespective of origin. This is not teleology; the DNA fragments do not *know* that they are breaking new biological ground. The chemistry takes over.

In practice, the annealing joints made between fragments by such 'passive' chemistry are weak; they are simply 'hydrogen bonds', brought about by electrostatic attraction between adjacent, active molecules. Stronger bonds must be created by yet another enzyme known as *ligase*, one of those whose normal function in the cell is to keep the DNA in good repair. As was discussed in Chapter 2, earthly life can be seen largely as a dialogue between nucleic acids and proteins: the former constructing the latter, the latter (acting as enzymes or in other capacities) attending (and to some extent directing) the former. Genetic engineers merely break in upon this dialogue.

So, DNA can be broken into bits, at prescribed places, in orderly fashion; and different bits, including bits from different organisms, can be joined up again. So long as all this takes place merely in a test-tube, it is interesting but not very useful. Usefulness requires a little more.

CLONING

A crucial undertaking in the analysis and exploitation of DNA is to produce the pieces concerned in worthwhile quantities. The standard way to do this is to introduce the relevant piece of DNA into a single-cell organism and then to allow (a) the DNA to multiply in the organism and (b) the organism itself to multiply. The organisms that play host to the new genes are simply required to act as store-houses and multipliers. If the host organism is a bacterium (which more often than not is the case), a clone of bacteria is soon produced which contains the DNA in relatively huge amounts.

However, the desired DNA will not be multiplied in the host bacterium unless it is first properly introduced. The standard way to introduce the new DNA is initially to attach it to another piece of DNA which is recognised by the host as being acceptable, and which it can deal with. This accompanying DNA is then known as a cloning vector. Cloning vectors can take several forms – they may, for example, be λ viruses, or artificial constructs known as 'cosmids' – but the standard vector is a plasmid. Plasmids, as was seen in Chapter 2, are circular pieces of DNA that reside in the 'protoplasm' of bacteria. Plasmids pass from bacterium to bacterium in their 'primitive' version of sex; thus they at least effect an exchange of genetic information. Indeed, plasmids have in part evolved as agents for breaking and entering, and thus are equipped by nature to act as Trojan horses, and bearers of foreign DNA. Once again we see that genetic engineers merely exploit the existing processes of nature, albeit processes that are somewhat arcane.

In principle, then – the details are somewhat more exacting – the appropriate plasmid is broken open with an appropriate restriction enzyme, leaving sticky ends; the new 'foreign' piece of DNA is allowed to anneal to the two cut ends of the severed plasmid, thus completing the circle again, and the plasmids are then added to a suspension of the host bacterium, whereupon, on contact, they are 'taken up' by the hosts in the process known as *transformation*. Organisms *transformed* by introducing novel DNA are said to be *transgenic*. This kind of process was first achieved in 1973, by Herbert Boyer and Stanley Cohen, again in California.

One last step completes the basic picture. The DNA separated from the tissues of any organism contains samples of all of its genes; its entire genome. This entire genome can therefore be broken into fragments with restriction enzymes, and each fragment can then be introduced separately into some convenient host organism, usually a bacterium. Each host bacterium, with its own piece of the donor's genome, will then develop into a clone. Thus it is possible – indeed, relatively straightforward – to produce a series of clones of bacteria which between them contain, nurture, and continue to multiply the entire genome of the original donor. This is a *genomic library*.

These, then, are the four very basic components of genetic engineering: first, the ability to break DNA into orderly fragments of varying lengths, at predictable places; second, to join different samples of DNA together; third, *via* plasmids (or some other 'cloning vector') to introduce samples of DNA – samples of our choice! – into some compliant organism, such as a bacterium, in which it will be multiplied to provide useful amounts in an accessible form; and finally to establish a genomic library containing all the genes of the organism of interest.

After that, the possibilities multiply; but the ever expanding plethora of goals and techniques divides into two clear categories. The techniques are employed for analysis, which we will discuss in Chapters 8 and 9, and to provide genetically *transformed* organisms, to serve the needs of agriculture, medicine, environmental protection, and virtually all branches of industry, as discussed in the rest of this chapter.

Before they can press genes into service, however, genetic engineers must undertake a few more preliminaries.

FIRST FIND YOUR GENE . . .

Classical breeding is a wonderful, essentially teutonic exercise in observation and logic. By noting the characters of millions of offspring of thousands of crosses, modern breeders build a picture of the underlying genetics of each required character: how many genes contribute to that character; what other effects those genes have; what other genes they are 'linked' to, and hence their positions on the chromosomes. The 'classical' geneticist's picture of a crop or animal's genome that he or she knows well can in fact be fairly impressive, a map of what does what and where it is.

Yet each individual 'gene', as far as the classical geneticist is concerned, remains an abstraction. He may represent it simply as a point on a line, where the line represents the chromosome. He is likely to think of each gene as a bead on a string. For the practical purposes of breeding by crossing and selecting, the 'beads-on-a-string' model is perfectly adequate; its efficacy is demonstrated by the wonderful array of modern crops that have been produced by applying the theory that depends upon that model.

But such abstractions are of limited use to genetic engineers. They deal in chemistry. They need to know precisely which stretch of DNA in the donor organism *corresponds* to the particular bead on the geneticist's hypothetical string. As Mrs Beeton would say, 'First find your gene.' The route by which genetic engineers do find their gene varies, depending on the nature of the problem.

In general, it is relatively simple to find genes that code for proteins that are produced in large amounts in particular cells. Suppose, for example, the genetic engineer wanted to produce a strain of millet that produces seed proteins similar to those of wheat. This could in fact be a very good thing to do, because millet grows in very dry places where wheat cannot, but wheat flour on the whole is more versatile and generally more pleasant than millet flour; at least, people tend to gravitate towards wheat flour when given a choice. And it is the proteins that make most of the difference.

The engineer could begin by identifying a particularly desirable seed protein in the wheat; one of the *glutenins*. He could then pick out the mRNA that produces that particular protein, which is not as difficult as it may sound, because a cell that specialises in producing a great deal of a particular protein will contain a correspondingly large amount of mRNA. He then might 'label' this mRNA with a radioactive isotope. The labelled mRNA is then 'introduced' to the wheat genome library, and it attaches to the part of the library that contains the corresponding DNA. The fact that it has attached to one clone and not to others is revealed by placing the clones in contact with a photographic plate. The clones with the mRNA attached will produce a shadow. This is *autoradiography*.

Thus the engineer knows where, in the clone, the relevant DNA is hiding. He knows the structure of the protein, and hence can infer (or directly work out) the structure of the corresponding mRNA; so he knows the rough structure of the corresponding DNA, and hence he has identified the gene he needs, and has it to hand in a cultured form.

This, however, is a very simple example. A plant storage protein is a definite, identifiable product, so in seeking the relevant gene, the engineer knows precisely what he is looking for. The protein itself is the 'character' that he seeks to transfer or to improve. But most 'characters' are not perceived as simple entities. They are *qualities*, or *abilities*: the ability to resist a particular disease, or to grow more quickly. Behind each such ability must lie a particular metabolic pathway, dominated by crucial enzymes. But plant physiology is not a particularly advanced science, and most of the time no-one knows precisely which metabolic mechanism results in which ability, and which particular enzymes are crucial. If the engineer does not know what enzymes – what 'gene products' – he should be looking for, he cannot easily home in upon the relevant gene. There are about 250,000 known species of flowering plant, and each one has about 30,000 genes; and although many plants obviously have many of the same genes in common, that amounts to a lot of genes. But of all the genes in the plant kingdom, only a few hundred have been identified, and they for the most part are the easy ones – those, for example, which produce storage proteins.

In short, identification of the actual genes that lie behind desired qualities may prove to be the greatest hurdle among several that need to be overcome. Now, however, a new technique is coming on line whose powers of analysis promise to be stupendous. This is the use of *antisense genes*, or, more generally, of *reverse genetics*.

ANTISENSE GENES; REVERSE GENETICS

The underlying logic of reverse genetics – at least for the purposes of analysis; we can discuss production later – is to nullify specific stretches of DNA to see what difference it makes to the organism. This is a time-honoured method of analysis in all physiological research. Often specific mechanisms or organs have been knocked out with drugs or by surgery – to see what difference it makes. In genetics, pioneers such as Sir Archibald Garrod explored the ill effects of particular mutations in human beings; classical geneticists in the Thomas Morgan school provided their own, random, mutations by X-rays.

The moderns employ particular stretches of DNA that are designed exactly to complement particular stretches within the genome. These complementary stretches are stitched into the organism's own genome and interfere with the expression of the 'real' gene. A very nice example was described by A.J. Hamilton, G.W. Lycett and Don Grierson from the University of Nottingham in *Nature* in 1990 (vol. 346, pp. 284–7). Their research involved ethylene, the gas which in plants serves as an all-controlling hormone, influencing the ripening of fruit, the abscission (shedding) and senescence of leaves, and responses to wounding. Whoever controls ethylene controls much of the life of the plant, and has a huge and obvious contribution to make to horticulture. Hamilton and his colleagues sought the gene that controls ethylene output.

They had previously identified an extremely interesting piece of mRNA whose synthesis correlated with ethylene production when fruit ripened and leaves were wounded. They then prepared a piece of complementary DNA (cDNA) which corresponded to that mRNA, and which they called pTOM13. They also showed that this cDNA would code for a protein with a molecular mass of around 35,000.

The idea was to prepare a piece of DNA that was complementary to, and would nullify the expression of, pTOM13; in other words, an 'antisense' pTOM13. They prepared such an antisense region, and introduced it into Ailsa Craig tomatoes with the aid of a vector called *Agrobacterium tumefaciens*, of which more later. These tomatoes, thus 'transformed', duly produced less ethylene than normal, thereby confirming that pTOM13 is indeed involved in ethylene production and also suggesting ways of influencing ethylene production, for purposes of crop improvement.

The next question is: what does pTOM13 actually *do*? What is this protein of molecular weight 35,000 that is produced? Well, the enzymes responsible for ethylene production are known to be ACC synthase, and ACC oxidase. Evidence from cucumbers shows that ACC synthase has a molecular weight of around 53,000, so the pTOM13 protein does not seem to be that. Presumably, then, it could be part of the ACC oxidase system – which, by further studies, Hamilton and his Nottingham colleagues showed is probably the case.

Note in this example how much depends upon inference, and how vital it is to bring data from many different kinds of experiment together. But note, too, the simple logic of the antisense gene (one example of reverse genetics in action) in pinning down the precise piece of DNA that does the precise job. Hence the opinion of Paul Berg of Stanford University, a pioneer of genetic engineering and winner of a Nobel Prize: 'Reverse genetics has been, and will continue to be, the most powerful experimental approach to understanding the molecular details and physiological roles of gene function.' (*Biotechnology*, April 1991, vol. 9, pp. 342–4.)

These, then, are the kinds of ways in which relevant genes are being identified. Once identified, they can be isolated as outlined above. The next task is to introduce them into a new host.

. . . THEN GET YOUR GENE INTO THE HOST . . .

Genes stored in libraries are not generally supposed to function, that is, to be 'expressed'. They are simply chemical entities, being nurtured and multiplied. To be truly of use, they have to be put into a context – in general this means a new host organism – in which they do express. If the new host is a bacterium, as it would be in any one of a hundred industrial microbial enterprises, this is not too complicated. It can readily be put into the new bacterial host as part of a plasmid. However, the plasmid that carries a foreign gene into a bacterium in a library has merely to contain other genes that will help the foreign gene to 'relate' to its host, and be multiplied within it. The plasmid that transfers the foreign gene into an industrial microbe for productive purposes has also to contain stretches of DNA that turn the gene on (allow it to express) and provide the 'full stop' at the end.

Animal cells can sometimes be induced to take up foreign genes simply by injecting directly into their cell nuclei. Provided these foreign genes are suitably attached to regions of DNA that the host enzymes can recognise, they can be stitched into the host genome as if by the normal processes of repair.

Technically the most difficult is to introduce new genes into plants: given that plant cells begin at least with a thick cell wall, and are very big and watery, with the nucleus being only a small part of the whole. Natural plant parasites are widely employed as vectors. Chief among these are the bacterium *Agrobacterium tumefaciens*, already introduced above, and its relative, *A. rhizogenes*, and various kinds of virus.

A. tumefaciens, first described about fifty years ago, attacks wounds of plants, and shows a particular but by no means exclusive preference for the rose family (including raspberries, of which it can be a severe pest) and vines. Somewhat like a virus, the invading *A. tumefaciens* takes over and re-directs the genome of its host cells. It induces them to multiply, to form a tumour known as a 'crown gall', and it induces the cells of the tumour to produce novel amino acids called 'opines'. This is a smart move. The bacterium can make use of the opines, and the host plant cannot; thus does the parasite organise its own exclusive food-store. The whole reminds me of a line of Christ's: 'Make ready wherein I may sup.' *Agrobacterium* invades its host via a special 'tumour-inducing' plasmid – 'Ti plasmid' for short. The plasmid contains genes to carry out three functions: causing the host cells to proliferate; causing the host to produce opines; and a third group, the 'virulence' or 'vir' genes, which breaks into a host chromosome so that the plasmid becomes part of the genome.

Once modified, Ti plasmids are excellent vectors of foreign genes. The vir genes are left to insinuate the foreign genes into the host. The genes that provoke tumours are removed. So too are *most* of the genes that provoke the production of opines, but not all; the stop-start regions, which delineate the part of the DNA that has to be transcribed, are left. It is between these two punctuation regions that the foreign gene is inserted. Thus is the novel gene insinuated into its new host genome, complete with the accompaniments – the means of getting in, and punctuation marks – that can enable it to function.

Invasion by *Agrobacterium* results in cells containing five to six copies of the novel gene, which should be an advantage, because more copies generally means more expression. The main snag with *A. tumefaciens* is its preference for dicotyledenous plants (such as raspberries) rather than monocots (which include cereals, the most important crops of all). But further modifications are increasing the host range. *A. rhizogenes* is less commonly employed than *A. tumefaciens*, but an intriguing and potentially important example is given later.

Viruses are also used as vectors. Cauliflower mosaic virus, CaMV, suitably modified to prevent it causing disease, is an established favourite which migrates rapidly through the whole plant. But CaMV will infect only a few families of dicots, and Robert Hayes and his colleagues from Imperial College London have proffered tomato golden mosaic virus (TGMV) as an alternative. (*Nature*, 1988, vol. 334, p. 179.) TGMV infects a wide range of dicots and monocots, and infects most cell types; it is very good at self-replication, and produces 200 to 500 copies of the foreign gene in each host cell, which results in strong expression. First evidence of foreign gene expression is seen in about ten days, and invasion of the whole plant takes only twenty-one days. Viruses often insert the foreign genes not in the host's nucleus, but among its chloroplast genes, which has both advantages and disadvantages. Their main snag, perhaps, is that they are small, and can carry only a limited amount of DNA at a time.

Then there are a number of mechanical, and sometimes bizarre techniques, techniques which are proving to be of special use in monocots, which *A. tumefaciens* is reluctant to invade. *Electroporosis* is one method: naked protoplasts of the intended host are exposed to the foreign DNA and then subjected to pulses of electricity. These pulses render the outer membrane porous, briefly – just long enough to allow the DNA to enter, whereupon, obligingly, the cell's own DNA repair mechanisms stitch it into place. In the mid-1980s scientists from Stanford University used electroporation to get maize plants to take up DNA directly – but they did not manage to grow the protoplasts into entire plants; as discussed in Chapter 6, cereals are reluctant to regenerate from protoplasts. But in 1988 Carol Rhodes and her colleagues from Sandoz Crop Protection Corporation in Palo Alto did manage to regenerate transformed maize protoplasts by nurturing them on a layer of 'feeder cells'. This was the first time ever that a foreign, introduced gene was shown to be expressed in a cereal plant. (*Science*, 1988, vol. 240, p. 204.) Since cereals are the world's most important crops, by far, this was significant indeed!

Finally, we may note one very surprising way of getting DNA into wheat, which has been explored at the Agricultural and Food Research Council's Institute of Plant Science Research at Cambridge. DNA is shot into the cell nuclei with a 'particle gun'.

The particle gun, produced by the American company Biolistics, fires tungsten particles which are coated in DNA: that is, with the gene that needs to be inserted. These pellets are fired from the gun barrel by a plastic macroprojectile (a bullet) propelled by a .22 cartridge. The macroprojectile strikes a plastic screen, which stops it dead; but the DNA-coated particles carry on. They fly through a tiny hole in the centre of the screen, and will drive through the walls of plant cells placed in their path. Some of the DNA arrives in the nucleus, is attached to chromosomes, and – the IPSR scientists have shown – is expressed. This is the kind of technique that leaves me astonished. I cannot quite believe that it works, but it does. (AFRC Annual Report, 1988/89, p. 13.)

. . . THEN GET YOUR GENE TO EXPRESS

Getting a gene into a plant or animal can be an uphill struggle, and yet it is only *half* the struggle. The other half is to bring the gene under control: to ensure that it expresses only in the appropriate tissues at the appropriate times. It would be silly, for example, if a gene that was supposed to produce better quality proteins in a seed began to produce those proteins in the roots. It takes energy to produce proteins, and such profligacy would enormously reduce total yields – and could, perhaps, make the root more susceptible to pests by making it more nutritious.

The general way to bring a foreign gene under control is to create a hybrid gene; that is, to attach the functional part of the gene you wish to transfer to the control region from some gene that performs an equivalent function in the appropriate tissues. An example is given below: the functional part of a human blood protein gene attached to the control region of a sheep milk protein gene. Again, though, it can be seen that genetic engineering does not provide an instant 'fix'. Genes cannot simply be transferred between organisms in the way that it was once possible to move carburettors between Ford Model Ts.

Introns, which we first encountered in Chapter 2, further complicate the issue of gene expression. These are stretches of DNA *within* the genes of eukaryotes, which have to be excised ('spliced') from the messenger RNA that the gene produces, before the message on that mRNA can be translated into protein. When they were first discovered they were thought to be 'nonsense' DNA: very puzzling, and with no obvious function. It now seems that they help to regulate the extent to which any one gene is expressed. For example: mRNA cannot function until the intron copies are spliced from it, and the ease or difficulty of this splicing determines the readiness with which the mRNA comes on line.

Research in the early 1980s revealed that the *mechanism* of splicing differs between plants, yeast and animals; this in itself means that an animal gene transferred into plants, or a plant gene into yeast, may not be expressed as freely as we might suppose. It might remain unspliced. Furthermore, it now seems that the means of splicing differs even between monocots and dicots. This clearly complicates attempts to transfer dicot genes (e.g. from beans) into monocots (e.g. wheat), or at least to transfer them in a useful, functional form. Thus it is that at the Scottish Crop Research Institute at Dundee, Robbie Waugh of Dundee University has transferred various monocot and dicot introns into the gene of zein: zein is one of the chief storage proteins of maize, and its gene (as it happens) normally contains no introns. Waugh transfers these and other modified genes into naked protoplasts and sees how splicing is affected. Research such as this will eventually pay handsome dividends, for, by modifying introns, genetic engineers of the future will be able to introduce the genes of their choice with precisely the right degree of expression. For the time being, however, the need to get the introns right is another hurdle to be overcome.

These, then, with a few examples *en passant*, are the basics. Now we should look at some work in progress, concentrating, for the rest of this chapter, on projects that have to do with production. Other kinds of endeavour can wait for later chapters. I will begin with one of the most basic endeavours: to place new genes into microbes, and employ those microbes as agents of manufacture. The possibilities are endless, and increasing all the time. I will pick out just a few.

THE QUEST FOR NEW MICROBES

Microbes – a colloquial term that generally embraces bacteria and some unicellular eukaryotes such as yeast – have served humankind since long before we had a written history; they are not simply a scourge, a source of disease. When it was that yeasts were first employed for bread, beer and wine can only be guessed at; probably since before the dawn of agriculture; perhaps for more than 30,000 years. They and other microbes such as the lactobacilli have also been employed to ferment milk to produce endless variations on a theme of cheese and yoghurt; and these and many others have also provided a vast range of pickles. *Aspergillus* fungi are involved in traditional production of tempeh, miso and soysauce (a vast industry!) in the Far East. In short, microbes have underpinned some of the world's largest and most important industries.

Yet, though yeast in bulk is a perfectly tangible cheesy mass, the separate existence of microbes as discrete organisms was not established until the mid-nineteenth century, by Louis Pasteur. Progress followed rapidly, in studies of disease (pioneered by Pasteur again, and by Robert Koch), and for production purposes.

The early decades of the twentieth century saw significant additions to the traditional roles of microbes in food production; indeed, they saw the start of what for several decades more was called (and in some circles still is), *industrial microbiology*. The first came in World War I when David Lloyd George, then prime minister of Britain, asked the young Chaim Weizmann (later President of Israel) to find a way to increase output of acetone, vital for the production of explosive. Within a few weeks Weizmann showed that strains of *Aspergillus* would produce acetone in endless quantity. Then in the 1920s came the first intimation of penicillin, from the *Penicillium* mould, which in World War II became the basis of a new generation of anti-bacterial antibiotics that changed the face of medicine and the significance of bacterial disease worldwide.

The range and versatility of microbes has now been transformed yet again by the superimposition of genetic engineering; and modern industrial microbiology, with genetic engineering thrown in as an option, is now generally known by the fancier soubriquet of *biotechnology*. To be sure, 'biotechnology' is used broadly, in many contexts. 'Industrial microbiology' is still a respectable expression; biotechnology does not have to involve genetic engineering; and it is legitimately applied to techniques (such as tissue culture) which do not involve microbes. In general, biotechnology may be defined as 'the creative use of enzymes', which may or may not be inside the cells that produce them. However (just to nail the point), 'biotechnology' is most commonly applied to microbial production, and often involves genetic engineering.

Microbes, whether simply bred for the purpose by conventional means or – as now is more likely – transformed by foreign genes, seem able to manufacture almost anything. Their chief drawback is their reluctance to produce very big proteins of the kind that may feature in animal or plant physiology: the bacterial cells are simply not big enough.

Yet one truly wonderful new example of bacteria in action, which seems to encapsulate the industrial promise of industrial genetics, *does* require a bacterium to produce an animal protein. Nick Ashley, at PA Technology in Cambridge, England, is now persuading *E. coli* to produce spider silk. (*New Scientist*, 29 September 1988, p. 39.)

The silk of Britain's common orb-web spider, *Araneus diadematus*, is truly miraculous. It is wonderfully flexible and, weight for weight, five times stronger than steel. Its strength comes from an ordered structure of crystalline proteins, of molecular weight around 300,000; its flexibility from disorderly protein molecules between the serried crystals. The principle (a combination of molecular order and disorder) is the same as that exploited by traditional Japanese and Saracen sword-makers who order the inner structure of the steel by tempering and sintering. Furthermore, the orb-web produces five distinct kinds of silk, for different purposes:

for example for making the web itself (two types), for cocooning its prey, and so on. Ashley studied the kind on which the spider dangles: the dragline.

The spider produces the silk in glands in the abdomen, with the protein dissolved in salt solution. It hardens on exposure to the air and is sheared into shape by the spinnerets. Ashley has analysed the amino acid sequence in the two kinds of protein, and from this has inferred the corresponding DNA. A synthetic version of the gene is then introduced into *E. coli* which obligingly produces 'spider silk' granules. These are then dissolved out in lithium bromide – similar in principle to the spider's own gland – and then spun through artificial spinnerets.

One of the buzz-words in materials science these days is 'composites': materials compounded from two or more materials that all lend different qualities. Tennis racquets and jet aircraft are among the many devices already transformed by composites. Spider silk is itself a composite (with more than one molecular structure), and it lends itself perfectly to integration in other composites, together with carbon fibres, ceramics, and all the rest. Furthermore, by modifying the gene structure (and the method of processing and spinning) Ashley could modify the properties of the silk *ad infinitum*.

Thus in this tiny example are all the main principles encompassed. The original source is an animal: emphasising the economic potential of 'biodiversity' – for what other splendours do the living Kingdoms contain? The two principal players – the spider and the bacterium – are infinitely renewable: one feeds on flies, the other on decay. The manufacture will be carried out by factories that operate at modest temperatures and pressures: no pollution, no profligacy. The factory could if necessary be small-scale: high tech could operate from a village, in England or India or anywhere that people care to live. The product should be truly fabulous, as revolutionary and valuable as carbon fibres. Yet this is only the beginning. What might be achieved in 100 years?

For the moment, however, genetic engineering is serving mainly to beef up the traditional microbial industries: in these, after all, the techniques for handling microbes are already well advanced, and the industries collectively are among the biggest and most important in the world. Prominent among them is antibiotics.

NEW AND BETTER ANTIBIOTICS

Many fungi and bacteria produce chemical agents specifically to inhibit or destroy their rivals, and these agents are 'antibiotics', which means 'anti-life'. The most famous is the fungus *Penicillium*, which produced the first known microbial antibiotic, penicillin (and whose relatives manifest in many a blue cheese). But the most versatile producers of antibiotics are the *Streptomyces*, which produce streptomycin, erythromycin and the tetracyclines, as well as a herbicide, an anti eelworm compound, and feed additives.

A dozen or more genes, working in clusters, collaborate to produce an antibiotic. By combining genes from different clusters, David Hopwood and his colleagues at the John Innes Institute at Norwich were able to produce the world's first hybrid antibiotic in 1984. It is called mederrhodin, and combines part of the molecule of actinorhodin with part of the medermycin molecule.

Now, Professor Hopwood and colleagues from various British universities are producing quite new combinations of the antibiotic genes and thus can in principle produce an effectively infinite range of new molecules. Not all of them will actually be useful, but some certainly will be. Once an active molecule is identified, it will be possible to make small, controlled modifications of its structure (by altering the order of genes in the cluster) and so make a range of variants, to refine the action of the most effective types. It has long been possible to modify natural antibiotics by chemical means, and to create novel, synthetic antibiotics. But microbes are much better chemists than human beings. Thus, in antibiotics, genetic engineering again takes us into a new age.

We can attack disease organisms with drugs (of which antibiotics are an example), or we can simply help the body to mount its own defensive responses. Here genetic engineering at present offers two main possibilities: to provide antibodies (the body's natural agents of defence), or to provide vaccines (to provoke an innate response). Again, there are many intriguing examples.

ENGINEERING ANTIBODIES

Antibodies are proteins produced by vertebrate animals which (a) recognise foreign invaders, which are liable to be parasites, and (b) help to mount an attack upon the foreign invaders. They do both these things by latching on to specific recognition points (which usually include proteins) on the surface of the invader, known as *antigens*. Parasites do not carry their antigens specifically to invite attack; it's just that every living thing has to have a surface, and antibodies (or, rather, the cells that produce antibodies) can differentiate between the surface of the body's own cells ('self'), and that of the invader ('not-self').

Antibodies that you and I produce every day in our bone marrows are wonderfully varied, even when intended to attack one particular invader, such as flu virus. They are produced on the basis that there is more than one way to skin a cat, and there is more than one antigen per pathogen for antibodies to latch on to. But antibodies are also employed in biological analysis, to pick out particular (antigenic) proteins, and for this purpose the broadside approach of the natural antibodies is less appropriate. But in the early 1970s César Milstein of the Molecular Biology Laboratory in Cambridge, England, produced antibodies in pure form that attacked one antigen and one antigen only, and in indefinite amounts, by combining particular kinds of antibody-producing cells with tumour cells, to produce antibody-producing cell cultures with indefinite life-span. The resulting *monoclonal antibodies* (mAbs), specifically pinpointing minute quantities of esoteric chemicals in mixtures more complex than minestrone, have transformed vast areas of biological analysis, from cancer to plant taxonomy.

Monoclonal antibodies are already being attached to drugs, particularly drugs of the very powerful but potentially destructive kind, including anti-cancer agents. The mAb would be designed to latch specifically to the target cells, and could in principle latch *only* to those cells, even in the middle of a complex tissue such as the liver or pancreas. One problem is that the mAbs that are most conveniently employed are from mice; when they are injected into a human patient the recipient treats the mAbs themselves as foreign invaders and mounts an immune attack upon them. For this reason J.D. Rodwell reported in *Nature* (1989, vol. 342, pp. 99–100) endeavours to create synthetic mAbs that were more human-like. One solution is to create hybrid mAbs, from mouse and human antibodies.

Another approach – the one he was advocating – is to create 'Fabs': 'antigen-binding fragments, which contain *only* that small fraction of the total mAb that actually binds to the antigen. These small fractions provoke less of an immune response in the human recipient precisely because they are small; for the same reason they penetrate the target tissues far more easily. They could for example be attached to radioactive isotopes and then used to target thrombi (blood clots) simply to find out where they are, or to carry cytotoxic drugs into tumours. Again, production of Fabs – modified natural proteins – is a task for the genetic engineer. Again, this is only one example among many, of relatively early work in progress. Again, the eventual possibilities – delivering extremely powerful and highly specific drugs not only to precise cells but to exact systems within those cells – portend a quite new age of therapeutics: essentially, chemotherapy with absolutely minimum side-effects. Later, I will discuss another way of producing antibodies – from plants!

VACCINES BY GENETIC ENGINEERING

Production of antibodies is half of the body's defence against parasites. The other half is production of various kinds of white blood cell which attack the parasite directly, but generally in combination with antibodies. It costs the body dear to mount a concerted antibody and/or cellular defence; the amount of protein mobilised in a heavy infection is extraordinary. This is one reason why undernutrition predisposes to infection; the body simply cannot spare the response. So the body does not in general manufacture antibodies and immune cells in large amounts *until* the infection has struck. In fact, the first time it is attacked by a new parasite it is generally taken by surprise: it may be several weeks before the body can mount a convincing defence – unless of course it is already dead. If it survives the first attack, however, then it is ready and waiting the second time. To be sure, it does not maintain a vast amount of antibody in store, but it does retain just enough to pounce as soon as the parasite appears again. Hence, if an animal (including a human animal) survives a first infection, it is effectively 'immune' from then on.

Vaccines are designed to imitate the first infection but without causing disease. They present the body with antigens, as contained in the live, virulent parasite, and so provoke the state of immune alertness. But the vaccine itself has been deprived of the physiological wherewithal to follow the infection through. Traditionally, vaccines are made either from killed parasites, or from parasites that are alive but weakened (attenuated), or from parasites of an antigenically similar but pathologically distinct disease (so that exposure to cowpox viruses – vaccinia – immunises against subsequent exposure to smallpox – variola.

There are, however, numerous difficulties. Live vaccines are in general likely to be more effective than killed ones, but they can also cause disease in their own right. Vaccines, particularly if impure, can sometimes provoke inappropriate immune responses that rebound upon the recipient (responses analogous to allergies) which can be highly dangerous. Worst, perhaps, is that some of the world's most troublesome parasites – such as the plasmodium parasite that causes malaria and the blood-fluke that causes bilharzia, known as schistosoma – at best provoke only a weak immune response in their host, even though they are so destructive. Both plasmodium and schistosoma have ways of fooling the host. Both, by different tactics, can change their 'antigenic profile', so that the immune system is constantly faced with a moving target; and the malaria parasites spend much of their time hiding within blood cells or liver cells, where the immune system cannot get at them. Finally, vaccines are made from living organisms (or from killed organisms that have been living) and are delicate. Delivery to where they are most needed (for example to remote villages in tropical countries) is often difficult. The back of a jeep or a bicycle is no place for delicate biological material. I think in this context particularly of a new, robust form of foot-and-mouth vaccine, which I saw being developed in India in the late 1980s. Traditional Indian village life depends upon cattle (for milk, transport, ploughing and dung) but Indian cattle, like cattle in much of the poor world, are permanently debilitated by foot-and-mouth. The new robust vaccine was high tech designed not to obliterate traditional life, but to reinforce it; a proper use, indeed.

In commercial, governmental and academic laboratories worldwide, therefore, the present aim is to produce a new generation of vaccines that provoke *precise* immune attacks that will hit the parasite most effectively without provoking inappropriate immune responses; to provide vaccines that are robust, and can be delivered from the back of a jeep; and – most important – which will provoke immune responses against parasites that naturally induce only a weak response, including schistosoma but also – by far the biggest prize of all – malaria. The new vaccines do not contain whole parasites, dead or alive. They are cocktails of parasite antigens, and, more particularly, of the particular antigens that have been shown to provoke the most damaging (to the parasite) immune response. The production of isolated antigens, which invariably consist largely if not exclusively of proteins, is an obvious task for genetic engineering.

Genetically engineered vaccines are already in use: the genes that produce the surface antigens of *Eimeria*, a serious protozoan parasite of chickens, are expressed in *E. coli*, which thus provides an effective vaccine of synthetic antigen.

The bilharzia worm might soon yield, too. In 1987 J.M. Balloul and his colleagues from the Institut Pasteur in Lille, France, reported that one of the key antigens of *Schistosoma mansoni* had been cloned within *E. coli*, and the proteins thus produced provoked a strong antibody response in rats, hamsters and monkeys, and led to 'significant protection' against subsequent invasion by the parasite larvae. (*Nature*, 1987, vol. 326, pp. 149–53.) Already there are safe and effective drugs against schistosoma, but delivery is all. Reinfection is a perpetual threat, and the parasite still infects 200–300 million people in seventy countries, mostly poor countries. Vaccination, conferring permanent or at least long-term immunity, would be a boon indeed.

But malaria, which kills perhaps a million people a year in Africa and remains a scourge throughout the tropics – optimistic plans to wipe it out have long since been shelved – remains the Big One. To be sure, its impact has been overshadowed somewhat in recent years by AIDS. But each human being dies only once, and malaria does a very good job on its own. To the immune system – and hence to the creator of vaccines – it poses numerous problems. It is injected by a mosquito into the human (or animal) blood stream in a mobile form known as a sporozoite. But these soon flee to the liver cells, where they incarcerate themselves and multiply to produce up to 10,000 each of a quite new form known as a merozoite. The merozoites are then released, but quickly invade red blood cells, where they multiply again. One obvious problem is the constant change from one form to another: the moving target. Another is the hiding away in liver cells or blood cells, where they are hard to attack. A third is that no stage naturally seems to provoke much of an immune response, so a vaccine that slavishly imitates any one stage would not provoke much of a response either.

Thus vaccines aimed at invading sporozoites – the obvious target – have had only a modest impact. Those aimed at the merozoites after they escape from the liver have also had only mixed success. There is evidence, however, that when the sporozoites first invade the liver cell they produce antigenic changes on the surface of the cell itself – at least in some genetically favoured individuals. These antigenic changes then provoke an antibody response against the newly infected liver cells, so the individuals whose liver cells respond by producing altered antigens of their own show extra resistance to malaria, because their newly infected liver cells (which in any case are doomed) are weeded out before the parasite can multiply within them.

Thus, present research aims to provide antigens that imitate the newly infected liver cell of the favoured, resistant individuals. Here is a strange vaccine indeed: designed not to imitate the parasite itself, but the cells that have been blighted by the parasite. Undoubtedly, the successful malaria vaccine that is finally produced (assuming it ever is) will be a cocktail: of antigens from as many as possible of the various parasite forms, so as to attack it at all possible stages. (See the account by F.E.G. Cox, *Nature*, 1992, vol. 360, pp. 417–18.)

Yet my favourite example of work so far involves rabies. Parasites of all kinds are now known to play an enormous role in regulating the populations of wild animals throughout the world. Ecologists traditionally have overlooked them, but they are a key part of any ecosystem. Ecology is the sum of rival survival strategies, and the survival strategy of parasites is as valid as any other.

The survival strategy of the rabies virus (if you can look at it objectively) is truly wonderful. Like the AIDS virus, it cannot invade a new host except subcutaneously. The AIDS virus can rely upon the natural behaviour of the host, for all animals, sooner or later, will undertake sexual intercourse. But the rabies virus is injected by biting, or from saliva laid on an open wound, and to achieve that it must, in general, change its host's behaviour because (contrary perhaps to popular belief) wild animals do not go around biting each other. Rabies effects the change by invading the host's nervous system and evoking two different kinds of behaviour that between them maximise transmission. It does indeed make some individuals vicious, in the archetypal 'mad dog' mode, and they rush about and bite all and sundry, often far afield. Others it induces to behave in an over-friendly fashion, approaching and licking complete strangers; there is nothing so dangerous in a tropical country as a friendly wild animal. These sloppy creatures ensure that all the litter-mates and neighbours are exposed to the virus. The virus wins both ways.

Human beings are interested in rabies not primarily because it plays such a part in the lives of animals (and it is at present – 1993 – threatening to wipe out the East African subspecies of hunting dog) but because it affects humans. The infection is rare – very low on the list of life's hazards – but it is extremely nasty, with agony and madness preceding inevitable death. Bats sometimes transmit the infection to people, but the commonest source in mainland Europe is the fox. Eliminate rabies from the fox, then, and rabies in Europe should cease to be a problem. The British would like the disease eliminated because at present they exclude it only by quarantine laws, which will be very hard to enforce once the Channel Tunnel is complete.

The wrong, indeed potentially disastrous, way to eliminate rabies from foxes is to eliminate the foxes. Clear foxes from one area and they re-invade from another, thus potentially spreading any disease more than before. Besides, foxes are fine creatures. The better way is to vaccinate the foxes in the field by baiting their territories with chicken heads loaded with oral vaccine. Astonishingly, the tactic works. The principal problem, until recently, has been to provide a vaccine that is truly effective and safe, and this has now been produced by various laboratories in France. The vaccine contains antigen from the rabies' virus surface which have been introduced into the vaccinia (cowpox) virus; the virus which formed the basis of human smallpox vaccine (before smallpox was eliminated). Note the principle: placing the crucial (but not dangerous) part of an extremely dangerous parasite in another parasite that is extremely good at invading a host but which is not itself dangerous. Now 25,000 baits dropped from helicopters over 2,000 square kilometres of southern Belgium have produced 81 per cent immunity among the foxes – high enough to ensure that the rabies parasite can no longer sustain itself. If it had not been for the war in Yugoslavia, the EEC might indeed have achieved its target of eliminating rabies from all European foxes by the end of 1992. As it is, we will have to wait. But there is no doubt, now, that we have the means.

These, then, are just a few possibilities offered by transformed, transgenic microbes: from industrial composites to safe and super-efficient vaccines. Other transgenic hosts offer other possibilities. We should look now at plants.

NEW GENES IN PLANTS

Sometimes organisms are given a new gene not as it were to *improve* them, but simply to turn them into vehicles of some *ad hoc* material which they would not normally produce. Sometimes they are fitted with a new gene (or genes) so as to improve (in the opinion of the person doing the transforming!) their ability to do the things they normally do: apples that last longer in store, wheat more proteinous, Brussels sprouts more sprout-like. Here are examples of plants in both capacities.

PLANTS AS 'VEHICLES'

One of the most intriguing examples I have come across of 'plants as vehicles' was provided by Andrew Hiatt and his colleagues at the Research Institute of Scripps Clinic, in La Jolla, California. (*Nature*, 1989, vol. 342, pp. 76–8.)

They employed *Agrobacterium* to carry genes into tobacco plants that coded for mouse antibodies. But there was more to it than that. The proteins of antibodies consist of two distinct kinds of chain, which in the case of the commonest kind of antibody (those known as IgGs) are called gamma and kappa chains. In practice, then, the La Jolla scientists transferred the gene for the gamma chain into one tobacco plant, the gene for kappa into another, then crossed the two plants sexually – and then showed that the offspring produced sizable amounts of the entire antibody, which worked just as efficiently as the original!

All this has several striking implications. First, it suggests that one way to get the genes for complex entities into new hosts is to do as Hiatt did: put some of the required genes in one host, some in another, and then cross the two hosts. Conventional plant breeders do analogous things. More generally, this work illustrates how obliging plants can be. In most of the present world – particularly the poor world, which most of it is – there is an abundance of sunshine and human labour. The two together are the prime ingredients of horticulture. In theory, transgenic plants could produce virtually anything (if animal antibodies are possible, what isn't?), and we can envisage the present market gardens of India, say, which now provide spices and perfumes and dyes, extending their range in a thousand different directions. Here again, we see very high tech working for the economies of poor people (depending on who owns the gardens).

However, Hiatt and his colleagues envisage specific reasons for putting animal antibodies into plants, which in fact would push them from the 'hosts as vehicles' category and into the class of plants that are improved 'for their own good'. Once produced inside the plant cells, antibodies will not get out again, because they are too big to pass through the cell walls. But they can (because that is what they evolved to do) serve to *bind* small organic molecules of a kind that might normally pass in and out: such as herbicides or plant hormones. Thus crops fitted with antibodies might be far more (or less!) sensitive to any chemical applied to them (depending on whether the effect of binding was to enhance or reduce the chemical's action), which would give the farmer far more control over it.

Furthermore, plants fitted with antibodies might, says Hiatt, 'provide new options for the recovery of an array of environmental contaminants, as well as other biologically significant organics'. I love the idea of transgenic city trees busily mopping up soot and traffic fumes, and pondweed scouring spoiled rivers – and perhaps, with a little more engineering, feeding on the filth they remove.

Finally, says Hiatt, antibodies linked to suitable catalysts could influence the metabolism of the plant directly, effectively producing new metabolic pathways. Again, though these are early days, one feels that the only limits are imposed by human imagination.

PLANTS IMPROVED

Small genetic changes to plants can make huge differences. The genetic basis of the alteration that produced the semi-dwarf varieties of wheat and rice, which underpinned the Green Revolution, was extremely simple: a single gene that altered the response to the 'hormone' giberellin. Less visible but still genetically simple impositions can transform a plant's nutritional value or flavour, or its resistance to pests or herbicides, and hence its agricultural and economic potential. One conceptually simple example with huge economic and nutritional implications was described in *Science* in 1992 (vol. 257, p. 1480): to transfer a gene from *E. coli* that produces starch, into a potato. The transgenic potato produces 20 per cent more starch than any existing type. This increases its energy value, and also – more importantly – reduces the water content. When potatoes are fried to make crisps (which the Americans call 'chips') or chips (which in America are 'French fries') the water is replaced by fat. So crisps made with the new high-starch potatoes contain only 30 per cent fat instead of 36 per cent, and they cook more economically because less heat is needed to remove water.

More universal, however, are the many attempts worldwide to provide crops with greater resistance to pests and diseases. As discussed in the last chapter, specific genes for specific disease resistance can be added without upsetting the qualities for which the plant is prized. The danger of compromising the original qualities by providing a new and fancy crop is thus avoided.

Just as many bacteria and fungi produce antibiotics to outdo their fellow microbes, so there are some that produce toxins (which could reasonably be called 'antibiotics') that kill insects. Most notable is *Bacillus thuringiensis*, which produces a polypeptide (small protein) which, when ingested by an insect, forms a toxin that poisons it. In fact, different strains of *B. thuringiensis* produce different toxins, active against different insect groups: *israelensis* against two-winged insects, which include blackflies and mosquitoes; *kurstaki* and *berliner* against butterflies and moths; *tenebrionis* and *san diego* against beetles. Many two-winged insects, moths and beetles (which include weevils) are extremely important commercially, and for more than twenty years farmers have employed soups of *B. thuringiensis* as a pesticide.

With genetic engineering, however, it becomes possible to transfer toxin-producing genes – or relevant bits of those genes – from *B. thuringiensis* into the crops themselves, so that they can produce their own insecticides. This was achieved in the 1980s by Mark Vaeck and his colleagues of Plant Genetic Systems NV, Gent, Belgium (*Nature*, 1987, vol. 328, pp. 33–7). They employed *A. tumefaciens* as a vector to carry *berliner* genes into tobacco, which duly expressed the gene – and indeed produced enough insecticidal protein to protect against the larvae of tobacco hornworms. Furthermore, the transferred gene was passed on in proper Mendelian fashion to further generations. To be sure, tobacco is not the world's most important crop, even if it does make some people a lot of money, but it is obliging, and what can be achieved with tobacco today can be done with other plants later. Potatoes and tomatoes are close relatives of tobacco.

Virtually all synthetic pesticides provoke resistance in the pests sooner or later, but no insect resistant to the toxins of *B. thuringiensis* has yet emerged in the field, despite wide use. In part this is probably a property of the toxin: insects may just find it hard to evolve the necessary defence. But in part it results also from the way they are used, for *B. thuringiensis* toxins rapidly disappear after application, and it is difficult for insect populations to evolve resistance if they are exposed to a toxin for only a short time. If plants produce the toxin all the time, so that exposure is constant, this could favour the development of resistant insects.

But these are early days. The trick (as I observed in the last chapter) will be to introduce different genes – or variations of genes, perhaps artificially altered – into different plants, and then combine different genes in different plants by conventional breeding. *B. thuringiensis* is by no means the only source: many plants already produce their own insecticides, often *in response* to insect attack, and these too can be transferred from plant to plant. Such potential is encouraging, but also cautionary. We can immediately see that genetic engineering is unlikely to produce new, usable crops instantly, as at first seems possible. Instead it will provide new breeding 'lines' – experimental strains, containing particular genes – which must then be integrated into conventional (and lengthy) breeding programmes. Thus we see the power of genetic engineering in agriculture – its theoretical ability to seize upon virtually any gene, from any source – but also perceive that it absolutely does not supersede conventional breeding. The true role of the genetic engineers is to provide the breeders with genetic raw material.

I cannot resist mentioning one variation on the *B. thuringiensis* theme, from the AFRC's Welsh Plant Breeding Station at Aberystwyth. As was mentioned in the last chapter and will be addressed again later, many plants harbour bacteria in or around their roots that 'fix' atmospheric nitrogen: that is, turn nitrogen gas in the air into soluble ions which the plant can absorb as nutrients. Most notably, nitrogen-fixing bacteria of the genus *Rhizobium* live in nodules in the roots of leguminous plants such as clovers, beans and acacia trees. The relationship is symbiotic: the plant provides its bacterial guests with sugars, and the bacteria provide soluble nitrogen.

But nodules on the roots of legumes are attacked by weevils of the genus *Sitona*. Scientists from WPBS have transferred toxin-producing genes from *B.t.tenebrionis* into *Rhizobia*, and so produced leguminous plants that resist weevil attack. The task at the time of writing is to enhance the expression of the gene, and so increase the toxicity, which could be done by introducing a 'promoter' sequence of DNA 'above' the toxin-producing gene. Such a promoter must, of course, be able to function within the *Rhizobium* host, and a suitable candidate is the promoter that controls the nitrogen-fixing genes themselves. Such a promoter has been isolated, from *Rhizobium trifolium*, which operates specifically in the roots of clover. It may have to be modified before it will operate in the rhizobia of other legumes, such as peas and beans. When this is achieved, the task will be to produce 'transformed' rhizobia that can actually compete with 'wild' rhizobia in the field. In general, wild types are more robust than transformed types, because they do not waste energy doing things that they do not strictly have to do.

The WPBS scientists are also seeking to transfer *B. thuringiensis* genes directly into plants (birdsfoot trefoil – a clover-like plant – is the experimental host), using genes from cauliflower mosaic virus as a promoter. Here again we see the opportunism of applied science and the sheer scope of genetic engineering: for lotus, *B. thuringiensis* and cauliflower mosaic virus seem an odd trio. But as the future unfolds, we will see many more bizarre combinations than this. It is better in some ways to put the *B. thuringiensis* resistance genes directly into the plant, for once they are in they are in. But there are also theoretical advantages in leaving the gene in the rhizobia. In theory, it would be easier to spread to the transgenic rhizobium technology from farm to farm (in particular into the Third World), and the bacterial insecticide would not then be put in the crop itself, which at least in theory might be better for consumers.

Crops resistant to *herbicide* are also attracting enormous interest. The logic here is that crops are, in practice, protected from weeds by herbicides, and although this can be (and often has been) overdone, there are advantages. Herbicides adroitly applied need not be environmentally damaging; there is some environmental advantage in farming as intensively as possible (that is, producing maximum crops per unit area) because this (in theory) releases more land for wildlife; and weeding, traditionally, has been and remains a fearsome task. But, in general, whatever kills the weed is likely to kill the crop. Farmers must overcome this by timing and cultivation (encouraging weeds to grow and then hitting them before the crop gets above ground) or by careful dosing (for example, some herbicides affect dicots – which many weeds are – more than monocots, which include the important cereals).

Yet many problems remain. Maize, for instance – a monocot – can detoxify the herbicide atrazine. But dicots cannot. So atrazine is good for clearing maize, but it stays in the soil and so would affect any subsequent crop of soya. Yet soya-following-maize is a good rotation. An atrazine-resistant soya, then, would clearly be good news. Glyphosate, on the other hand, does not persist in the soil, and so raises no such problem. But glyphosate is indiscriminate: it kills *all* kinds of plant. So farmers would like *all* their crops to be glyphosate-resistant.

In the mid-1980s L. Comai and his colleagues reported early steps to glyphosate-resistant crops (*Nature*, 1985, vol. 317, p. 741). They found that in bacteria, glyphosate attacks a particular enzyme known as EPSP synthase – an enzyme involved in the synthesis of some amino acids. They also found glyphosate-resistant strains of bacteria; specifically, as it happens, of *Salmonella typhimurium*, which is a food-poisoning bacterium. These resistant bacteria produced a version of EPSP synthase that was *not* susceptible to glyphosate. Comai and his colleagues identified and isolated the gene that made the resistant EPSP synthase, and employed *Agrobacterium rhizogenes* to transfer this gene (plus a few controlling stretches of DNA) into tobacco. The tobacco proved to be glyphosate-resistant.

Again it seems that with this first step achieved, the world is the farmer's oyster. There is of course an important conceptual difference between the herbicide-resistant crop and the pesticide-resistant crop. The pesticide-resistant crop should encourage the farmer to apply less chemical, while the herbicide-resistant crop in theory enables him to spray herbicides like never before. This is one of those nice dilemmas (we will meet many others as this book unfolds) where a potentially advantageous technology must be tempered by responsibility and restraint.

But it isn't only the conventional crops that may thus be transformed. Trevor Fenning and his colleagues at Horticultural Research International, at Wellesbourne in the English Midlands, are seeking genes that will confer specific resistance to Dutch elm disease, for transfer into the English elm.

For hundreds of years the elm had been the glory of England's open farmland, punctuating the hedgerows. It was of great economic value – a source of much fine furniture and big, solid boards – and, as a native broadleaf, it supported and sheltered entire ecosystems of attendant creatures. Then, twenty years ago, came 'Dutch elm disease', caused by a fungus, *Ophiostoma novo-ulmi*, which is inoculated into the trunks by *Scolytus* beetles and stifles the trees by spreading through their vessels. In the late 1970s and early 80s, 30 million English elms succumbed.

The elms, however, are a versatile lot, and many are resistant to Dutch elm disease, including species from Siberia, Japan and China. But they do not look the same as the English elm, or support the same pyramids of wild creatures, so they cannot simply be brought in as replacements. Thus does Dr Fenning seek, rather, to transform the native trees.

Late in 1992 the work was still at an early stage, but the Wellesbourne scientists have already taken several crucial steps. They have regenerated elm trees from single cells, which is no mean feat. When the required genes are identified, they will of course be introduced into single cells. The scientists have cause to believe that *Agrobacterium* will provide a suitable vector, and they are already screening genes in plants other than elms: for example, one which produces a toxin in cabbage that poisons the *Scolytus* beetle.

Purists would argue – and the point should be made – that a disease-free plant growing wild sets a somewhat equivocal ecological precedent. All wild organisms are held in check by their pests and diseases. We could of course argue that Dutch elm disease is just a bolt from the blue, and not part of the 'natural English order', but that would be spurious science indeed, for it is clear now that comparable bolts from the blue, devastating entire ecosystems at intervals of perhaps centuries or millennia, are very much a feature of 'natural' ecosystems, although they do not seem so because (obviously) we do not see them very often. On the other hand, we have to admit that the English agricultural landscape is almost entirely artificial, and that if we are going to retain any 'wild' nature at all then we have actively to manage what there is. A pristine landscape could no doubt endure periodic wipe-outs of major tree species, but the fragments left to us in Britain seem far too sparse and fragile to be left to time and chance. On balance, then, I suspect that most conservation scientists would favour the disease-free English elm. (Report by Robin McKie in the *Observer*, 18 October 1992, p. 3.)

Equally problematical ecologically, but of enormous agricultural significance and on balance surely benign, are attempts to manipulate *nitrogen fixation*: the process by which some plants co-opt the help of bacteria to obtain nutrient from the nitrogen of the atmosphere.

NITROGEN FIXATION

The growth of many crops, worldwide, is limited by lack of water. Other crops – those of the temperate North – find themselves running out of sunlight, or of warmth. When plants have enough water, light and warmth, their growth may still be limited by lack of nutrients, and of all nutrients, the one they need in greatest amounts is nitrogen, an essential component of proteins and of nucleic acids. Farmers, accordingly, seek in general to 'feed' their crops with nitrogen.

They obtain this nitrogen from various sources. Many rocks contain nitrogen, and although most are useless because they are insoluble, potassium nitrate has been an important nitrogen source, particularly in the nineteenth century. Living things contain nitrogen, and in a tropical forest there is far more nitrogen in the trees than in the soil, so that if the trees are removed, the underlying soil may appear remarkably infertile. When the living things die and return to the ground, bacteria release the nitrogen in the form of ammonia, of which some is immediately absorbed by plants (in solution), some disappears into the atmosphere, and some is oxidised by other bacteria into nitrates. Then follows a race: the plants must absorb the nitrate before it is washed away in the soil water, or acted upon by other bacteria which convert it into oxides of nitrogen, in which form it disappears into the atmosphere. In a good summer the plants win the race and take up most of the nitrate before it is dissipated. But any nitrate left in the soil at the start of a European winter is liable to be washed away – 'run-off' – and thus to pollute the ground water.

However, the biggest source of available nitrogen worldwide is the atmosphere. This contains about 80 per cent nitrogen gas; billions and billions of tonnes of it. Plants cannot absorb nitrogen gas and could not do anything useful with it if they did absorb it. But nitrogen can be combined with hydrogen to provide ammonia, which they can absorb and utilise as it is, or after it has been converted in the soil to nitrate. This combination of nitrogen with hydrogen is called *nitrogen fixation*, and it occurs in three main ways. First, nitrogen and hydrogen gas are combined directly in an industrial process involving catalysts which was invented by Fritz Haber in the early part of this century; this forms the basis of modern 'artificial' fertilisers. Second, a comparable process takes place in nature to a remarkable extent, for example during lightning strikes, with clays acting as catalysts. Third, various bacteria practise nitrogen fixation.

In fact, many bacteria of very different kinds fix nitrogen. Archaebacteria, which are some of the most ancient organisms on Earth and live in marshes producing methane gas that manifests as 'Will o' Wisp', fix nitrogen. They survive only in dank marshes because they date from times before there was oxygen in the atmosphere, and they remain oxygen-sensitive. Many cyanobacteria – previously known, erroneously, as 'blue-green algae' – also fix nitrogen; lying as they often do on the soil surface, they are an important source of fertility. One particular kind of cyanobacterium, *Anabaena*, also lives inside the leaves of an aquatic fern, *Azolla*, which floats on the water in paddy fields; it is vital to the traditional rice crop. Many 'ordinary' bacteria – 'Eubacteria' – also fix nitrogen. These, in general, are the ones that have attracted plant breeders and the genetic engineers.

Many nitrogen-fixing bacteria live free in the soil, and many plants produce carbohydrates from their roots that encourage such bacteria to live in the region around their roots (the 'rhizosphere'). Many tropical grasses do this, including those of economic importance, such as millet and sorghum. Indeed, breeders at ICRISAT have shown that one strain of millet grew 17 per cent faster than controls when the soil around its roots was inoculated with the bacterium *Azospirillum lipoferum*. The relationship between the grasses and the bacteria is symbiotic: the plant provides carbohydrate, and the bacteria provide fixed nitrogen.

However, the symbiotic nitrogen-fixing bacteria that have caused most excitement are those that actually live inside the roots of plants, in special nodules that the plant creates to house them. Thus do bacteria of the genus *Frankia* live in the roots of wetland plants such as alder and bog myrtle, for waterlogged soil tends to be nitrogen deficient. The extremely important tropical tree *Casuarina* – one of the world's best firewoods – also fixes nitrogen with the aid of *Frankia*. Most important, however, are the *Rhizobium* nitrogen-fixing bacteria that live in the roots of leguminous plants: pulses such as peas and beans; trees such as the acacias and the tropical fodder crop *Leucaena*; and forages of worldwide significance, such as clovers, vetches, and lucerne (alfalfa).

Research here has several foci. One is simply to develop more efficient strains of rhizobia; some that will operate at lower temperatures than usual, or at higher soil pH, or which will continue to fix nitrogen even though the soil already contains significant amounts of nitrogen, for most rhizobia stop working if the soil is nitrogen-rich.

But the most exciting prospect is the one first suggested in the 1970s by scientists at the Unit of Nitrogen Fixation at the University of Sussex. In 1971 Ray Dixon transferred genes for nitrogen fixation from the free-living *Klebsiella* into *E.*

coli, the first transfer of such genes. This opened the floodgates. For example, if legumes contain genes that enable them to form nodules to contain N-fixing rhizobia, why not transfer those genes into cereals? What, after all, could be more alluring than a wheat plant that provides its own fertiliser?

In practice, of course, there are several difficulties: the genes that enable legumes to form rhizobium-containing nodules have not been identified; it is not easy to get any kind of genes into cereals; and genes from legumes (which are dicots) may not easily be controlled within wheats (which are monocots). Ideas like this, however, are too good to abandon, and sooner or later the technical problems will be solved. There is still one conceptual drawback, which is that nitrogen fixation is an energy-consuming process, so a nitrogen-fixing wheat would yield less than an ordinary wheat that was fertilised artificially. But in places where fertiliser is too dear, or run-off a severe problem, N-fixing cereals would be a boon.

We could of course go even further. Instead of transferring the ability to house rhizobia into wheat, we could transfer the bacterial, nitrogen-fixing genes themselves. Here there are different snags. The N-fixing – or *nif* – genes have been identified, but there are seventeen of them, and they would have to be transferred as a functional group, like a concert party. Second, bacterial genes have the same code as eukaryotic genes, but they do not necessarily respond to the same control signals (and of course lack introns). Control could therefore be a problem. They could, however, be joined not to the plant's nuclear DNA, but to the DNA that resides in its chloroplasts: DNA that is very similar to that of prokaryotes. N-fixation would then be carried out in the leaves, more or less as in *Azolla*. The problem here is that nitrogen fixation is a 'reducing' process: nitrogen is made into ammonia by attaching it to hydrogen. The presence of oxygen knocks this reaction off course. But chloroplasts generate oxygen as they conduct photosynthesis. Another problem sent to try us. But problems are there to be overcome, and who can tell what the next 100 years will bring?

Most radical of all, however, is the suggestion made by Ralph W. Hardy of the E.I. du Pont de Nemours & Company in 1983 (at the CIBA conference on 'Better Crops for Food'). He pointed out that although nature chose to fix nitrogen by joining it to hydrogen, it would be much more economical and practical to join it instead to oxygen, and thus generate nitrate almost directly. This process did not evolve in nature, but then, at the time natural fixation evolved, there was no oxygen in the atmosphere, because free oxygen did not become common until photosynthesis had evolved and was well established. Joining nitrogen to hydrogen requires energy. But when nitrogen is joined to oxygen, energy is generated, and as oxygen is a major component of the modern atmosphere, a process that used it (instead of being inhibited by it), would clearly be a tremendous bonus. But as nitrogen fixation by oxygen does not occur in nature, enzymes able to effect it do not exist. They would therefore have to be invented, synthesised from scratch. This may seem like a very tall order indeed, but again, we must see what the next century may bring.

In short, the potential of the transgenic plants seems endless. Transgenic animals have many possibilities too.

TRANSGENIC ANIMALS

Technically, it can be easier to produce transgenic animals than plants: there are no cell walls to breach, no watery vacuoles to negotiate, and there is more nucleus to aim at per unit volume of cell. Direct injection of the novel DNA is often all that is required. The key problem, however – which, as always seems to be the way, has not yet been properly acknowledged – is welfare.

As has already been shown in this book, it is all too easy to produce genetic changes in animals that cause them to suffer: chickens and turkeys which grow too fast for their skeletons and become crippled; dogs which are mechanically unstable (such as dachshunds with their extra-long backs), or which, like pugs and pekes, have such short noses that they suffer permanent respiratory distress. It simply is not true, as some breeders seek to maintain, that any creature capable of survival is bound to be physiologically stable and psychologically content. If that were true, there would be no such thing as a genetic disease, but there are in practice a great many such diseases.

Altering an animal by genetic engineering does not, I believe, raise any *radically* new problems of welfare, or of ethics. I do not believe that an animal that is given a gene from another species has *ipso facto* suffered a bigger insult than one that has simply been inbred, or bred to contain mutant genes that are known to be harmful. Yet genetic engineering does raise a specific danger, which we will discuss again in Chapter 11, that of *pleiotropy*. This is the phenomenon (first discussed in Chapter 1) that a gene which does one obvious thing may also have other, quite unrelated effects, and may indeed interfere with the behaviour of other genes. *Inter alia*, this has obvious welfare implications in the context of animals, and could in theory endanger consumers, whether they are eating transgenic animals or transgenic plants. Potatoes, for example, contain genes that produce toxins, and those genes are turned on when the potato is exposed to light (so that green potatoes are poisonous). Suppose, now, we introduce a gene for some quite different purpose which, inadvertently, causes the toxin genes to be turned on even when the tubers are not exposed to light (as indeed is the case in many wild potatoes). Then we would have produced a toxic potato, while trying to do something quite different. In fact, toxic strains *have* occasionally turned up in normal breeding programmes, not because new genes have been introduced, but simply because existing ones have been allowed to express. Pleiotropy must be taken seriously. With animals, genetic engineering could also raise extra welfare problems precisely because it is intended to produce radical change more quickly than conventional breeding can achieve – and indeed to produce changes that are more radical than conventional breeding can achieve at all.

In short, for these and other reasons which I will discuss in Chapter 11, when we create transgenic animals we tread on ethical egg-shells. Minimum requirements should surely be that the added genes impose *no* additional stresses on the animal, and that transgenic animals should be produced *only* for very serious purposes.

An instance which I believe is justified, and where good easily outweighs the caveats, has been pioneered at the AFRC's Institute of Animal Physiology and Genetics in Edinburgh, and is now a commercial enterprise. Genes that produce human proteins that are of enormous value in human medicine are attached to the control regions of genes which produce milk proteins in sheep. These hybrid genes are then introduced into single-cell sheep embryos, and the resulting animals obligingly produce the human protein in their milk. Proteins produced in this way include various blood-clotting factors which are vital for treating haemophiliacs. Obtained from this source, there is no danger of contamination from hepatitis virus – or from the HIV virus of AIDS which, tragically, has now infected many thousands of haemophiliacs following conventional blood transfusions. To produce milk with slightly different protein does the sheep no harm, and the sheep that do the producing are very well looked after. The benefit – the saving of human life – is immense. Such technology as this must be justified.

All welfare problems can be overcome by culturing animal cells which, deprived of any kind of nervous system or organisation, must be considered insentient. One of the earliest examples of a useful gene in action was provided in 1985 by Sharon Busby and her colleagues at ZymoGenetics Inc., Seattle. (*Nature*, 1985, vol. 316, pp. 271–3.) They put the gene for one of the human blood-clotting proteins – Factor IX, whose absence leads to haemophilia B, or Christmas disease – into cultured hamster cells. Some animal cell lines (particularly embryonic cells, or cells hybridised with tumour cells) can be sustained indefinitely, and for production of valuable proteins which, for example, may simply be too big for microbes to produce, they will prove invaluable.

At the other end of the acceptability scale are the attempts by companies worldwide to introduce genes into various forms of livestock to enhance their growth rate, or to increase their milk yield. The chief of these is the gene for bovine somatotrophin, or BST, sometimes known as growth hormone. At present BST is produced from transgenic microbes, and when injected or implanted into cows can increase milk yield up to 25 per cent (over the whole lactation). It achieves this in two ways: by increasing the cow's appetite, and hence its energy intake; and then by causing the cow to divert a greater proportion of that energy intake into milk. But extra production implies extra strain which, in some animals at least, has led to increased udder infection, *alias* mastitis. This can be overcome by better husbandry, but on real, modern, cut-to-the-bone economy farms this often is not possible.

The general point is that further increase of livestock production outside the Third World should be considered a luxury; it is designed to increase the profits of the producers, and not to solve essential nutritional problems. So attempts to raise production by transgenics do not meet the ethical requirement: that they should be of outstanding human benefit. Gene transfer could conceivably be employed to improve various qualities of Third World livestock (for example, resistance to heat stress) but I know of no present attempts to do this, and would lay large bets that in the case of livestock such attempts would not be cost effective because there are so many other more obvious shortcomings to be corrected first.

Yet I have heard rumours – perfectly plausible, the way things are – of plans to produce transgenic chickens that do not mind having their beaks cut off. Battery chickens are often 'de-beaked' – the tip of the beak removed with a hot guillotine – to stop them pecking each other, which they do from sheer frustration in their congested cages. But evidence gathered by monitoring the signals from the nerves in their beaks indicates that de-beaked birds suffer permanent pain, in the same way that human amputees often suffer from the 'phantom' limb. The plan is, then, to produce chickens that lack nerves to the beak. This, surely, is technology gone mad: an astonishingly ingenious procedure to ameliorate a mutilation that is intended merely to facilitate a quite unacceptable form of husbandry, while the whole sorry exercise is driven by greed.

In short, livestock and livestock farmers might benefit from genetically engineered vaccines, and probably from genetically engineered crops; and some animals might, as a favour to humankind, lend their udders or their cells for the production of vital medical agents. But I can envisage no immediate justification for any transgenic livestock that is designed simply to 'improve' the livestock. The faults in present livestock production lie in husbandry. This should be the target for improvement. Of course, genetic alteration of the human animal to correct frank disorder is another issue entirely, and is discussed in Chapters 9 and 11.

Finally, although it is still obviously difficult even to transfer ready-made genes into new hosts in forms in which they will function, scientists are already working on projects that seem conceptually even more advanced. They are for example seeking to modify genes to produce 'designer' proteins with prescribed functions.

'DESIGNER PROTEINS'

Genes make proteins, or, rather, provide the code that determines the protein structure. Proteins are the principal functionaries of the body, serving as enzymes and sometimes as hormones; as haemoglobin; as structural proteins such as collagen; and as food stores in plant seeds. The way a protein functions depends in part upon its overall shape, and in part upon the exact positioning of particular parts of particular amino acids within the sculptural whole.*

But the overall shape of the protein molecule and the positioning of particular groups ultimately depend upon the sequence of amino acids in the protein chain. In other words, the *tertiary structure* depends upon the *primary structure*. Thus, if you want to fabricate a particular protein that carries out a particular job, all you really need to do is put the amino acids together in the right order. To achieve this, you merely have to create a piece of DNA with the appropriate structure, and insert it into an appropriate host.

Some of all this is already possible. The genetic code is known, so it is possible to create DNA with the appropriate nucleotide sequence. Knowledge is still deficient in two areas, however. The exact way in which an amino acid chain of a particular sequence will fold is not yet understood exhaustively. Neither can we predict precisely how a protein of any particular shape will behave. Yet the science of 'protein engineering', leading to 'designer proteins', is already advancing rapidly. For example, scientists at the AFRC's Institute of Food Research are collaborating with several university groups to produce better, or at least more versatile, protein-digesting enzymes from papaya.

The latex of papaya – surprisingly, perhaps – contains several

*In practice, in living cells, this simple statement clearly needs two modifications. First, complicated proteins do not necessarily fold into their tertiary, functional form all by themselves. Many have to be assisted to fold correctly by 'chaperone' proteins ('chaperonins'), which are made by yet more genes, and the study of which is a new science. Second, the function of proteins is further modified after they are made, by addition of sugars, which in general are added by the activity of enzymes, each of which in turn has its own gene to produce it. This fact is liable to prove supremely important in our understanding of biological functions as a whole, and in the creation of new drugs and vaccines. But the above generalisations hold true none the less: protein function depends, obviously enough, on protein structure.

proteases (protein-digesting enzymes) of which the best known is papain, and the most abundant is papaya proteinase omega, or pp omega. Papaya for this reason is employed in traditional cooking, to tenderise meat. In practice, the two enzymes do not break down protein into single amino acids, but into short chains of amino acids known as peptides, or polypeptides. Peptides and polypeptides (which effectively are 'subproteins' in a very digestible form) are used in snack foods and special diets, and so papain and pp omega are both used widely in the food-processing industry. Proteins can be broken down into polypeptides by chemical means, but if this is done then the resulting mixture is too salty.

The IFR scientists have pursued several lines of study. They have shown, for example, that papain molecules contain 212 amino acids, which are preceded by a 'pro-sequence' of 133 acids that may be essential for correct folding. They have also isolated the gene that produces the entire structure, and cloned it within the filamentous fungus *Neurospora crassa*. The bacterium *Escherichia coli* is a more usual host for foreign genes, but papain produced by the gene kills *E. coli*.

The scientists have also revealed the three-dimensional structure of papain and pp omega by computerised X-ray studies, and studied the particular 'active site' of the enzymes which actually break protein chains in the course of digestion. In the active site, the amino acid cysteine, carrying a negative electric charge, is positioned opposite a histidine, which carries a positive electric charge; and the electrostatic 'tension' between the two is what makes the site 'active'. However, in conventional papain and pp omega, this essential electric charge is maintained only when the conditions are neither too acid, nor too alkaline; that is, at pH 4.0 to 8.0. It would be extremely desirable to produce a papain that could operate in very acid conditions – that is, at low pH – because in acid conditions there is little fear of bacterial contamination. It seems very likely that, by altering the sequence of amino acids around the active site, it should be possible to produce a structure in which the essential electric charge is stable at low pH. Since the structure of papain is known, these alterations can be made and tested in an orderly fashion. Since the genetic code is understood, it will then be possible to modify accordingly the gene that produces the papain. Thus will the scientists have produced a brand new protease, to carry out an important industrial role.

In practice, of course, the technologies that produce the end result are extremely complex: the X-ray machine itself, which reveals the three-dimensional structure of a protein in a few weeks – a task that hitherto took years – is quite staggering. But the underlying logic behind these astonishing processes is straightforward, and the advance is inexorable.

This, then, is a taste of some of the enterprises now under way. Where might we go from here?

A GLIMPSE OF THE FUTURE

We can start by thinking trivially, and work up to grand generalisations. On the trivial front, and as suggested in the Prologue, I like the notion of clothing all buildings – or all, at least, that are not architecturally distinguished! – with vegetation. But we could make the vegetation earn its living: besides providing insulation and shelter for birds, it could provide us with food. We might for example produce a Virginia creeper that did not respond to seasons, and was tolerant of cold, and could clothe domestic houses throughout the winter. This should not be too difficult, for it is the roots rather than the stems of plants that need to be kept warm, and basic ground warmth could be supplied from the house. With more genetic engineering, we could add some provender: Virginia creeper that produced strawberries, or passion fruit, or runner beans. I like the idea of harvesting the back of the house for Christmas dinner, and see no reason at all why this should not be possible within 100 years. There are huge difficulties to overcome (as has been shown), but 100 years is a long time in this business.

Somewhat less trivially, I envisage plants in which modern notions of biological pest control have been taken a few steps further. For example, scientists at the former Plant Breeding Institute in Cambridge and at the Rothamsted Experimental Station in Harpenden sought to produce potatoes with hairy, sticky leaves, by crossing domestic potatoes with the wild *Solanum berthaultii*. It has been known for many years that these sticky hairs repel and generally make life difficult for aphids, which are a severe pest of potatoes because, as they suck sap from the leaves, they also spread virus diseases. However, R.W. Fibson and John

Pickett at Rothamsted have shown that the hairs' effect on aphids is not merely mechanical. They exude a substance called (*E*)-beta-farnesene which is very similar to an 'alarm pheromone' produced by aphids; an alarm pheromone being a chemical that they emit when they are attacked, and which warns other aphids that there is danger. (The evolutionary advantage in doing this is explained in Chapter 4. Aphids reproduce asexually as well as sexually, so a group together may be a clone. As they all contain the same genes, it is advantageous to any one to help the others.) Thus, when one aphid lands on a hairy-leaved potato, an 'odour' is emitted that frightens it, and all the others, away.

John Pickett is taking the general idea – of using insect pheromones to control insect pests – in all kinds of directions: for example, to produce sex pheromone sprays that lure pests away from crops, or from the succulent bits to the less succulent bits, whereupon they can be zapped with an inoffensive pesticide, such as one of the pyrethrins.

But with genetic engineering we could take the idea further still: for example, produce hedgerow plants (such as transformed hawthorn) that would produce, say, aphid sex pheromones, and lure the insects from the crop. Then again, everyone knows that a range of plants – Venus' fly-trap, sundew, pitcher plants – in various ways practise insectivory: digesting hapless insects that land on their treacherous leaves, and thus deriving nitrogen. Less appreciated is that a great many plants practise insectivory to a minor degree. Teasels, for example, also seem to digest the insects trapped on their sticky, bristly leaves. So our alluring transformed hawthorn could also be fitted with teasel genes – or, better still, with pitcher-plant genes – and feed itself on the insects it entices to itself.

Alternatively, we could equip the crops themselves with the powers of insectivory. These would be of especial value in the Third World. Imagine indeed a final flourish of Third World cropdom: groups of plants (for mixed cropping is always liable to be a good idea in harsh climates) that endure drought, provide the proteins, flavours and textures desired by the local people, fix their own nitrogen, are specifically resistant to local pathogens, but supplement their diet by gobbling up passing insects. Such a vision is possible. As ever, the real challenges, and the real possibilities, are in poor countries. Never mind the fact that the crops themselves would be enormously high-tech. First, all Third World countries produce excellent scientists, perfectly able to keep control, once they are given the opportunity; second, the seeds from which the super-plants grow would themselves be very simple to use. The bicycle, after all, is the universal Third World transport, generally considered to be the friendliest technology of all. But bicycles – bicycles that actually work, that is – are products of high tech, from their synthetic rubber tyres to their nylon bushes. Complicated things can be simple to use. We will discuss such issues further in Chapter 11.

More broadly, there has been much talk lately of domesticating exotic or wild species to create brand new crops. At a trivial level, British farmers are toying with sunflowers for oil, once the prerogative of southern Europe, and are now contemplating big-seeded lupins for oil and protein (though not, emphatically, the toxic Russell lupins of the suburban garden!). Of more significance is Third World interest in *Leucaena*, a leguminous tree that provides excellent fodder; or *Echinochloa*, an Australian grass that will mature and set seed after a single watering; or *Zostera marina*, 'eelgrass', which grows and sets seed under the sea; or various saltbushes of the genus *Atriplex*, which can grow through over-irrigated soil that is thick with salt, to provide fodder. All these and many others have tremendous promise, but all brand new species have to be bred to adjust to the cultivations the farmer is able to provide. To take a comparable example from animal husbandry, eland antelope from Africa have often been tamed for drought and milk, and show tremendous promise to replace cattle in arid countries because they have a quite extraordinary ability to survive drought. However, as soon as eland are husbanded – kept under control – many of their advantages disappear. If, for example, they are corralled at night (as they would be, so that they are available for milking in the morning), they cannot forage at night; and if they cannot forage when it is cool and relatively damp, they are obliged to feed when it is hot and dry. Under these circumstances, they do not out-perform conventional cattle quite so convincingly. The same kind of principle applies to promising wild plants: once restraints are imposed upon them, their excellent qualities can be compromised.

With genetic engineering, however, we can think laterally.

Undoubtedly it is worth trying to turn *Echinochloa* into a respectable crop. But we could also, instead, try to produce wheat – which is already a respectable crop – that can do the things that *Echinochloa* does. If we could identify the genes that give *Echinochloa* its magical properties, we could perhaps transfer them. In the distant future (100 years) we need not think in terms of 'species', or 'new species'. We could simply regard each crop plant as a kind of Christmas tree on which to hang the particular qualities we require for a particular circumstance. My super-resistant, self-fertilising, insectivorous cereal is just one example.

We could, of course, envisage comparable transformations in animals: sows, for example, that were just a mega-uterus, turning out piglets like a queen termite. We could, but I prefer not to. Animals are sentient beings. For the time being we are obliged to farm them, yet the task here is not to exploit them further, but to devise husbandry that gives them the most agreeable life, and us the food that we require. So far the world has deployed its technology clumsily. We (and other creatures) suffer as much from technology's side-effects as we benefit. True sophistication is not a matter simply of ingenuity, but of restraint.

As for transformed bacteria, and other single-celled organisms – yeasts, isolated animal or plant tissue – the sky seems to be the limit. We can envisage soups of transformed bacteria that would make short work of oil-spills and other pollutants (mainly with an eye to cleaning up the sins of the past, for the future world should be cleaner!). Oil-consuming bacterial cultures are already employed to disperse oil-slicks, as in the Exxon-Valdez disaster in Alaska. It surely will not be long before the human species takes active steps to regulate and maintain the chemistry of the atmosphere: in particular, in the short term, to keep carbon dioxide within bounds, and slow the destruction of the ozone layer. It would be surprising if transgenic organisms had no part to play in this: for example, super-efficient photosynthesising soups of blue-green algae to mop up surplus CO_2.

Arcane chemicals, from paints to solvents, that are now produced in high-temperature, high-energy factories could be turned out by the thousand tonne by quiet, compliant bacteria that operate at room temperatures – and perhaps energise themselves by photosynthesis. Fungi could be fitted with the genes that give flavour to beef (or lamb or chicken or what you will) to provide steaks by the kilometre that could be distinguished from the real thing only by minor variations in texture, and have a fat content precisely adjusted to the nutritional theory of the day. They would take the heat out of factory farming for ever. Perfect microbial replicas of milk (cows', goats', human – or any one of a thousand exotic species to feed to animals in zoos) should be relatively easy to provide.

Add to all this super-safe vaccines, provoking enormous and enhanced immune responses to pathogens but with absolutely no possibility of producing disease; vaccines that could, when circumstances are difficult, be self-transmitting, from animal to animal or person to person; human genetic diseases reversed, and cancers stopped in their tracks; all these and more are among the prizes – and perils! – to come. I will discuss further implications in later chapters.

Thinking far ahead, we can envisage various end points. The first is to separate DNA from the organism entirely, and to create structures in which it operates *ex situ*. In the same way, many enzymes already operate with surprising efficiency when removed from living cells, and indeed are standard components of soap-powders. Unfortunately, I cannot think of an application for *ex situ* DNA that is not trivial. I can envisage, for example, kitchen disposers that consist, effectively, of giant artificial cells, with DNA annealed to the surface of a giant synthetic sponge. Throw in household waste at the top, and out of the bottom will come a beautiful protein soup which, with a few onions, would be the perfect meal. Of course there would be little point in this (even if it were not so trivial) because cultures of transformed yeast cells, say, could probably achieve this end. But the existing example of isolated industrial enzymes suggests that isolated DNA may have its uses too.

More boldly and far more interestingly, we can envisage the creation of entire new *kingdoms* of organism. 'Kingdom' is the largest recognised grouping of living things. All plants collectively form a kingdom; so do all animals; so, too, all fungi; and so do all bacteria (though I believe that bacteria should be divided into more than one kingdom). Suppose, however, we combined qualities of different kingdoms in one organism – not just in a small way (as scientists do when they place animal genes in bacteria, say), but to create qualitatively different creatures? In particular, we might envisage the merger of plants (which are big, and use sunlight as a source of energy) with various bacteria. Those bacteria could include the ones that employ a form of photosynthesis, but employ light energy to split hydrogen sulphide, H_2S, rather than water, H_2O; and could also include the thermophilic bacteria which survive in hot springs. Then we would have plants able to survive in an acrid and fiercely hot environment: the kind of conditions we might theoretically envisage in a few million years, as the atmosphere continues to evolve. Thus, as life becomes intolerable for human beings and for most present-day life forms, we might none the less leave big and noble organisms behind us.

Or we might eventually create an organism from scratch; beginning with the raw materials from which DNA is made (two purines, and two pyrimidines, all of which can be put together by organic chemistry) and ending up with super-crops, or dinosaurs, or replicate human beings, or what you will. This, too, is discussed in a later chapter.

CHAPTER 8

Conservation, Evolution and Murder Most Foul

The world needs genes – or, more precisely, it needs allelic variation. Populations die out, which means that species die out, unless they contain within them sufficient genetic (allelic) variation to avoid death by inbreeding, or wipe-out by epidemic. Allelic variation is thus the raw material and the *sine qua non* of that fashionable concept 'Biodiversity', where 'Biodiversity' means 'variety of species'. Allelic variation is also a *form* of biodiversity; indeed, if you admit (*vide* Chapter 4) that the 'selfish' gene rather than the individual or the species is the true unit of life, then we might argue that allelic variation *is* biodiversity.

I would not argue myself, as many do these days, that the *purpose* of maintaining biodiversity (and hence of maintaining allelic variation) is simply to provide human beings with another resource – as if genes were the same as coal or oil or tin. I maintain, rather (as in *Last Animals at the Zoo*, Hutchinson Radius, London, 1991) that we human beings should strive to maintain biodiversity for no better or other reason than that it is *right* to do so; that the only proper attitude for us to adopt to the world is one of reverence and guardianship. None the less, the fact is that biodiversity – whether we mean variety of species or variety of alleles – *is* one of the world's most valuable resources; far more valuable, in the long run, than coal or oil or tin. But whether you regard the conservation of biodiversity primarily as a moral issue, or as an economic one, its importance is paramount. In any book about modern genetics, it deserves at least a chapter.

However, the conservation of genetic diversity is a very complex business, one that since the 1970s has given rise to a whole new branch of genetics (essentially a form of 'population genetics') and also occupies an entire group of scientists under the auspices of the World Conservation Union (IUCN). All serious conservationists in charge of all the world's wildlife reserves now recognise the necessity of the new genetics; and so, too, do all the world's zoos and botanic gardens, whose role in preserving animals and plants by 'captive breeding' is now universally acknowledged in serious circles.

In practice, the conservation of allelic diversity cannot succeed unless the geneticist knows which individual animal (or plant) is which; who belongs to what species, or to which population within that species; and who, within each population, is related to whom. If the geneticist has perfect records – which in zoos would mean a studbook of long standing – he can work all this out from pedigree and family tree. But perfect records of long standing are rare even in the best-run zoos, and, with a few modern exceptions, are completely lacking for animals in reserves. So modern conservation geneticists have to work out for themselves who is related to whom, who belongs to what population and species, just by looking at the animals that now exist.

That is a tall order. If you were a stranger in a city, for example, how could you hope to establish with certainty who precisely was related to whom among all the residents, just by looking at them? Nowadays, however, such relationships *can* be established – by looking directly at DNA. Furthermore, the same techniques that enable conservation biologists to identify relationships within populations can also (albeit applied slightly differently) help to reveal evolutionary relationships between species: to show, for example, whether pandas are 'really' bears or are racoons, or indeed are neither; and whether cheetahs are really as different from the rest of the cats as they look. At the other end of the scale, these same techniques (with modifications) can also reveal which *individual* is related to which; who is whose sister; who is whose father. These same techniques can also show whether a particular hair in a particular nest belongs to this individual or that, and hence can help field biologists to trace the day-to-day activities and interactions of animals in the wild. In fact, the techniques for exploring family relationships and identifying individuals were originally developed not for conservational purposes at all, but by *forensic* biologists, for the pressing of paternity suits and the identification of murderers and rapists. Here is a rare example indeed of benefit to animals from human-orientated research!

One final bonus is the discovery – broadly speaking, and with many caveats! – that different kinds of DNA mutate at particular rates: the degree of difference between DNA from different organisms shows how long ago it was that they shared a common ancestor. Thus does mutation serve as a built-in *molecular clock*.

So what are these techniques for exploring DNA?

DNA EXPLORED

Geneticists interested in relationships look not only at DNA, but also at the way the DNA is packaged – that is, at the chromosomes – and at the proteins which are its immediate products. Different aspects provide complementary information.

Molecular biologists employ two main kinds of methods to compare and analyse DNA from different organisms. The first – DNA *hybridisation* – merely gauges, in a crude but none the less informative way, the *difference* between two or more samples. The idea is to heat DNA so as to separate the strands, and then to mix the single DNA strands from different species together. Separated but complementary strands from the same species cleave together precisely, and single strands from different species cling together to different degrees, depending on how similar they are. The extent to which they cling together can be measured by changes in overall viscosity. Such measurement is of limited value for *absolute* measurement, but it is useful for comparative purposes. Thus if DNA from animal A clings more firmly to B than to C, then it is presumed to be more closely related to B. By many such cross-matchings, it is possible to build up family trees.

The second principal approach employs the enzymes confronted in Chapter 5: the restriction enzymes. Their general task is to cut molecules of DNA, which they do by latching on to particular 'restriction sites'; and the DNA from different organisms will differ somewhat in the frequency and spacing of any one kind of restriction site. So if DNA from different organisms is exposed to the same restriction enzyme, it cuts those samples into different lengths. These different lengths are known as *restriction fragment length polymorphisms* or RFLPs: an accurate expression, if cumbersome.

The fragments of DNA – RFLPs – in any one sample can be separated by electrophoresis. Each fragment carries a small electric charge: if they are placed in a gel through which they can move, and an electric current is run through the gel, they will move across the gel in tune with the current. The lighter (smaller) pieces move faster than the heavier pieces. Hence they separate.

Then the analyst has merely to reveal which fragment has got where. To do this he first transfers the RFLPs from the gel to something more permanent: in one standard technique, by blotting on to nylon. He then applies 'probes', which are short sequences of DNA that are complementary to known sequences within the DNA he is exploring. These probes attach to the particular RFLPs that they match, and they are 'labelled', commonly with radioactive isotopes. Thus the position of the radioactive probe (and hence of the RFLP to which it is clinging) is revealed by exposure to a photographic plate: each place where a probe attaches is revealed as a dark band. Hence if a particular sample of DNA is first chopped with a particular enzyme, and then exposed to a particular probe, it produces a characteristic 'banding pattern', while DNA from a different organism will give a different banding pattern. Molecular biologists seeking to compare DNA from different organisms treat both with the same enzyme, and then put the two (or more) samples side by side on the same gel, and expose them to the same probes simultaneously. Then the contrast between the two banding patterns can be seen directly.

Chopping DNA into RFLPs and then exposing the pieces to probes is the general method applied in all the different kinds of DNA study discussed below. It is a job for experts, there is much to go wrong, and the comparisons can be very difficult to carry out. But in principle (and generally in practice) the method is extremely powerful, and the information is of the kind that makes you wonder how we ever did without it.

Now we should look at the technique in action in testing different degrees of relationship.

INDIVIDUAL DIFFERENCES: 'GENETIC FINGERPRINTING'

First, as was discussed in Chapter 2, the DNA in the nuclei of eukaryotes (plants, animals and fungi) does not consist entirely of genes: that is, of DNA that codes for specific proteins. Much of it has a function that is largely unknown, and although some of this 'non-genetic' DNA may in fact serve to regulate gene expression, we should not be surprised if much of it actually had no function at all. DNA that simply hung around without getting in the way would escape the notice of natural selection. In Chapter 2 I looked briefly at 'jumping genes', or, more broadly, at the notion that sections of DNA may move from one part of the genome to another, replicate, and spark more changes wherever they land. Chapter 5 showed that genes are 'selfish', and a gene (or any stretch of DNA) is perfectly capable of behaving selfishly *vis-a-vis* its companion genes within the genome. In short, a piece of DNA within the genome that has acquired the trick of replication will continue to replicate, and spread, just as long as its activities do not bring the entire organism crashing down and cause self-immolation.

DNA has a great tendency to alter, both by mutation and through the mobility and replication of particular sections; if the alterations do no immediate harm, and are not selected against, we should be totally unsurprised to find that 'non-genetic' DNA can indeed vary enormously from individual to individual. So it is not surprising, in general terms, that each individual's DNA, at least in the apparently non-functional regions, is as individualistic as his or her fingerprints.

This fact alone, however, is not as immediately helpful as it may seem. After all, it would take hundreds of scientist-years to analyse the non-functional DNA of any one creature base by base, to reveal its full individuality. We need a simple guide, an easily pinpointable quality that differs in a simple way from person to person. And Nature, so often obliging, has obliged us here as well. For within the apparently non-functioning DNA are discrete stretches of nucleotides known as *minisatellites*. Within any one genome, minisatellites hop and spread from place to place, like 'jumping genes'; thus the same (or very similar) minisatellites are scattered throughout the genome. Each minisatellite consists of a distinct sequence, or 'set' of nucleotides that are repeated again and again within the minisatellite. Part of each set is variable, but part – typically about twenty nucleotides – is the same within all the sets in any one satellite, and this consistent stretch is called the *core sequence*. Different minisatellites differ in the nature of core sequence, and in the number of repeats. Because minisatellites have these two sources of variability – length and nature of core sequence – and because they have an innate tendency to alter, the total minisatellite pattern is unique to any one individual (although patterns of *particular* minisatellites may be replicated in different individuals). Minisatellites, then, form the basis of *individual* identification: that is, of *DNA fingerprinting*. To carry out a DNA fingerprint, the DNA is chopped into RFLPs and then exposed to particular probes that latch on to particular core sequences within the minisatellites.

DNA fingerprinting was originally developed for forensic purposes in the mid-1980s; it has since been used to match signs of crime – hairs on jackets, semen, or blood stains – with the DNA (as revealed in white blood cells) of suspects. It has also been used widely to settle paternity cases, for although it is impossible to say on the basis of DNA fingerprinting that a particular man *is* definitely the father of a particular child, it is certainly possible to show on such evidence that he is *not*. In 1992 some doubts were thrown on the uniqueness of DNA fingerprint patterns in humans, and in December 1992, for the first time in Britain, the Crown dropped charges against a man who, one month previously, had been convicted of sexual assault on the basis of DNA fingerprint evidence. But these doubts seem to be based on a statistical nicety: whether there is only a one in ten million (or so) chance that two different people would have the same genetic fingerprints, or whether the chances of two people sharing the same DNA minisatellite pattern are in fact as high as one in one million. Most people would surely argue that one million to one is pretty good odds, and DNA fingerprinting seems set fair to occupy a key role in forensic medicine in the future.

Conservation biologists have been quick to adopt DNA fingerprinting, as nicely shown by Mike Bruford's studies of Mauritius pink pigeons at London's Institute of Zoology. Mauritius pigeons are very endangered: by the 1950s only sixty remained, and half of them were wiped out by Cyclone Carol in 1960. The wild population now numbers sixteen, but since the 1970s they have been bred in captivity by the Jersey Wildlife Preservation Trust, which now has about fifty birds, and by the Government of Mauritius, which also now has about fifty. There is also a reintroduced population in Mauritius – that is, reintroduced to the wild after breeding in captivity – which stands at around twenty birds, and has been kept separate from the original wild birds. There are also a few flocks in zoos elsewhere. The total world population is around 200.

But the captive flock at Jersey has been showing signs of inbreeding depression, including a loss of fertility: after all, all those birds, plus all the other captive birds in the world and the reintroduced flock, are descended from just eleven original *founders*. But inbreeding depression can be reduced. If an inbred and therefore homozygous bird is crossed with another *to which it is not related*, the offspring will be more heterozygous, and therefore less depressed, than either parent. This is the principle outlined in Chapter 6, whereby homozygous and perhaps somewhat depressed parent plants (maize, for example) are crossed to give vigorous, heterozygous offspring.

But there is a snag. No-one knows at present which Jersey pigeon is related to which, because the relationships of the original founding birds – who were caught from the wild – were unknown. Birds which the studbook says are only distantly related may in fact have descended from the same founder. Mike Bruford's DNA fingerprinting studies are sensitive enough to enable him to distinguish between birds that are siblings and birds that are merely cousins; and of course, to distinguish between siblings and non-related birds. This knowledge enables his Institute of Zoology colleagues to devise breeding plans for the captive birds to ensure that inbreeding is kept to a minimum in the future – in the manner described later in this chapter.

Larger scale differences can be explored by looking at DNA that is not so variable as minisatellites: indeed, by looking at differences in the DNA of genes.

MORE DISTANT RELATIONSHIPS: BONA-FIDE GENES

Variations – mutations – within functional genes may well affect the life of the animal; if they are at all harmful they will be punished by natural selection. Thus DNA that actually codes for proteins tends to be less variable than DNA that does not. But natural selection does permit some degree of variation. Not every sequence of amino acids in every protein is critical; some variation in protein structure is possible and yet the protein may still do its job. Hence the genes that code for those proteins can vary a little as well, and, in practice, so they do. We would not expect the variations in the proteins (or the corresponding genes) to be enormous, as is the case with minisatellites. We would expect the same allele in two different members of the same population to have effectively the same structure. But we might expect identifiable and consistent differences between individuals from *different* populations which had not been in contact for some time; or indeed between different species. Furthermore, variations within genes accumulate with time: there will be more difference between the genes of organisms that have not shared a common ancestor for a long time, than between organisms who recently shared a common ancestor. Indeed, when the difference in DNA is correlated with independent evidence, for example from the fossil record, it can provide a measure of the *actual* time, measured in thousands or millions of years ago, at which two organisms last shared a common ancestor.

Yet again, too, Nature has proved even more helpful than we might have hoped. Different *kinds* of genetic DNA within any one organism vary to different degrees. Mitochondrial DNA is on the whole more variable than nuclear DNA. And within nuclear DNA, the genes that are concerned with aspects of immunity are more variable than other genes: they need to be, because they need to be versatile to cope with foreign invasions. So differences between 'ordinary' genes coding for 'ordinary' proteins can in general give information on relationships of species, but differences between the DNA that codes for the immune system, or for mitochondrial DNA, in general provides finer grained information on the relationships between populations within species.

The cat family, Felidae, provides a fine example of molecular classification at work. In general, the relationships between cats have been very difficult to pin down. Some taxonomists (the 'lumpers') have at times placed virtually all thirty-seven living species into the same genus (*Felis*), while others (the 'splitters') have broken the family down into ten genera or more. Most taxonomists are happy now with five: *Felis* (including small cats, such as domestic cats); *Panthera* (the big cats, such as lions); *Neofelis* (the clouded leopard); *Lynx* (the lynxes and bobcat), and *Acinonyx* (the cheetah). But it is difficult to say whether any two species of cat look similar because they have the same recent ancestor, or whether they have different recent ancestors but have been shaped in similar ways by natural selection. Contrariwise, cheetahs look different from other cats and so are placed in their own genus, or even in their own sub-family. But did they have a different ancestor from other cats, or was it just that natural selection has shaped them – uniquely among cats – to chase rather than to ambush?

However, biologists such as Stephen O'Brien of the National Cancer Institute in Maryland, USA, have now studied the molecules of cats in various ways, including DNA hybridisation and analysis of DNA genes. In the main, molecular studies support and take their lead from the classical studies, which are based on fossils, anatomy and behaviour. But as O'Brien records in *Great Cats* (Merehurst, London, 1991), the molecular studies can also produce some surprises.

For example, fossil studies indicate that the Felidae split from the Canidae, the dog family, about 50 million years ago. No argument there. The molecular studies now suggest that the cat family – or at least, the relatively few living members of it – first began to divide up into major sub-divisions about twelve million years ago. But this split divides the small South American wild cats – ocelot, margay, oncilla, Geoffroy's, etc – from the rest. In other words, it drives a wedge through the middle of the genus *Felis*, and also means that half the *Felis* (wild cats, jungle cat, sand cat, golden cats, etc) are more closely related to big cats, lynxes and cheetahs than they are to ocelots and their kind.

The next big split, among the cats that were left after the South Americans had split off, occurred about 8–10 million years ago. The wild cats, jungle cat, sand cat, black-footed cat, etc – that is, various *Felis* of the domestic cat type – split off from the rest, where 'the rest' includes the puma and golden cats (which are both *Felis*) plus *Panthera*, *Lynx* and *Acinonyx*. So, according to the molecules, lions, tigers, leopards and jaguars are all much of a muchness, which is not surprising (they are all *Panthera*); *Lynx* are also much of a muchness, closely related to *Panthera* – no surprise there; and pumas, cheetahs, and golden cats are all a bit different from *Lynx* and *Panthera*. The surprise, though, is that cheetahs are no *more* different from *Panthera* and *Lynx* than are pumas or golden cats.

In short, the classical classification of cats is beginning to look a bit of a mess. The genus *Felis* seems to contain four groups that each seem to deserve the status of genus – that is, ocelot *et al*; wild cats *et al*; pumas; and golden cats. Cheetahs, though they too deserve their own genus, are not widely separate from the rest. They almost certainly did not acquire their present peculiarity through any particularly odd ancestral route, but have simply become adapted to life on the open plain. In the Felidae, then, direct studies of DNA offer some sharp and surprising insights. Many other groups are now being looked at in this way, including the entire class of birds, the classification of which is clearly most unsatisfactory. The next twenty years of taxonomy should be very exciting indeed!

Studies of mitochondrial DNA clarify degrees of difference that are intermediate between species differences and individual differences. Thus they are particularly appropriate for exploring the relationships between populations *within* species. Mitochondrial studies have thus been employed to great effect in recent years to study the ancestry of human beings.

To go back a little: in 1967 Vincent Sarich of the University of California at Berkeley compared the blood proteins of human beings with those of chimpanzees. Blood proteins are, of course, direct products of genes, and more will be said about them later. In the 1960s (though not now) it was easier to analyse protein primary structure than to analyse DNA directly.

By various intellectual manoeuvres Sarich was able to use the differences between the chimp and human blood proteins as a molecular clock, and he concluded that chimps and humans had shared a common ancestor as recently as 5 to 7 million years ago. This caused shock and consternation, partly because some modern biologists were as anxious as Darwin's contemporaries were to emphasise the gap between men and apes – and such recent common ancestry implied an uncomfortably close relationship – but mainly because palaeontologists had concluded from fossil evidence that the last common ancestor of humans and apes lived at least 25 million years ago.

Since then, the weight of evidence – including fossil evidence – has firmly supported Sarich, and partly because of that success, the molecular approach to human ancestry has been placed squarely on the map.

After 1980, analysis through RFLPs made it possible to study and compare DNA structure directly, and Allan Wilson, Rebecca Cann and Mark Stoneking at Berkeley began to study the mitochondrial DNA of different people from all around the world.

As was shown in Chapter 2, mitochondrial DNA is (in general) inherited only through the maternal line, and this produces an odd effect. Suppose you begin with a population that contains ten unrelated women, and suppose that population remains at the same size for many generations (as human populations did before agriculture). In any one generation, some women will fail to breed, while others will have more than their share of offspring. Those who die without issue will take their mitochondrial genes with them. Thus, after a certain number of years – that number being the generation time *times* the population size *times* 2 – all the mitochondrial genes left in the population will have derived from just one of the founder females. Note that because of sex, this does not apply to the nuclear genes: the genes in a female who fails to breed may already have been distributed through the population, and be shared by other females of her generation. So it is not correct to say that the female who provides all the mitochondrial genes is the 'mother' of the entire population, like the Biblical Eve. She is merely the source of all the remaining mitochondrial genes.

From their studies of DNA from 241 different individuals from different races, later augmented by more studies of 120 Africans, Wilson and his colleagues concluded that all human beings now on Earth derived their mitochondrial DNA from a single female who lived just 200,000 years ago. As we have just said, this does not make this single female the mother of all humankind: she is not equivalent to the Eve of Genesis. In fact, demographic studies suggest that the human population of 200,000 years ago could have numbered 10,000. Nevertheless, this hypothetical source of all modern human DNA has become known as 'the mitochondrial Eve'.

Wilson and his colleagues also used the differences between mitochondrial DNA of different human populations to construct an evolutionary tree of humankind. The principle is the same we have seen adopted in the Felidae: individuals A and B are more similar to each other than either is to C, but all three are more similar to each other than they are to D, and so on. In practice, however, the picture is not at all clear when you are dealing with samples of DNA from hundreds of individuals. For example, A might differ from B in two nucleotides, but both might differ from C in two different nucleotides. Who then is more closely related to whom? Nevertheless, with the aid of powerful computers, the Berkeley biologists did produce a convincing tree, and this showed that 'the mitochondrial Eve' lived in Africa. Independent evidence supported this: for example, the fact that the mitochondrial DNA of Africans is more variable than that of all other human beings, which suggests that they have been around longest. Many biologists, including Chris Stringer of London's Natural History Museum, also believe that the fossil data support an 'out of Africa' origin for humans.

Thus the matter rested until 1992. Then, David Maddison of Harvard University and his colleagues argued in *Systematic Biology* (in press at the time of writing) that the statistical methods used by Wilson and his colleagues were at fault. They pointed out that a computer can produce literally millions of different but equally convincing family trees out of Wilson and his colleagues' data. The longer the computer runs, the more versions it produces, and the Berkeley group simply did not do enough runs. The Harvard group produced a great many trees from the Berkeley data that did *not* suggest an African origin.

There, at present, the matter stands. The 'out of Africa' hypothesis is still strong because of the fossil data, but the molecular support for it is shaky. The moral, perhaps, is that although molecules give enormous evolutionary insights, they do not provide the Royal road to unequivocal wisdom. There are inconsistencies and uncertainties within the molecular data, and in general, when judging relationships between animals, analyses of DNA must be used to complement data based on fossil and other records; they cannot be assumed simply to be superior. That said, however, there is no doubt that molecular studies of evolutionary relationships are going to make huge inroads into conventional wisdom over the next few decades.

I mentioned that molecular studies of human relationships began with blood proteins. Blood protein studies still have much to offer, and it is worth looking at some modern studies.

BLOOD PROTEINS: MESOPOTAMIAN FALLOW DEER

The Mesopotamian fallow deer, *Dama dama mesopotamica*, is a subspecies of the 'ordinary' fallow deer of Europe which was thought to be extinct in the 1940s but was rediscovered in Western Iran in the 1950s. It is still extremely rare. There are only about 200 in the world, largely in zoos in Germany and parks in the Middle East.

But there are doubts about some of the existing Mesopotamians. Have they been cross-bred in the past with European fallow deer? Are they in fact inter-subspecies hybrids? Paul Sunnuchs, again at the Institute of Zoology, set out to establish this by exploring an enzyme, glucose phosphate isomerase (gpi), which occurs in fallow deer in two different forms. The Mesopotamians contain one form of the enzyme, and the Europeans contain the other, while hybrids contain a mixture of the two.

Enzymes can again be identified by gel electrophoresis. They are placed on the gel, and their position is pinpointed by exposing them to their natural substrate: the enzyme is present where the substrate is broken down. For somewhat complicated chemical reasons, gpi enzymes show a three-banded pattern on an electrophoretic gel. But whereas the gpi of Mesopotamian fallow deer produces a pattern of three bands high on the gel, the European gpi produces a three-band pattern near the bottom. Hybrids produce a mixed pattern, with seven bands more in the middle. At the time of writing Paul Sunnuchs has found that 17 out of 20 fallow deer that he examined in a Middle Eastern herd were pure Mesopotamian, while the other 3 were European. No doubt there are hybrids in the herd as well. The work continues.

You may of course ask, 'Why bother to differentiate the Mesopotamians from the Europeans? Why not let all fallow deer run together?' In fact, there are two answers to this. The first is aesthetic. Much of the aim of conservation is simply to conserve biodiversity, because biodiversity is wonderful. Mesopotamians do look a little different from Europeans – they are slightly bigger – and their obliteration would be an impoverishment. Second, although different subspecies may interbreed perfectly happily, with no drawbacks whatever, such interbreeding can at times cause difficulties. Sometimes there is 'outbreeding depression':

particular, advantageous groups of genes in each parent type are disrupted by the cross-breeding, to produce slightly enfeebled offspring who, for example, may be somewhat subfertile. Sometimes the two subspecies live in slightly different habitats, and the cross-bred hybrids are not well adapted to either. This could theoretically be a problem if big, shaggy Siberian tigers which are adapted to the snow were crossed with smaller, smoother Sumatrans, which prefer tropical forest. In any case, since there are *possible* snags in cross-breeding and few theoretical advantages, it is best to keep the different subspecies apart.

Finally, biologists explore relationships between animals and plants by examining the packaging of DNA: the chromosomes.

CHROMOSOME STUDIES: ORANGS, OWL MONKEYS AND GAZELLES

The technique of examining chromosomes – *karyotyping* – has been with us for decades. The chromosomes are revealed by staining when the cells are dividing, and they are coiled up and compact, in preparation for mitosis. The number and shape of chromosomes is characteristic for each species, although in some species some chromosomes are sometimes joined up with others and sometimes not, which results in variable chromosome patterns that can be sorted out only by experience. A huge number of karyotypes have now been carried out both in animals and plants. Just a few are outlined below.

First, there are two distinct populations of orang utan: the Bornean and the Sumatran.* It is becoming important to breed orangs in zoos as a back-up for the wild, and the two subspecies must be kept apart. It is important, too, not to reintroduce captive Borneans to Sumatra or *vice versa*. The two subspecies look somewhat different: for example, the Borneans tend to be slightly darker, and the enormous cheek-pouches of the males are even more pronounced and forward-pointing than those of the Sumatrans. But there is considerable overlap in these features:

*In fact there are probably three orang subspecies, since modern studies suggest that the East Borneans are just as different from the West as either is from the Sumatran; after all, the mountains in Borneo are just as divisive as the sea between Borneo and Sumatra – and a great deal older! But the two Bornean groups have so far generally been lumped together.

many Sumatrans are darker than many Borneans, for example. Fortunately, there is a clear difference in the chromosomes. Like chimps and gorillas, orangs have 48 chromosomes. In Sumatrans, one of the arms of chromosome number 2 (the second largest) is turned head to tail ('inverted'), relative to the same arm in the equivalent chromosome in Borneans. Nowadays, no responsible zoo would breed orangs without first ascertaining their subspecies by looking at their chromosomes. Karyotyping has revealed that about 200 of the world's 900 or so captive orangs are hybrids, and these animals should not be allowed to breed.

Many species of tropical monkey are divided into separate subspecies, even when they all live in the same forest; what looks like a continuous forest to us may not look like a continuous forest to them. The classification of South America's spider monkeys, for example, is proving very complicated. Woolly monkeys are difficult too. Dourouroulis – owl monkeys – all look much of a muchness, but there are clearly nine different 'karyomorphs' (chromosome types) and it seems that hybrids between the types are often infertile or subfertile. Nowadays, responsible zoos co-operate to ensure that between them they breed all the different types, and that no-one mixes the types. The whole operation is coordinated from London Zoo. In the wild, it is important to work out where one population ends and another begins, for newly designated reserves should not be placed across the borders of different populations: this would create two subpopulations, each of which might be too small to be viable, yet would run into trouble if they hybridised.

Gazelles are an extraordinarily difficult group. Different types may look almost identical, but breeding failures in the past suggest there may be more differences between them than meet the eye. Modern chromosome studies support this view. Indeed, karyotype studies by scientists of the Institute of Zoology working at the King Khalid Wildlife Research Centre at Thumamah, Saudi Arabia, have shown that Saudi gazelles, which previously were thought simply to be a minor variant of Dorcas gazelles, are in fact a distinct species. There are only about forty or so Saudi gazelles left in the world, and dedicated conservation is urgent. If it were not for the Institute's chromosome studies, the separateness of Saudi gazelles might never have been realised, and if they had simply been mixed in with Dorcas gazelles their hybrid offspring could have failed and the entire species would have disappeared.

Such stories could be proliferated almost *ad libitum*: this is a very exciting and fast-growing area. But I must press on to look at yet another modern analytical technique that is transforming vast areas of both palaeontology and archaeology. This is the *polymerase chain reaction*, or PCR.

MINUTE SAMPLES AND CREATURES LONG DEAD: PCR

DNA is an extraordinarily robust molecule, much more so than most proteins. Minute traces can therefore survive in tissues long since dead: Egyptian mummies, the frozen flesh of mammoths, or even on ancient weapons in museum cupboards (if they have not been too conscientiously scrubbed). There are many animals, too – or even human beings – from which it is possible to take tissue samples in minute amounts, but not enough to carry out straightforward DNA analysis. It is here that PCR comes in, for PCR is a means for multiplying minute quantities of DNA many thousands of times – to produce as much as the scientist requires – in double quick time.

The technique is based upon polymerase enzymes. First, the normal, double-stranded DNA is placed in a medium containing all the nucleotides needed for DNA construction. Then the DNA is heated to split it into single strands. The mixture is cooled slightly and polymerase is added, which then builds a complementary strand of DNA from the loose nucleotides, as a match for each single strand. Hence the original sample is doubled. Then the whole process is repeated.

Until a few years ago, this was all very tedious. As the DNA mixture was heated to separate the strands, so the polymerase enzyme from the previous doubling was destroyed. Each repeat doubling had to be carried out separately, with fresh enzyme added each time.

Then scientists began to take a serious interest in the 'thermophilic' (heat-loving) bacteria that live, miraculously, in the near-boiling water of hot springs. Their polymerases are extremely heat resistant, and they are perfect for PCR studies because they survive the surges of heat needed to separate the DNA strands. So

PCR can be carried out simply by mixing DNA and thermophilic polymerase in a suitable soup of spare nucleotides and then heating and cooling, every few minutes, to just above and just below the temperature at which DNA strands separate. In a matter of hours the scientist can produce huge batches of DNA – enough for all the analysis required – from the most minute samples.

PCR finds its most immediate application in field studies: for example, for examining DNA from the tips of feathers, and thus establishing which particular population the original owner came from; and thus showing whether two neighbouring populations really do intermix, or remain within their own territories. Information of this kind has enormous genetic and ecological implications, which must be understood if the wild populations are to be managed and conserved.

The technique also has enormous implications in archaeology and palaeontology: so much, indeed, that some people have wondered whether it will be possible to analyse the entire genomes of people or animals that are long since dead, and even to 're-create' those animals by introducing their DNA into existing cells. Such matters will be looked at again in Chapter 10. Suffice to say here that although DNA is robust, it is rare to find complete genes, or anything like complete genes, in ancient samples. Sometimes there can be enough to reveal some particular polymorphism, which may indicate that one particular individual is more likely to be related to this modern race or to that, but not much more than that. These are early days, however. Introducing a new technique of this power is like lighting the blue touch paper of a firework. We can only stand back and see what transpires.

Finally, in the matter of analysis, there is a related, though different set of questions: where, in any one genome, do all the genes lie? And what is the nucleotide sequence of all those genes, and of all the alleles and common mutants within the gene pool? These are the issues of *gene mapping* and *DNA sequencing* which are of enormous relevance in human medicine and potentially in agriculture, as well as in basic biology.

GENE MAPS AND GENE SEQUENCES

Gene mapping – finding out the relative positions of all the genes on the chromosomes – dates back to the classical studies of Thomas Morgan in the 1920s. The method is based on the observation that genes which are not on the same chromosome, or are far apart on the same chromosome, have no or virtually no tendency to be inherited together. They follow Mendel's rule: any gene can be inherited independently of any other. But genes that are close together on the same chromosome are physically *linked*, and are very likely to be inherited together. Genes give rise to discernible phenotypic characters. So whether or not genes are on the same chromosome, and how far apart they are, can be gauged by seeing whether or not, and to what extent, the characters they give rise to are inherited together more often than chance would dictate. By such 'linkage experiments' the chromosomes of a fair number of organisms have now been precisely 'mapped'.

By measuring the sequence of nucleotides in overlapping fragments of DNA – that is, in different sets of RFLPs created by employing different restriction enzymes that cut the DNA at different known points – it is also possible to establish the sequence of nucleotides in entire genes, or indeed in entire chromosomes: given that each chromosome consists of a single, though huge, DNA macromolecule. Nowadays such sequencing is routine, and indeed is largely automated, which has made it possible to contemplate the sequencing of entire genomes. Organisms whose genomes are now being mapped and sequenced include the fruit fly *Drosophila*, great stand-by of the geneticist; the thale cress, *Arabidopsis thaliana*, a diminutive relative of the wallflower; the roundworm *Chaenorabditis elegans*, another biologists' favourite; the mouse; the pig, and the human being. Some of this mapping and sequencing is being done in the interests of pure research (for example in *Chaenorabditis*). But some has enormous and almost immediate practical spin-offs. The European programme to map the pig genome for example (code-named PiGMaP and coordinated by the AFRC Institute of Animal and Genetics Research at Edinburgh) is intended to accelerate the breeding of pigs. If the position of genes that code for valuable characters was known precisely, breeding programmes could be refined.

The worldwide plan to map the human genome – 'HUGO' – is coordinated from Britain, and the immediate aim is not to sequence every last nucleotide (there are three billion in the entire human genome!) but at least to sequence the regions of interest, once these regions are identified by mapping. Regions of interest of course include the particular alleles that cause genetic disorders, including such obvious and genetically simple disasters as cystic fibrosis on the one hand, and more complex and subtle predispositions on the other – for example to heart disease. More follows in Chapter 9.

This, then, is the range and power of modern DNA analysis. Its implications are obviously far-reaching, and some of them will be discussed in Chapters 10 and 11. For the rest of this chapter, the applications in animal conservation will be explored. But first, we must ask what it is we are trying to conserve.

WHAT ARE WE TRYING TO CONSERVE?

Biologists have now named and described between one and two million different species of living creatures. Most of those species are animals; most of those animals are insects; and most of those insects are beetles. Being a modest lot, biologists concede that they certainly do not know *all* the species on Earth. Fish are more conspicuous than beetles or worms but among teleosts alone (just one of several classes of bony fish) 100 new species are named each year. So perhaps, biologists used modestly to say, there would be as many as five million different species. But most species live in tropical forests, and recent studies in the forest of Panama suggest that biologists have not been nearly modest enough. By counting the insects living in the canopy of just one species of tree, and then multiplying by the number of species of tree in the forest, Terry Erwin of the Smithsonian Institution in Washington has calculated that the total number of species on Earth could be anywhere between 30 million and 100 million. To be sure, some biologists now feel that Erwin got carried away, and that the true number may be less than 10 million, but whatever the truth, we must conclude that (a) the variety of life on Earth is truly staggering and (b) after 300 years of serious and ever-accelerating natural history we have come nowhere near a complete description of that variety.

But the tropical forests, which harbour most of the species, are being felled; so rapidly that conservationists now soberly calculate that half of all species may disappear in less than a century. There have been several 'mass extinctions' in the past, including the one which, about 65 million years ago, brought about the end of the dinosaurs. But the natural extinctions of those long-gone times took thousands or millions of years to run their course. The present wipe-out will be achieved in decades. The man-made extinction is occurring thousands of times faster than those of the past, and Andrew Dobson of Princeton calculates that genes are disappearing a million times more quickly than mutation is able to supply new ones. Very obviously the conservation of existing life forms is among the most pressing issues of our age. We cannot hope to save more than a token proportion of those species without applying good science, and a large chunk of that science is genetics.

GENETICS FOR CONSERVATION: POPULATIONS, COMPUTERS AND NOAH

Noah, according to Genesis, saved all the animals on Earth by marshalling two of each kind on to his ark – which, apart from the matter of numbers (did he really gather together 2 × 30 million-plus individuals?) may seem plausible enough in principle. After all, a healthy breeding female and a healthy young male ought to be able to found a lineage. In practice, however, studies with hundreds of populations of animals in the laboratory, on farms, in zoos, and in the wild show that Noah must have been very lucky indeed; an impression that is amply reinforced by modern computer models.

To begin with, animals do not always stay fit. Apparently, no animal on board the ark slipped over in a storm and broke its leg. There was no disease. No individual was infertile. Yet in practice, disease and accident are commonplace and, in large groups, over time, they are *inevitable*. Computer models (and many a practical study) suggest that unless a population contains at least dozens of individuals, then accidents alone – a run of bad luck wiping out or compromising key individuals – can drive a population to extinction.

We can add to that the phenomenon of *demographic stochasticity*. 'Demographic' of course refers to populations, and 'stochasticity' refers to the kind of peculiar things that happen for purely statistical reasons. The main point here is that of sex ratio. Most large populations of animals tend to finish up with 50 per cent males, 50 per cent females, just as a spinning coin will come down heads or tails with roughly equal frequency. But there is no *mechanism* to ensure that males and females, or heads and tails, even each other out. That is just the way things tend to turn out, after a time. If you toss a coin twice, then it might come down heads both times. Less often, but sometimes, you will get six heads out of six, and there is no theoretical reason why you should not get 100 heads out of 100 or 1,000 out of 1,000 – although, of course, this is rare. The same considerations apply to populations of animals. If 100 babies are born in any one generation, then, most probably, there will be a fair representation of both sexes. But as the population diminishes, and numbers grow smaller, the chances are that the babies of any one generation will all be of one sex. If this is so, then the population shudders to a halt. Thus it was that the first efforts to breed Arabian oryx in captivity in the early 1960s (after heroic rescue from under the guns of hunters in the wild) at first produced a succession of males. The last six pure-bred dusky seaside sparrows on this Earth were all males. For this kind of reason alone, computer models show that small populations – less than a few score individuals – are liable to go extinct sooner or later.

Even if the population rises above a few score, however, the population cannot be considered safe. If there are fewer than a few hundred individuals then inbreeding depression is liable to rear its head. 'A few hundred' may seem a lot, but at any one time, some will be immature, some will be sick, some infertile, and some too old. So the number able to breed – the *effective population* – will always be lower than the real population, and often considerably so. The smaller the population, too, the faster genes are lost by genetic drift. Finally, many animals in a natural state tend to exacerbate inbreeding problems by practising polygamy – one male with many females, or one female with several males – so that in any one generation a proportion of individuals may fail to breed, even if able to do so.

Put all these reasons together – accident and disease, demographic stochasticity, the presence at any one time of non-breeding individuals, and inbreeding – and conservation biologists now calculate as a rough rule of thumb that unless a wild population contains around 500 individuals, it is liable to go extinct, sooner or later. Yet even 500 is only enough to allow the population to tick over, with the rate of loss of genes by drift roughly compensated by new mutation. Five hundred is not enough to allow the population to evolve, and hence to adapt to the changing environment. Neither – as will be shown later – is it enough to enable populations to resist systematic depletion by poachers. Five hundred, then, is a very conservative figure.

But big animals – especially big carnivores – need a huge amount of room. Siberian tigers living in their hostile plains typically require 500 square km (200 square miles) *each*. A 'safe' population of 500 individuals would therefore need 250,000 square km (40,000 square miles), a very significant slice of land. In parts of India the terrain is richer: individuals may need as little as 20 square km (8 square miles). But in general a truly viable population of tigers needs as much land as there is in a large English county, or a small country, and India, unlike most of Siberia, is dense with human beings and their cattle, who do not get along with tigers at all. In practice, then, although there are five living subspecies of tiger, there seems now to be only one viable population in the whole world: in Corbett National Park. Unsurprisingly, three subspecies of tiger – the Caspian, the Javan and the Bali – have gone extinct within the past few decades.

All this has several implications. The first is that genetic considerations – lack of genetic variability leading both to inbreeding and to a lack of ability to evolve – play a large part in driving small-ish populations towards extinction, at which point accident, disease and statistics may finish off the job. The second is that endangered animals are even more endangered than most people realise, because populations that may seem healthy enough at first glance may be near, or already over, the brink, while the areas needed to restore them may be far greater than are available. The third is that breeding in captivity at present is very often *essential*, if animal populations are to be raised to safe levels.

BREEDING IN CAPTIVITY

Populations of animals in captivity do not need to be as large as those in the wild. Thus Tom Foose of the World Conservation Union has calculated that wild populations of rhinos should contain at least 2,500 individuals – which very few of them do – partly for all of the general reasons given above, and partly because virtually all wild rhino populations are still being poached. But captive populations could tick along without significant genetic decline with only 150 individuals. Note, however, that 150 rhinos is still a great many: far more than any one breeding centre – which generally means any one zoo – could possibly manage alone. Point one, then, is that conservation by captive breeding is doomed unless zoos co-operate, which they are beginning to do, sometimes on a region-wide basis and sometimes (as with very rare animals such as the giant panda) on a worldwide basis.

There are several reasons why zoo populations can safely be smaller than wild populations. For one thing, captive animals should be safer than wild ones. Second, they can breed faster, so whereas wild female orang utans average only one baby every seven or eight years, in zoos they can produce a baby every two years or so. But the chief reason is that zoo animals can be genetically *managed*. This has two main connotations. First, and always, the conservation breeder (as opposed to the breeder of domestic livestock) must seek to minimise the inbreeding which compromises the welfare of individuals. Second, he must seek to ensure that his captive population contains the greatest possible genetic variety: that is, that each gene in the gene pool is represented by as many different alleles as possible and ideally contains all the alleles that are present in the wild population. In principle these two aims – reducing individual inbreeding and maintaining overall diversity – are not identical, but the techniques that promote either one also tend to abet the other.

To ensure maximum possible diversity within his gene pool, the captive breeder should strive to begin with the largest possible *founder* population, and all the founders preferably should be unrelated to each other, although they should all of course belong to the same subspecies. Thus in an ideal world the prospective captive breeder would first select an equal number of young, healthy males and females from different parts of the animal's range. Of course, he could not expect to trawl *all* the allelic variations in the wild population unless he took the entire wild population into captivity, but if he just took one outbred (heterozygous) wild male and one unrelated heterozygous wild female then he would (so genetic theory suggests) thereby capture as much as 70 per cent of the variation within the wild population. If he began with six pairs of unrelated, heterozygous individuals, he could capture 90 per cent of all the allelic variants. Most conservationists feel that to retain 90 per cent of existing variation in a captive population is as good as one can reasonably expect. So, half a dozen pairs of unrelated animals is the generally accepted desirable minimum.

This might not sound too much, but a little contemplation suggests how rarely this modest ideal is liable to be reached. Conservation biologists – as opposed to old-style menagerie managers! – have not so far felt pressed to begin captive populations until animals are already rare. If animals are already rare, then the capture even of a dozen young, wild, healthy individuals may seem an intolerable burden, and not one likely to be countenanced by field biologists or governments. To gather particular animals from different parts of the range, and to demonstrate their unrelatedness to others, would compound the problem.

Founding populations tend to arise in one of three main ways, or some messy combination of the three. First, biologists may simply decide that a wild population is doomed, and round up all the animals that are left. This was attempted with the Arabian oryx in the early 1960s, and achieved with the California condor, the black-footed ferret and the red wolf in the 1980s. Second, as a compromise, conservationists seek to gather up a fair proportion of wild individuals that seem for some reason to be doomed in the wild, but hope none the less to leave a wild population large enough to recover (at least when closely managed). Thus it was that a proportion of golden lion tamarins – small monkeys from Eastern Brazil – were taken into captivity by biologists from the Smithsonian Institution in the 1970s, and half a dozen Sumatran rhinoceroses that were wandering aimlessly in forests that were due to be felled have been taken into American zoos (New York, San Diego and Cincinnati) and into Port Lympne, England. In neither kind of case can the biologist guarantee that the animals he picks up are unrelated to each other, or that they are all healthy and able to breed. He just gets what he can. In fact, the individuals are quite likely to be related, since, for reasons both of cost and logistics, he is liable to gather individuals from just one or a few locations.

The third way in which serious zoos attempt to found breeding groups of endangered animals is simply to identify all the animals already existing in zoos, and then persuade the different zoo directors to join in a co-operative breeding scheme. The problem here is that many zoos kept very poor records before the 1970s, so the origin of the wild ancestors of the existing zoo animals is often unknown. So it turns out, first, that many zoo animals are in fact closely related to each other – even if they are now in different zoos on different continents – because, in fact, they may all have descended from a single litter some time in the past. Genetic fingerprinting is helping to sort this out, as described earlier with the Mauritius pink pigeon. Second, it has transpired in recent years that many zoo animals are sub-specific hybrids, but again, modern techniques are coming to the rescue. I have already cited the 200 'zoo hybrid' orangs, and a genetic study of Asian lions in the mid-1980s showed that all but four of the individuals in European and American zoos were contaminated with African lion 'blood'. Captive breeding of Asian lions is essential: there are only a few hundred left in one forest in north-west India and there is no room for expansion. But the breeding programme has had to be re-started.

Once the founders are in place, the curator must then induce them to breed as rapidly as possible, in accord with the excellent advice that God gave to the creatures who left the ark to 'Go forth and multiply'. The reason (though not alluded to by the authors of Genesis) has to do with genetic drift. The founders obviously contain all the alleles that the future generation will contain, but each founder parent passes on only half of his or her alleles to each offspring. If each pair of parents has two offspring that grow to adulthood, this will ensure that the F1 generation is the same size as the founder population, which on the face of things seems reasonable because at least there is no loss of individuals. But it is extremely unlikely – in fact it is virtually impossible – that any one parent will manage to pass on *all* of his or her alleles to just two offspring, so the genetic loss in the first generation in such a case would be enormous. If the founders have huge numbers of offspring, however – enough to give every parent allele a fair chance of finding a home – loss by drift can be kept to a minimum. It is in this situation that the new reproductive technologies could be used to best advantage: storage of sperm; maturation of ova outside the ovary; *in vitro* fertilisation (IVF); embryo storage, and embryo transfer, presumably (if the target animal is already very rare) into a closely related species or subspecies. At London Zoo, gaur calves have already been born to Friesian cows, and zebra foals to ponies.

However, the art of conservation breeding in zoos is not simply to produce as many of any one endangered animal as possible. After all, if zoos set out to produce 1,000 Siberian tigers, they will have no room to keep Sumatrans, or Bengals, or Chinese or Indo-Chinese, each of which has as much or a greater need for zoo protection. Instead, the zoos that are co-operating in breeding must work out exactly how many they really *need* to keep. The number required depends upon two main factors. First, if a population begins with very few founders, then the population required to keep all the alleles going has to be *larger* than it would be if there were more founders at the beginning. The maths shows this clearly, but the point in words is that large numbers of founders contain more alleles to begin with, so the individual descendants of those founders are more heterozygous than they would be if there were fewer founders, and a small number of highly heterozygous individuals can contain as much genetic variation as a larger number of more homozygous individuals.

Second, the optimum size for a breeding population depends upon the *generation interval*: the time between each generation. The reason is the one already given: genes are lost when the animals reproduce. It follows, then, that more alleles will be lost per year or per decade if animals breed frequently in that time than if they breed less frequently. Thus mice, which breed every nine months or so, lose alleles by drift many times more quickly than, say, Caribbean flamingoes, in whom there may be twenty years between generations.

In practice, zoo conservation breeders recognise that some loss by drift is inevitable, and the general rule of thumb is to keep enough animals to ensure that 90 per cent of the alleles that were in the founding population can be retained for 200 years. The hope is that in 200 years it may be possible to release at least some of the animals back into the wild; or, perhaps, that genetic technology will be advanced enough to create some genetic variation artificially. It transpires, from all the above considerations, that you need to keep more than 1,000 mice if you are to retain 90 per cent of founder alleles for 200 years (because there would be more than 250 generations in that time), but you could meet this genetic criterion with only 37 flamingoes. Note that these are *minimum* figures, which assume that every individual breeds and none falls ill; the real figure would have to be several times higher than the minimum.

Thus we can define the overall numerical strategy for zoo conservation breeders. First, from a knowledge of the animal's biology and taking into account the number of founders, they should calculate how many of each kind of animal they really need to keep to maintain 90 per cent of founder alleles for 200 years. Then they should encourage the first few generations of founders to breed as quickly as possible to ensure that this required population is reached as soon as possible, to minimise drift during the early, vulnerable period. Then they should try to maintain the breeding population at the required level: certainly never below, and not too far above because space allocated to one animal cannot be used for another.

Some of the consequences of this strategy – though they must be followed if captive breeding is to succeed! – seem odd. First, serious zoo breeders should try to *lengthen* the generation interval, once the required population is reached: which means they should prevent their animals from breeding too young. But although lions, say, can and therefore should breed only at 10–12-year intervals, they *can* breed at four years or so. So we find that serious zoos may keep rare animals on contraceptives for the first few years of their reproductive life, which seems a strange thing to do when the animal is rare and the priority (apparently) is to increase numbers. Less responsible institutions simply encourage their animals to produce babies, and receive public plaudits for doing so. A second oddity springs from the need to keep some animals in far greater numbers than others. Thus, William Conway of the New York Zoological Society has pointed out that it costs several times as much to keep a viable population of striped mice, as of Caribbean flamingoes!

When the conservation breeder does allow the animals to breed, he must again follow several genetic guidelines. First, he must ensure that *all* the founder animals are encouraged to breed, and indeed that the genes of all founders are equally represented in future generations. If you think in terms of conserving overall genetic diversity, then this rule is obvious, but it did not become obvious to conservation breeders until the 1970s. After all, until recent decades, directors thought they were doing well if they got their animals to breed at all, and it seemed sensible to breed from the ones that bred most freely, and ignore the ones that did not. This idea still prevailed in the 1960s, when the first Arabian oryx of the modern population were first rescued from the wild. In consequence, a few individuals from the original founders have left a great many descendants – or as biologists say, they are *over-represented* – while others did not breed at all, and their alleles have died with them. This is a tragedy. Arabian oryx have bred well in zoos, and have now been returned to the wild in Oman and other parts of the Middle East, but geneticists from London's Institute of Zoology are now finding signs of inbreeding within them, largely because of lack of overall genetic diversity. Equally to the point, animals that lack genetic diversity are less able to continue their evolution and adapt to changing circumstance after returning to the wild. A failure to breed in captivity is no criterion by which to judge an animal's ability to thrive in the wild, and even an apparently 'sickly' individual may contain genes that prove of great value in future generations.

Just as we must encourage the reluctant breeders to breed in captivity, so we must discourage the most fecund individuals from breeding too much, which would ensure that the future population contained only their descendants. The trick here is to equalise family sizes. Over-fecund individuals must be given contraceptives, and if one lioness (say) produces a litter of five cubs while another produces only two, then three of the larger litter must be prevented from breeding in the future. Wherever possible, the three surplus cubs should go to good homes, to be kept on standby. But culling is sometimes necessary. This raises another paradox: people seriously interested in the conservation of endangered animals sometimes have to cull some of those animals, albeit with extreme reluctance and heart-searching. The alternative, often, would be to fill all the available spaces with inbred animals and bring the entire programme crashing down.

Finally, the serious conservation breeder should ideally seek to ensure that every male in every generation mates every unrelated female. In practice, this is almost never achievable, and instead matings are arranged as far as possible between particular unrelated individuals, to ensure the greatest possible mix and equal representation of each type. These mating programmes rapidly become extremely complicated. The coordinators of regional conservation breeding programmes must first become adept in the use of computers. I have described all this in my recent book *Last Animals at the Zoo* (Hutchinson Radius, London, 1991).

This, then, in outline, is how captive breeding works in the conservation of animals, and how it is underpinned both by 'population genetics', which provide the grand strategic background, and by techniques of analysis, from DNA fingerprinting to karyotyping. Note that true 'conservation breeding' differs from the kind of undisciplined production of babies that menagerie-keepers indulge in in several crucial ways, including the fact that serious breeders, once they have established populations of a required size, do *not* seek to maximise reproduction. Crucially, too, conservation breeding differs from the breeding of domestic livestock, as discussed in Chapter 6, although not everyone apparently understands the difference.

CONSERVATION BREEDING VERSUS LIVESTOCK 'IMPROVEMENT'

It was not until the 1970s that zoo directors finally realised that they had a serious and necessary role to play in conservation, by breeding animals. Until then, there seemed to be so many animals in the wild, and many of them bred so badly in zoos, that the notion of 'conservation breeding' generally seemed absurd. In consequence, most zoo directors before the 1970s had a very different attitude to breeding from what they do now. Commonsensically – or so it seemed at the time – they bred simply from the individuals that showed the greatest willingness to breed in captivity. Both the Père David's deer and Przewalski's horse have been saved from extinction by breeding programmes that started in the last decades of the nineteenth century, or the first decades of the twentieth. In both cases, the good breeders were encouraged, and the rest allowed to go by the wall. Père David's came through this well: so far as I know, they have shown no particular signs of inbreeding. Przewalski's did less well; so much so that earlier breeders felt it necessary to 'enrich' the gene pool by cross-breeding with domestic horses, which do breed freely with Przewalski's even though they are generally acknowledged to be a different species. Modern breeders of Przewalski's now seek to eliminate signs of domestic horse, including the 'foxy' colour, otherwise known as 'chestnut'.

Second, early zoo directors did not perceive a clear difference between breeding for conservation and breeding for livestock improvement. The aim of conservation breeding is to preserve the *maximum* genetic variation within the population that it is logistically possible to keep. The conservation breeder has no preconceptions whatever about how the captive wild animal should look (except in the case of Przewalski's, where the population is known to be 'contaminated' with 'foreign' genes). He is just as anxious to breed from the runt as from the most macho of studs; more anxious, indeed, because the macho stud will generally make sure that he breeds without any help from keepers. Thus a few years ago zoologists at San Diego zoo had to decide whether to continue breeding from a black-and-white lemur that had a sunken chest. Old-fashioned breeders would have had no doubt: throw the 'defective' animal out! But the San Diego team observed that the sunken chest, though unsightly, did no apparent harm to its possessor, and they reasoned that that animal would also carry other alleles that should not be lost. So, in the true spirit of modern conservation, they kept the animal in the breeding programme.

In sharp contrast, the improver of domestic livestock has a very clear preconception of what it is he wants to achieve: a cow that produces 4,000 gallons per year; a dog that will fetch wounded rabbits; a ram with a 'challenging expression'. He seeks not to maximise genetic diversity in his population, but to narrow it until he produces creatures that conform absolutely to the preconception and are alike each other as peas in a pod, so that they *all* conform. In fact, of course, the competent domestic livestock breeder seeks to avoid too much inbreeding. More generally, however, he seeks to eliminate all deleterious alleles completely so that his animals do not suffer too much from homozygosity and hence can tolerate inbreeding.

Unfortunately, some curators still seem to confuse conservation breeding with livestock breeding. We still hear directors who should know better saying that tropical deer kept in northern parks have to be 'toughened up', which means they allow the ones that are least cold-tolerant to fall by the wayside. Thus they are not conserving genetic variation but breeding a new strain of domestic animal suited to northern climes. We still find zoos worldwide that should know better – particularly in India – specifically breeding white tigers, and attracting the crowds because people at large do not realise that this is pernicious, a narrowing of an already narrow gene pool. It is of course still tempting in this commercial world for zoos to boast that they have the world's tallest giraffe or the world's biggest gorilla, and then to try to produce a super-race. But this temptation must be firmly resisted. Conservation needs zoo directors who understand genetics, and a discriminating public.

Fortunately, though some zoo directors still perpetuate the mistakes of the past, conservation biologists are already seeking to combine captive breeding with conservation in the wild.

ZOOS AND THE WILD IN CONCERT

We have seen that wild populations of animals are safe, physically and genetically, only if they are large: generally with thousands of individuals. But most wild populations these days are not large. Two thousand five hundred was the ball-park figure for a safe population of rhinos, but the only single population above that figure is that of the Southern white rhinos of South Africa. In only one other species is the total number of animals greater than 2,500. There are in fact about 3,500 black rhinos in Africa, but the blacks are divided into at least four major subdivisions – west, east, south and south-west – and each subdivision is extremely fragmented. In short, many wild populations of animals – particularly large animals – are already too small to be viable in the long term. Therefore, they must be genetically managed as if they were captive populations.

Such genetic management takes several forms. As mentioned, scientists from the Institute of Zoology are now testing the relationships between individual animals in the newly created wild herds of Arabian oryx in Saudi Arabia. In Kenya, once all the black rhinos have been safely rounded up into reserves, different individuals will have to be translocated between reserves to ensure genetic mixing. As the reserves are now separated by farms and cities, the rhinos can no longer simply walk from herd to herd, as they might have done in the past.

We can also expect to see increasing intercourse between wild populations and captive. Indeed, the two kinds of population should be managed in parallel. Sometimes particular individuals from either group, which contain alleles that are known to be lacking in the other group (or are known simply to be unrelated) could be brought from the wild into captivity (or *vice versa*). Theory shows that even a few such interchanges can make an enormous difference. There are precedents, of course; in particular, female Asian elephants in logging camps are commonly allowed to roam at night in the forest, to be impregnated by wild males (although the number of wild males is going down steadily). In fact, because some captive populations were gathered from far and wide in the past, while many modern populations are confined to small areas, today's captive populations sometimes contain *more* genetic variation than survives in the wild. This is the case with the golden lion tamarins. More generally, there are many alleles among captive animals that are already extinct in the wild.

With only minor development of present-day technologies it should be possible to arrange genetic exchange between captivity and the wild even without exchanging entire animals, with all the traumas and dangers that that could entail. For example, with modern techniques of pregnancy analysis (including sampling urine in the field) it is becoming possible to determine when wild females are in season. With modern techniques of anaesthesia, it should be possible to bring those animals down (safely) and then impregnate them with the sperm of a captive male, by AI. Then again, ova from wild females that have recently died for whatever reason could be 'rescued' and matured, then fertilised *in vitro* and transferred into the wombs of captive females. Indeed, there are many possible permutations.

Many scientists worldwide already feel that it is high time to take DNA storage more seriously. That is, samples of flesh from any creature now alive should be put into cold storage. This would cost very little indeed, but it needs to be done. Future generations of geneticists, more knowledgeable than those of today, could then engineer some of the stored genes back into tomorrow's animals, which will have been further impoverished by genetic drift. *In extremis*, tomorrow's conservation biologists might even increase genetic variation artificially by producing multiple copies of alleles within any one genome, and then mutating some of those alleles in a controlled manner. Finally, it should be possible to de-differentiate genes from differentiated tissue, and thus, perhaps, to regenerate entire animals from somatic DNA, as was discussed in Chapter 3. This might not ever be possible, but it is certainly worth employing today's technology – tissue storage – in case it is.

These, then, are the ways in which geneticists can serve the needs of other animals. Alexander Pope commented in the eighteenth century, however, that 'the proper study of mankind is man'. In the next chapter, I will act upon his observation.

CHAPTER 9

The Improvability of Man

The human gene pool is a wonder and yet we find reasons for discontent. Some of these reasons are legitimate, because the pool includes many thousands of mutations which between them produce 3,000 different named 'single gene disorders' that are usually distressing and are often life-shortening. But some of the discontent seems merely self-indulgent: we marvel at the speed of Carl Lewis, the elegance and power of Steffi Graf, and the genius of Shakespeare or Einstein or Mozart and ask (or at least some of us do), why aren't we all like that? Why do we think Lewis is wonderful, though he is slower than a well-trained collie-dog? Why do we think Einstein is a mathematical genius, when he could not do mental arithmetic as well as a £5.00 pocket calculator? If we have diverged as far as we have from the apes, why did we not advance even further? Has human evolution actually stopped, and if it has, can we give it a helping hand and breed super-humans?

In practice, human beings have been trying to dictate the course of their own biological evolution at least since the dawn of history, as witnessed by the self-centred inbreeding of the pharaohs and of all aristocracies since, and the prevalence, worldwide, of arranged marriages. Charles Darwin's cousin, Francis Galton, in a spirit of both typical nineteenth-century reform and, as he felt, modern hard-headed biology, advocated breeding programmes to produce superior humans; an approach he called *eugenics*. In fact, eugenics became a respectable notion in the early, idealistic decades of the twentieth century, and was embraced by, among others, the humanitarian biologist Julian Huxley, grandson of Thomas Henry, and H.G. Wells, who in 1901 suggested in *Anticipations* that right-thinking people should 'check the procreation of base and servile types . . . of all that is mean and ugly and bestial in the souls, bodies, and habits of men . . . [and] the method . . . is death'. Wells was praised to the skies by Arnold Bennett – 'absolutely overwhelmed by the sheer intellectual vigour' – and by the Fabians Sidney and Beatrice Webb ('luminous hypotheses'), but roundly and very properly condemned by G.K. Chesterton, while

Arthur Conan Doyle averred on moral grounds that *Anticipations* was 'vile and villainous . . . horrible' and on scientific grounds that it was 'muddle-headed'. Since Wells and the Webbs are generally regarded as forward-thinking social reformers, while Chesterton and Conan Doyle are popularly remembered as pleasant old fuddy-duddies, this example illustrates yet again how skilfully and remorselessly history is re-written, especially by those who believe that literary reputation must outweigh every other consideration.

But it fell finally to Adolf Hitler to demonstrate just how hideous eugenics could be in practice. He strove to eliminate the peoples he felt were inferior, and established breeding farms to multiply the Aryans whom he felt contained the seeds of the super-race.* As if Hitler had not made the point, various of the US states continued the endeavour to eliminate 'bad genes' after the war, and sterilised individuals whom they considered mentally defective or, sometimes, 'perverted'. In November 1992 there were Turks in Germany mourning the deaths of their relatives, killed by neo-Nazis, while the countries that once were Yugoslavia have added to the world's vocabulary the chill phrase 'ethnic cleansing'. The persistence of crude ideas is astonishing. Even the most evil never seem to go away.

Allied, of course, to this corruption of eugenics is the notion that some races are innately – genetically – superior to others. Most biologists in the nineteenth century (like most white people, indeed) felt this to be self-evidently the case; Wells and Hitler also took it as read. In the mid-twentieth century, intelligence tests carried out in the United States revealed differences in average IQ between peoples of different racial groups. In the 1960s in particular, various educational psychologists suggested that the reasons for this were not exclusively to do with background, but with genetics. To be sure, these psychologists were not in general motivated by racial hatred, although such ideas can obviously be used to fire racial prejudice. Rather did they argue that it was in the interests of the children from what they perceived to be the less

*A childhood acquaintance of a German friend of mine, who was a serving-maid, actually attended such a farm, and was duly impregnated by a young, blond, handsome German soldier, carefully selected for his compatibility.

The son who was born to them was beautiful, intelligent, and a very nice person. That's life.

favoured races that their innate inability should be acknowledged, so that they were not over-stressed by too much education. This is a strange argument, with obvious overtones of self-fulfilling prophecy.

Liberal opinion in the west nowadays does favour some measure of genetic intervention in human affairs. Most people evidently feel that to control or eliminate the genes that cause frank disease in human beings is good and desirable medicine. I will discuss these attempts later in this chapter.

But most of us also feel that eugenics is distasteful, and that the idea of 'racial intelligence' is hateful. The safe course, then, is simply to leave such notions out of this book, or hurl a few safe, liberal missiles at their protagonists. In matters like this, however, I favour the methods of Freud. The only way to deal with psychological skeletons is to drag them from their cupboards, and see what they are made of. If we discover biological facts that are uncomfortable – well, that becomes a test of ethics, for, as Immanuel Kant argued, there is no morality simply in complying slavishly to our own 'natures'. To be truly ethical, we must be prepared to override what is natural to us. If the biological facts are displeasing, they should not simply be suppressed, for such suppression is itself immoral. But biological 'facts' need not be slavishly acted upon, either.

So let us ask the dangerous questions, and see where they lead. To begin at the beginning – are we evolving still, into creatures even smarter than ourselves? Why did we not evolve further in the first place? If evolution will not improve us further, could we in fact breed superior people, as Galton first envisaged?

WHY DIDN'T SUPER-HUMANS EVOLVE?

Of all the features that we may feel distinguish us from other creatures, the ones that truly count, and have given us our ecological edge and our overweening conceit, are our individual intelligence, and an ability to communicate ideas precisely, an ability which enables each of us to operate as part of a collective intelligence. In indulging our quest for even greater excellence, then, we might reasonably focus on these two features: intelligence and communication. So, why aren't we brighter and more articulate than we are? Why aren't Einstein and Shakespeare the norm? Clearly those two are far from perfect, so why are the rest of us so feeble that we grant them the status of genius?

In practice, we can identify at least three reasons why we have not diverged further than we have from our fellow-apes. Overall, they leave us asking a quite different question: how on Earth did we ever get this far?

The first of these reasons (though I take them in no particular order) is genetic. There is very little direct information on the genetics of human intelligence, but we know enough to surmise that intelligence, genetically, is extremely complicated, and is likely to be a very difficult trick for natural selection to pull. The human genome as a whole is estimated to contain around 100,000 genes, divided (indeed scattered) among about 3 billion base-pairs. According to Christopher Wills, writing in *Exons, Introns, and Talking Genes* (Oxford University Press, Oxford, 1991), around 30,000–35,000 of these genes are expressed in the human brain, and of those, 15–20,000 operate *exclusively* in the brain. All of those 15–20,000, we may presume, are directly involved in brain structure and function, and all may therefore be said, in some measure, to influence intelligence.

We know that some genetic combinations in some people's heads result in greater intelligence (however that is measured) than the genetic combinations in other people's heads. Biological variation is the norm: only within clones are any two individuals the same, and even members of clones differ somewhat. All human beings have the same *number* of genes expressing in their brain (or at least, that is a fairly reasonable assumption). The difference lies, then, as ever, in polymorphism. At least 5,000 of the brain-exclusive genes seem to be polymorphic. Each of those 5,000 exists in more than one allelic variation.

As a crude first approximation, then, we might simply suggest that the way to increase intelligence (or at least, the genetic contribution to intelligence) is to breed out less helpful alleles in favour of more helpful alleles. With such a large number of genes to play with, this is a tall order numerically. But conceptually, it seems simple enough.

However, a little elementary genetic thinking, of the kind outlined even in Chapter 1 of this book, immediately suggests that such an approach would be too crude by half, and would at best bring about very limited improvements. A key concept is that of heterozygosity, leading in various ways to what Darwin called 'hybrid vigour'. This principle ought to apply in the human brain, as in virtually all other biological systems. But we cannot produce heterozygosity at any locus unless we cross two individuals with two different genes at that locus. If both parents are geniuses, it seems very likely that both are highly heterozygous, so their children will be a very unpredictable hotch-potch. Contrariwise, two relatively homozygous (and therefore 'ordinary') parents could, if well matched, produce an offspring who was extremely heterozygous in all the right loci, and hence be a genius. The front page of the Sunday newspaper that lies before me (*Observer*, 29 November 1992) carries an account of Nicholas MacMahon, who 'spoke fluently before he was one', but also taught himself to read *before* he could speak, and at the age of four 'can tell a Boeing 747 from a DC10, knows the Highway Code, devours encyclopaedias, reads the *Daily Telegraph* avidly and is well on the way to becoming a violin virtuoso'. Though his parents doubtless have hidden talents, they in fact work as gardener and housewife. I do not know (and neither does anyone else) whether the simple genetic notion of homozygosity and heterozygosity literally applies in Nicholas MacMahon's case, but such a model certainly fits the facts.

But heterozygosity is likely to be only a small part of the issue. As I have several times argued in this book, genes tend to operate in concert parties, and these concert parties tend to be broken up through sexual reproduction, unless the different genes are closely linked. No doubt many groups of genes associated with intelligence are closely linked. But, in general, the 5,000 or so we should probably be looking at in particular are spread throughout the genome. Again we see that a particularly felicitous combination will not be passed on intact to the next generation. Add that to the probable need for heterozygosity and we see that breeding for genius is likely to be very haphazard indeed.

But then, we might suggest that breeding for genius is a luxury. The main point, surely, would simply be to raise the *general* level of intelligence, and let genius take care of itself. Would it not be reasonable simply to try to reduce the proportion of less helpful alleles, and raise the proportion of more helpful ones?

Here, probably, we are on more hopeful ground, and such a process presumably did take place during human evolution.

But even here we have to be very careful, for the effect of any one gene depends very much on context. A gene that helps to produce genius in one individual might have no very helpful effect if present in a different combination of genes in a different individual. In fact, the gene that is helpful in one context could even be detrimental in another. So even if we knew a great deal more about the genes that underpin human intelligence, it would still be difficult to identify, unequivocally, those that are 'good' and those that are 'bad'. Goodness or badness probably depends largely upon context.

All these comments are pertinent both to the artificial breeding of super-intelligent human beings by eugenics, and to human evolution. Chapter 6 showed how extremely difficult it is to produce improvements in plants and animals that will persist, reliably, generation by generation, if the genetics of the desired character is at all complicated. But in human intelligence we seem to have a character of *maximal* complexity. Even when the character is relatively simple genetically (or may be presumed to be so), the breeding is extremely drawn out if the creature has a long generation time, or is not particularly fecund. Thus it is hard to breed better coconuts because there are at least twenty years between generations. But at least coconuts have plenty of offspring to choose between. Humans also take about twenty years to breed, and our output of offspring is niggardly.

Furthermore, when the genetics is complicated, it is extremely difficult to tell how the offspring would turn out just by looking at the parents. To be sure, intelligence and particular talents often run in families, as in the musician family of Bach, and in the polymathic Huxleys. But it is very hard in such cases to differentiate the 'breeding' from the up-bringing. Clearly, too, genius has a habit of springing from nowhere, as with the Beethovens and Nicholas MacMahons of this world.

It would be interesting – purely as an intellectual exercise – to set plant or animal breeders the task of breeding super-intelligent human beings. It would also be a huge mistake to expect very much from their efforts. Undoubtedly, the approach would be far closer to the 'mass selection' methods adopted for outbreeding plants such as millet, than to the straightforward crossing of outstanding types which is favoured for highly homozygous plants such as wheat. I imagine that to breed better humans, the initial gene pools would have to be very large. I would also imagine that, if only in the interests of heterozygosity, the breeders would want to begin with a wide variety. It would be quite wrong to do as Hitler tried to do, and concentrate upon one group only. In fact, I suggest, the ideal gene pool to begin the improvement process would be very little different from a random selection of human beings. I also suspect that the recommended breeding protocol would be very little different from random. It would probably be best just to let people choose their own partners, guided by nothing more accurate than Cupid's arrow. Indeed, the only significant departure from normal everyday life that a geneticist might reasonably suggest would be a little more infidelity. But even that – statistically speaking – would probably have only a marginal effect, given that human beings have so few children in any case.

If we apply the complex genetic model to the evolution of intelligence among real human beings in the wild, it becomes very surprising indeed that we have progressed as far as we have, and no surprise at all that we have not gone further. Only an immensity of time (and it is 5 to 8 million years since we diverged from chimpanzees: probably at least half a million generations, which is a huge number) could have brought us to this point, and the apes themselves (if indeed they were like modern chimps) were a pretty distinguished lot to begin with. If a character is complex genetically, and depends upon heterozygosity and a particular combination of genes, then it becomes very difficult for any specially favoured group to break away from the pack. A standoffish pair of prehistoric geniuses, who decided that they were too good for their companions and broke away from the herd to found their own colony, would probably find that their offspring were no better than ordinary. They would also find, contrariwise, that the pack they left behind had sprouted a new crop of geniuses of their own. So the whole pool has to grind on together: not so small that there is not enough variation for natural selection to work upon, and not so big as to preclude the possibility of change at all.

We also find that genius, far from being a target to which it is reasonable for the entire species to aim, is *bound* to be rare. It is in fact a side-effect: a phenomenon produced every now and again, by chance, when particular, perhaps fairly ordinary, individuals just happen to supply each other's gametes with appropriately complementary genes. According to this very plausible genetic model, then, genius *cannot* be prescribed.

We might also add, finally, that perhaps – just perhaps – genius is subject to frequency-dependent selection. Geniuses by their nature have a big effect in society. Some geniuses – by no means all, but it is a price that often has to be paid – can be very difficult people. Marx was extremely touchy, and Einstein, though amiable, was supremely absent-minded and needed a chaperon to prevent him being run over. Perhaps any one society can tolerate only so much genius. Perhaps primitive societies who produced too many geniuses, were selected against. Perhaps, taken all in all, 'ordinary' people just have more survivability. Perhaps we are all much more dependent on each other than we care to admit.

Where does such a view of the genetics of intelligence leave 'racial intelligence'? Rather barren, I suggest. Race to a very large extent is defined, in human beings, by skin colour. Consider, then, what the 'racial intelligence' hypothesis implies in genetic terms. It implies that the genes that produce, say, blue skin, are more likely to be linked to genetic combinations that result in high intelligence than are the genes that produce, say, green skin (where 'high intelligence' is defined as the ability to do well in intelligence – 'IQ' – tests). Yet we know, among real human beings, that the differences in average measured IQ between different skin-colour types are very small compared to the differences between different people of the *same* skin colour; we know that skin colour is complex genetically; we know – or can sensibly guess – that the genetics of intelligence is extremely complex; and we know (or can sensibly guess) that genius is certain to be rare in any community. In short, the hypothetical link between skin colour genes and intelligence genes, even if established, would still be no more than a statistical observation of great tenuousness, whose predictive power, when applied to any one *individual*, would be approximately zero. If ever there was an idea that was not worth the candle, this is it.

However, I do not belong to that ultra-liberal school which, in the 1970s in particular, wanted effectively to ban the word 'race' altogether, and refused to refer to people's skin colour, leaving us to guess that Mr Olroyd comes from Huddersfield and Mr Patel is from Karachi. To ignore skin colour seems to me as insulting as to deride it. In Mediaeval Italy people with blond hair were stigmatised because they came from the north and were therefore barbarians. In nineteenth-century Britain red hair was looked down upon because it was associated with Irish immigrant labourers. Now blondes and red-heads are openly admired. I look forward to the day – which in some circles is already with us – when people similarly admire each other's racial characteristics, on purely aesthetic grounds.

But to return to the general matter of human intelligence. There is a second, quite different reason why we should think ourselves lucky to have evolved such qualities as we have. Natural selection leads to evolutionary change if, in fact, individuals with particular characteristics survive better than those without. It is not too difficult to see, in a general, arm-waving way, that natural selection would have favoured intelligence. But it is not at all easy to perceive the natural selective pressures that could have produced individuals of the calibre of Shakespeare, or Einstein, or Mozart.

THE EVOLUTION OF EINSTEIN

Although Darwin argued that natural selection was bound to produce better and better adaptation to the prevailing conditions, he had no preconceptions whatever about the *kinds* of changes that would result in closer adaptation. Sometimes – more often than not, I would say – closer adaptation involves some measure of what might be called 'technological' elaboration. Early organisms that developed a double membrane around their cells were better protected, and so survived better, than those with a single layer. Amphibians finally liberated themselves from water, and began truly to exploit the extra resources of the land, when they developed a few extra membranes around their eggs, and became reptiles. And so on. But apart from a general tendency to elaborate structure, to gain independence from the vagaries of environment, and hence to exploit new niches, it is impossible to prescribe the qualities that will, in fact, prevail. Flight clearly benefits birds, but in practice, as soon as the selective pressure in favour of flight is relaxed, birds tend to re-evolve flightlessness, as demonstrated particularly among rails, water-fowl, and even pigeons and parrots. Mammals are warm-blooded, and warm-blooded creatures seem most likely to evolve the most complex nervous systems, and hence the greatest intelligence, and intelligence in general seems beneficial. Yet the first mammals appeared 200 million years ago, and for the next 120 million years did next to nothing because they were overshadowed and out-gunned by the dinosaurs which, despite some arguments to the contrary, were probably 'cold-blooded'.

Evolution by means of natural selection does not imply evolution in any particular *direction*. We may in retrospect feel, conceitedly, that somehow evolution was bound to produce super-intelligent creatures like us. But although some of Darwin's successors have suggested just this, Darwin himself argued no such thing. It was partly because he denied the possibility of direction in evolution that he offended some churchmen (and fellow scientists) so much. After all, evolution by natural selection would be perfectly compatible with the idea of Godly creation if we were allowed to suppose that God had prescribed the end-result, and was nudging the whole development towards the end-point that is us. But it is hard to reconcile such a view with the notion that we have developed the way we have only through time and chance.

In short, Darwin's notion of evolution by natural selection obliges us to seek reasons, within our own biological history, to explain how we evolved such an ability to think and to communicate. But it absolutely does *not* justify the notion that our superior intelligence and communicative powers were in any way inevitable. Their evolution is innately genetically difficult, as has been discussed, and there is no *a priori* reason to suppose that natural selection would, in practice, favour their development, even if that development had been genetically easier.

So what were the events that produced us? Why did they favour the kinds of qualities that we possess, but only so far and no further?

Some time around 5 million years ago, so the fossil record suggests, Africa became drier. The forests retreated and a group of chimp-like apes – perhaps virtually the same as modern chimps – were forced on to the open savannah. There they became more dependent upon a diet of meat (the fruits of the forest having gone). The ape-ancestors of humans must already have had a complex social life – as in modern chimps – and hunting would have encouraged further co-operativeness, for hunting in general is more efficient in groups. Co-operativeness encourages communication in general. And as Nicholas Humphrey argues in *Consciousness Regained* (Oxford University Press, Oxford, 1983), complex social living may have provided the selective pressure that encouraged consciousness to evolve: for consciousness, he suggests, began as self-consciousness, as our ancestors strove to 'read the thoughts' of their companions, so as to reduce the chances of conflict.

The general tendency to communicate was abetted by a change in the anatomy of the larynx. In virtually all mammals except human beings, the larynx sits very high in the throat, and as well as serving as an organ of sound, it functions as a valve: it enables the mammal to breathe and drink at the same time. In humans, the larynx has descended, and now sits half-way down the throat. It is useless as a valve, and we choke if we attempt to breathe and drink simultaneously, because the drink is liable to go down the wind-pipe. But now, above the larynx, there is a space which serves as a sound-chamber, and although the larynx has lost its value as a valve, it has gained enormously as a sound box. We can make a huge variety of sounds, and these, when further modified by our intricately muscled tongue and lips, give us the facility for speech. No-one knows why this laryngeal descent occurred – what selective pressures triggered it off – but for whatever reason, it has proved a tremendous boon, which far outweighs the occasional death by choking. A desire to communicate, a growing consciousness that caused us increasingly to monitor and direct our own thoughts, and the ability to articulate a wide variety of sounds to express those thoughts clearly – these are the components of human language. Human language on the one hand has become the hand-maiden and the director of conscious thought; on the other, it provides the means by which, as a species, we *pool* our thoughts, so that each of us can in principle partake of the thoughts of all other human beings who have ever lived (and many of our deepest thoughts, and our everyday inventions, have derived by word of mouth or by example from pre-literate times).

So there we have it: a perfectly good ape, adapted for the forest, very like a modern chimp, forced into open country by a change of climate; an apparently gratuitous change in the anatomy of the throat; time and chance, putting pressure on that ape to change, to become more intelligent, and more communicative. This brisk but objective overview puts quite a new slant on the question of human wondrousness. We may regret that the mathematical ability of human beings is so feeble; so bad, that we think Einstein is a genius, although he would be outclassed arithmetically by a child's pocket calculator. Yet surely we should ask a quite different question. We initially diverged from apes, it seems, when we became hunters, and it is easy to appreciate that a hunter should benefit from at least rudimentary arithmetic. Lose sight of one from a group of five buffalo and the fifth might creep up behind you. Fail to count your companions, and you may find yourself on your own. But what natural selective pressures could have given us the ability to perform calculus, or invent relativity? How could the need to hunt on the savannah have moulded the abilities of Isaac Newton, or Einstein, or Richard Feynman?

Indeed, the fact that the abilities of our ancestors – or of any animal, in different contexts – have apparently run so far ahead of the selective forces that are presumed to have shaped it gives encouragement to religious fundamentalists: for they might well say – 'Well, that proves it! Natural selection could not have produced creatures like us!' But orthodox biologists can still invoke two very plausible principles, both of which are known to operate in other contexts.

The first is simply the notion of *combination*. Vervet monkeys, for example, warn each other that predators are about. Furthermore, they make a different sound for 'snake' than they do for 'eagle'. There is clear selective advantage in this, for different predators require different escape strategies.

The earliest human beings were primates who lived exposed on the savannah, hunting some animals and hunted by others. Clearly the simple trick that was already developed in monkeys would serve them too: attaching different sounds to different animals. Selection would also surely favour an increase in versatility, to differentiate not simply between eagles and snakes, but between zebras, lions, tortoises, bees, leopards, antelope, and all the thousand and one other potential foodstuffs and hazards of the African plain. The descent of the larynx into the throat would clearly have assisted in this, by increasing the range of sounds possible.

But one further trick would have increased the vocabulary even further, that of combining different sounds. Natural selection (we may readily speculate) would have favoured a creature that could combine, say, the snake-sound with the eagle-sound to connote 'lion', or eagle-sound with snake-sound to mean 'elephant'. But once this process begins – 'eagle-eagle-snake' meaning baboon; 'snake-eagle-snake' meaning honey-badger – it becomes potentially infinite, and the path to infinity is smoothed if the animal can make more than two basic sounds, as early humans may well have been able to do. Thus a simple trick – in which there would have been obvious selective advantage – could have given rise to a potentially *infinite* vocabulary *en passant*. Vocabulary is not synonymous with language – or not, at least, in the human sense. Syntax must be added to provide order. But a potentially infinite vocabulary is at least the raw material of human language.

The second mechanism that can allow a creature's abilities apparently to outrun the obvious selective forces is that of the positive feedback loop, comparable with the Fisher's runaway, as described in Chapter 6. Thus in particular it seems that manual dexterity and human intelligence have fed upon each other. The ability to manipulate objects offers all kinds of possibilities, notably in tool-using and tool-making. There is clearly selective advantage in exploiting those possibilities. Thus manual dexterity imposes a selective pressure in favour of intelligence. Intelligence in turn suggests more possible tasks, which in turn increases the selective pressure that favours manual versatility. And so on. By the same token, the descended larynx made it possible to produce a greater range of sounds, and the increase in vocabulary that resulted from this would have given a selective advantage, which in turn would have favoured individuals with an even more descended larynx and an even greater range of sound.

Together, the positive feedback loop and the combination of existing faculties can produce leaps in evolutionary progress (in the technical sense) that far outstrip what otherwise seems possible. These two mechanisms are in turn driven by competition, for, as Darwin suggested, those creatures that achieve an edge over their fellows, by whatever means, are the ones most likely to produce most offspring. In this scenario, then, Einstein and Shakespeare emerge not as preconceived pinnacles of Creation, but as side-effects: the chance and inadvertent products of combination and feedback loop, acting on qualities that originally were shaped by the need to operate in collectives, to hunt, and to avoid being hunted.

But is this process continuing? Do the pressures that produced Einstein and Shakespeare still apply? Will they continue to the point where Einstein and Shakespeare are the norm, and genius is recognised only in people far more gifted than they?

The answer is surely 'No', for three reasons. First, natural selection can produce evolutionary change only if individuals with particular qualities achieve greater reproductive success than others. Is it the case that mathematical and literary geniuses commonly produce 2.4 children, while the rest of us are content with 2.3? Well, no, it is not. In short, the selective pressure that produced Einstein as a side-effect is no longer operating in a way that could turn Einstein into the norm.

We can broaden this point. Is it the case that natural selection is acting upon us at all? Well, it seems foolish to say 'No', for natural selection seems inescapable, just as gravity is inescapable. It is built into the fabric of the Universe. But as things are, natural selection is clearly not producing shifts in the human gene pool of the kind that neodarwinians recognise as evolution. To be sure, some groups of human beings are at present breeding faster than others: Moslems, Catholics in poor countries (though not necessarily in rich ones), Kenyans, and so on. But these fast breeders (for the time being) do not encompass any particular gene pool. In fact, both the Catholic and the Moslem communities are so widespread that each undoubtedly encompasses at least 99.9 (and we could probably add a few more 9s to that) per cent of the allelic variation in the human gene pool.

Neither should we regard every flux and hiccup in gene frequency as 'evolution'. It does not seem sensible to concede that evolution has taken place until there has been a *qualitative* shift in the pool; that is, until new alleles have come in, and others have dropped out. Undoubtedly the human gene pool is now acquiring new alleles faster than ever before, by mutation. There are more individuals for those mutations to occur within, and the selective pressures that in the past eliminated harmful mutations no longer operate so forcefully, so more mutations are surviving. On the other hand, there is very little loss of allelic variation at present. North Europeans are becoming less common relative to Asians, but they are certainly not becoming extinct, and the South Americans, who are presently multiplying rapidly, are largely the descendants of Europeans, and so contain European alleles. Rare alleles are presumably dying out as aboriginal groups of many kinds continue to dwindle, and this is immeasurably sad, but in terms of total allelic variation the loss is small. Overall, then, human beings seem to be in the same position that many animals have found themselves in in the past when they invaded new territories: the population is expanding rapidly, and there is an accompanying increase in allelic variation as more individuals (and the mutations they contain) survive. Such rapidly expanded populations tend either simply to find some new equilibrium level or else to radiate into new species.

In the future, the human population could diminish by voluntary means to some new, sustainable equilibrium, which of course would be the ideal. Equally plausibly (or perhaps more plausibly), it could enter a phase of unstable oscillation, as already observable in many wild animal populations, which sooner or later would lead to extinction (so computer models suggest), as one or other of the dips in population fails to bottom out. Plausibly, too, the human species could suffer catastrophic diminution but leave a few isolated populations here and there.

In the first of these three possible cases – the benign gradation into stability – it is difficult to see how the future human population could be genetically much different from the present one, except that the proportion of Asians and Africans will probably have increased by then, and there will be more overall genetic variation. In the second, of course, we would simply disappear. In the last plausible scenario, the isolated populations would presumably expand, and would be subject to selective pressures that would produce further qualitative change, perhaps leading to another round of speciation similar to that which brought human beings into existence in the first place. But there is no reason to suppose that any of the new, neo-human species that may develop in the future will be more Einstein-like or Shakespeare-like than present-day humans. Natural selection of the kind that produced our present gene pool seems – probably temporarily – to have come to a halt, although we are, in a random way, continuing to accumulate genetic variation. If and when the human species is again placed under strong selective pressure, there is no reason whatever to suppose that this pressure would move us down any of the paths that we might at present associate with super-humanness. Natural selection leads to greater adaptation, but super-people, as now defined by us, will not necessarily be especially well adapted to the world that is to come.

As far as Nature is concerned, then, we are as we are. Yet this is not a depressing thought. For as Marvin Minsky has pointed out, we do not realise – conceited though we may be – just how amazing we really are. Minsky has spent the last several decades at the Massachusetts Institute of Technology trying to devise machines that could out-think the human brain. But the more he has explored 'artificial intelligence' the more respect he has for what we might call the real thing. Modern computers can do things we find amazing, but in anything that requires true subtlety, they are simply not in the same league as us. They can calculate and remember stupendously, but their ability to do the things that we do all the time and without realising it is feeble in the extreme. They cannot easily read meaning into a mass of sensory data of the kind that constantly assails our eyes and ears. They are poor at making decisions of a practical kind and hopeless at those that require value judgement. They certainly cannot do all these things simultaneously – feats that we take in our mental stride. Furthermore, as Minsky argues in *The Society of Mind* (Heinemann, London, 1985), we are *all* of us amazing. The gap between Mozart and a 'tween-stairs maid is as nothing compared to the gap between the maid and the mightiest machine. Yet we remain modest. Instead of saying as Hamlet did, 'What a piece of work is a man!' we say, 'Mozart was a genius, but I am just an ordinary person.' 'Tween-stairs maids are brought up to be even more modest than that.

The real task, then, is now as always in the past simply to make proper use of the truly staggering talents that nature has already conferred upon us. We allow people to starve for no very good reason, which hardly exploits their abilities. Even in countries that take care of bodily needs (which, compared with most countries in the history of the world, is more or less true in modern-day Britain) we still create societies, and tolerate schools, that neither excite nor encourage. Even in countries that spend a fortune on education (as Britain does), the defects are stunning. In particular, it seems clear that small children have extraordinary faculties for learning which are rapidly lost and never re-acquired. Geniuses, perhaps, are those who lose those faculties more slowly than the rest; as Einstein said of himself, 'I never stopped thinking like a child.' In particular, virtually all of us learn our own language, almost perfectly and *en passant*, in the first few years of life, and never enjoy such facility again. In societies where hunters and gatherers still prevail, as in Australia and New Guinea, there is for some reason a commensurate plethora of different languages. It seems to me very likely that many if not most human beings in the history of the world have grown up to be multi-lingual; and that to speak only one language, as the English, Americans, Australians and French generally do, is anthropologically anomalous. Yet foreign languages are not taught in British schools until after the facility for language has long since died. There is evidence that music, too – including perfect pitch – is probably learned best in extreme youth, but music, too, is generally taught much later, and usually badly.

It is the case, too, that very few modern people have any worthwhile grasp either of quantum mechanics or of relativity, which arguably are the most significant intellectual contributions of the twentieth century, and were both formulated about eighty years ago. Both are difficult not because they are complex, but because they are counter-intuitive. Small children, though, I suggest, would have few problems with them. If electrons choose to be in two places at once – well, so do fairies. If people can stay at the same age by travelling quickly relative to others – well, so in a way did Sleeping Beauty. Only when such ideas are taken for granted can they be manipulated and improved upon. But so long as we delay the teaching of higher physics until the sixth form – when the possibilities for understanding are already limited – the ideas they contain will remain beyond our ken. Of course, those who did teach the ideas of physics to small children would be bound to be accused of indoctrination. But there is a world of difference between the mere introduction of ideas and the suggestion that those ideas – or any ideas – are a good or bad thing. Teaching *values* at a young age is indoctrination. But teaching ideas is merely striking while the iron is hot, which in many areas it does not seem to be for very long.

In summary, the idea that we should seek to improve genetically on the talents we already possess is biologically foolish and socially and politically misguided. Hope for the future lies, as it always did, upon the proper exploitation of the talents we already collectively possess. The scope for improvement there is truly prodigious.

Ethically, however, there is a world of difference between attempts to improve on a gene pool that already serves us beautifully – or would do, if we had the wit and humanity to exploit our own talents adequately – and the attempt to correct the defects in the gene pool that lead to frank disease. In western medicine, we make this distinction easily, and fairly clearly. Thus drugs that make us feel even better when, in fact, we were already feeling perfectly reasonable, we tend to take a dim view of. We allow alcohol but contrive to restrict its use. Marijuana, cocaine and opiates are quite simply banned. The point is not just one of health. There is within us, it seems, a strong Puritanical feeling (literally) that to take such drugs is *wrong*. But drugs that correct obvious illness are of course accepted, and their prescription is supported by taxes. To be sure, there are grey areas. Alcohol itself can be medicinal. With plastic surgery (not a drug, of course, but the principle applies) it is sometimes hard to draw the line between indulgent prettification and repair of frank ugliness that is truly psychologically damaging. In traditional Eastern medicine, too, the tonic – which improves on well-being that is already acceptable – is, in general, more highly favoured than straightforward therapy, possibly because traditional therapies are not in general very effective. But in Western medicine, at least, despite a few grey areas in practice, the distinction *in principle* is clear enough.

So although we can dismiss eugenics on grounds of taste and ethics, as well as on grounds of practicality, there are no such grounds for dismissing attempts to correct frank disorder that is brought about by damaging mutant genes. In the past, such therapy was impossible. Now – very properly – it is one of medicine's growth areas. More broadly, indeed, molecular genetics is causing a change of outlook through all of medicine that can properly be seen as a 'paradigm shift', as defined by Thomas Kuhn.

DISEASE IN THE GENES

The new medical paradigm is at present most evident in the

new approaches to the single-gene disorders. This is inevitable, because these are the simplest gene-based disorders, and is fully justified medically. But the work on single-gene disorders, preliminary though it still is, must be seen as the start of something far bigger, for successful attacks on single-gene disorders will lead on to studies of more complex gene-based conditions (such as cancers and coronary heart disease); and, as noted throughout this book, study of pathologies, when systems go wrong, provides crucial insights into normal metabolism.

Worldwide, the most common single-gene disorders are inherited forms of anaemia, which are of two main kinds. The *thalassaemias* – there are many different variants – are defects in the haemoglobin which can severely compromise the ability of the blood to carry oxygen, and affect tens of millions of people in South-east Asia and the Mediterranean. One form of thalassaemia previously affected 1 child in every 150 in coastal Sardinia, though genetic counselling since the mid-1970s has now reduced the incidence to 1 in 1,500. Astonishingly, 1 in 6 people carries the relevant allele (or at least one of the relevant alleles). *Sickle cell anaemia* affects millions of Africans (and their descendants). The red blood cells collapse when the oxygen concentration is low and become sickle-shaped. There is only one variant of sickle cell, which suggests that the mutation responsible has arisen only once in human history.

Cystic fibrosis is the commonest single-gene disorder among North Europeans. It is a disease of secretion: specifically, chloride ions cannot be carried properly across cell membranes. The result is variable but in general it leads to cysts and fibrous degeneration of the pancreas (hence the name of the disease); to male sterility (because of comparable effects in sperm-producing cells); and to the production of viscous, sticky mucus in the lungs. It is this last effect that proves fatal in severe cases, generally before adulthood. Sufferers have to be forced to cough (nowadays by first loosening the lung mucus with a vibrator) as often as they can stand it, but they generally succumb to lung infection sooner or later in any case. At worst (and many cases are 'at worst') cystic fibrosis is a horrendous disorder. Astonishingly, 1 in every 25 North Europeans carries the gene. This means that in 1 in every 625 marriages (25 × 25) both partners are carriers. One in 4 of the offspring of two carriers will have the disease (by the simple

Mendelian rules outlined in Chapter 1), so the incidence of the disease among North Europeans is 1 in 4 × 625 = 1 in 2,500.

Duchenne muscular dystrophy affects 1 in 3,000 male children. The sufferers appear normal at first, but then their muscles swell and then waste. After a time, the heart muscle is affected, and most victims die before the age of twenty from heart failure. Muscular dystrophy is variable; Becker's muscular dystrophy is a milder form of the same disorder.

Duchenne, then, is almost as common as cystic fibrosis, but for quite different reasons. First, the gene is X-linked, which means that when it occurs in males it is bound to make itself felt, since there is no homologous chromosome carrying a normal allele to compensate. Second, the gene that causes Duchenne has an extraordinary tendency to mutate: indeed, it mutates 10 to 100 times faster than the average. This implies that 1 in 10,000 X chromosomes in every generation mutate in a way that will lead to Duchenne muscular dystrophy. This very high mutation rate may be related to the vast size of the muscular dystrophy gene, for it is one of the largest known: occupying 2.5 million base-pairs on the X chromosome. In fact, only 11,000 bases are actually involved in coding for the protein that the gene produces, and the rest – the remaining 99.5 per cent – consists of no fewer than 65 vast introns. So this is a very untidy gene indeed. Whatever the reason, the stock of muscular dystrophy genes is being constantly replenished.

At the other end of the frequency spectrum are hundreds of distinct disorders that are very rare indeed, though some are of considerable practical interest for a variety of reasons. *Tay-Sachs* disease, a disorder of fat metabolism that leads to degeneration of the nerve cells, is an enormously distressing condition that typically leads to blindness, paralysis and death within the first year of life. Like many single-gene disorders, it is most common in particular populations, and in fact manifests mainly in Ashkenazi Jews. It is of enormous practical importance as the basis of a highly instructive and effective exercise in genetic counselling (of which more later) in North America.

Lesch-Nyan syndrome is a disorder of purine metabolism which again is caused by a single-gene disorder in the X chromosome (which means it is virtually confined to males). It leads, bizarrely and horribly, to self-mutilation: victims commonly bite their own fingers and, if restrained, bite their lips instead.

Very rare but again extremely distressing is *severe combined immune deficiency*, alias SCID, in which cells of the immune system are damaged so that the victims cannot fight off infection. Their fate is to live inside aseptic plastic bubbles. About a quarter of SCID cases are caused by lack of the enzyme adenosine transaminase, ADA, whose task is to help in the safe disposal of damaged DNA and RNA. In fact, like Lesch-Nyan syndrome, ADA deficiency is a disorder of purine metabolism. Both ADA deficiency and Lesch-Nyan syndrome are of particular interest right now because they lend themselves to the world's first ventures into gene therapy.

But although individual single-gene disorders are rare, there are so many different deleterious mutations – often, as in the thalassaemias, cystic fibrosis and muscular dystrophy, several or many different mutations in the same gene lead to similar disorders – that the overall *genetic load* of 'bad genes' in the human gene pool is enormous. In fact, each of us, on average, probably carries about five deleterious alleles which, if inherited in double dose, would cause frank and often very serious disease.

There are three main reasons why the load is so high. First, genes mutate all the time, while some, like that of Duchenne muscular dystrophy, have a particular tendency to mutate in particular ways. Second, most single-gene disorders are recessive, and since the homozygous sufferers are rare compared to the healthy (or in some cases slightly unhealthy) heterozygous carriers, there is no strong selective pressure to remove them. Those few bad genes that are dominant – Huntington's chorea is the most obvious example – do not generally make themselves felt until late in life; that is, until after their possessor has reproduced. As discussed in Chapter 3, selective pressure against genes that are expressed late in life is weak. Clearly, recessive alleles on the X chromosomes are the most likely to cause trouble, and indeed affect all males who possess them because males have only one X chromosome and so have no homologous, 'healthy' allele to counteract the bad one.

It has sometimes been suggested that by saving homozygous sufferers from an early death, and in some cases enabling them to reproduce, we are undermining the human species, by encouraging deleterious genes to become commoner. Contrariwise, I once sat through an extraordinary medical conference in which otherwise well-informed physicians discussed whether it was proper to sterilise sufferers from single-gene disorders so as, they said, 'to eliminate the gene'. The fatuousness of both lines of argument is obvious. Elimination or sterilisation of the rare homozygotes would have very little effect on the overall frequency of the bad gene, because the heterozygotes are so much more common. We have already seen that the cystic fibrosis carriers – 1 in 25 – are 100 times more common than the homozygous sufferers – 1 in 2,500. The rarer the allele, the greater the discrepancy between carriers and sufferers. Thus if an autosomal recessive bad gene occurred in single dose in 1 in 1,000 people, then 1 in 1,000,000 (1,000 × 1,000) marriages would bring two carriers together, and 1 in 4 million children would be sufferers. So in this case, the carriers would outnumber the sufferers by 4,000 to 1. Furthermore, natural selection has been snuffing out homozygous sufferers for as long as the bad genes have existed, for cystic fibrosis sufferers would rarely have reached adulthood until modern times. But this clearly has not eliminated the allele.

The third reason why 'bad' genes are so common is best illustrated by sickle cell anaemia and the thalassaemias. Both are far commoner than can reasonably be explained by mere recessiveness. After all, even the heterozygous sickle cell carriers suffer some degree of anaemia, and natural selection would surely have militated strongly against the sickling gene in our more exacting hunting-gathering days. It transpires, however, that these anaemia genes do not represent a simple pathology. They are an example of balanced polymorphism, as discussed in Chapter 1. The homozygotes are severely disadvantaged: in the case of sickling disease, the red blood cells actually collapse as the concentration of oxygen in the blood falls, because of a misplaced amino acid in the haemoglobin. But the heterozygotes both of sickle cell and of thalassaemia actually benefit, because their modest anaemia protects them against malaria. The malaria parasite lives for a time in the red blood cells, and prefers those cells to be big and healthy. In Chapter 5 I argued that parasites may have provided the selecting pressure that allowed sex to evolve, even though sex has obvious short-term disadvantages.

Here again, parasites provide selective pressure that has allowed the evolution of an otherwise disastrous mechanism.

This notion – that common deleterious alleles may sometimes indicate balanced polymorphisms – seems to apply far more widely than might be supposed. Emphysema, for example – the appalling disorder in which the lungs lose their elasticity, and which is seen most commonly in smokers – clearly has a genetic predisposition. But the genes that predispose to it seem to confer some protection against tuberculosis, which has been one of the principal scourges of most societies through most of history, and is still common in poor countries, and could yet return in force to the affluent ones.

Finally, emphysema introduces a broader concept: in addition to the genes that can lead straightforwardly and inevitably to disease, there is a whole class of genes that merely predispose to disorder. These disorders range from diabetes and coronary heart disease to various forms of schizophrenia, and to cancers: conditions that clearly are not merely passed on in a simple Mendelian way but which none the less seem to run in families.

The question is, what can be done about all these gene-based disorders?

PREVENTION: GENETIC COUNSELLING

Genetic counselling traditionally means the identification of carriers – or at least of people who are very likely to be carriers. Increasingly, it also implies detection of the homozygous state in babies *in utero*.

The most fundamental, ancient and universal way to identify carriers – or at least to narrow the field so you know who to examine further – is by pedigree: family history. Some populations may show such a high incidence of a particular disorder that all individuals should be examined, and thus some groups of Ashkenazi Jews should be screened for Tay-Sachs disease. More often, the counsellor is alerted to the possibility that an individual might be a carrier by his or her own family history (including the clinical state of the parents' previous children). In some cases, if enough family history is known (and the family is big enough), it is possible to assert on the basis of history alone that particular individuals *must* be carriers. Clearly parents who already have one child with an autosomal recessive disorder must both be carriers.

Sometimes, even using traditional techniques, it is possible to pinpoint carriers precisely because they, too, are slightly affected clinically. Thus heterozygous sickle cell and thalassaemia carriers both have anomalies of the blood.

The carriers of most recessive alleles are not clinically detectable, however, and it is here that the new molecular techniques come into their own. Here there are two requirements: first a probe, to attach itself to a known sequence within the damaged gene or to a region close to the damaged gene, and second a marker which indicates the presence of the damaged gene. The most convenient markers are RFLP sites: sites at which specific restriction enzymes will sever the DNA. Thus a deleterious allele is revealed when it is contained within a particular RFLP, as revealed by a specific (radioactive) probe. The production of probes and markers is now a growth industry.

In the very long term, it would in theory be possible simply to read the DNA: to do a printout of each individual and identify the stretches of DNA that (a) deviate from the normal (though accepting that the 'normal' is variable) and (b) correspond to known disorders. Present researches could be said to be laying the ground-work for this approach, but instant reading of everyone's genome may never be achievable, and certainly not for many decades to come.

However, the detection of carriers leads to all kinds of secondary problems. In the first instance, after all, the counsellor is able to establish only that one or both of the parents is a carrier. But that leaves a matter of odds. A baby born to two carriers of, say, cystic fibrosis would have a one in four chance of being homozygous for the bad gene, and therefore diseased, and a 50-50 chance of being a heterozygous carrier, like its parents. But it would also have a one in four chance of being homozygous for the healthy gene, and therefore neither diseased nor a carrier. Should two carriers decide not to have children, when they have a three-in-four chance of producing a child who is not diseased?

But for an increasing catalogue of disorders, it is becoming possible to assess the status of the fetus *in utero*; sometimes simply whether it is diseased (which is the most important question) but also (where molecular tests are available) whether it is a carrier.

Both the Duchenne muscular dystrophy and the cystic fibrosis genes can now be detected *in utero*. But of course the decision to abort can never be easy. Again, that decision falls to the parents.

The future, I believe, seems logically to lie with the methods of *in vitro* fertilisation, IVF, that have already been developed to assist childless couples. The mother would be induced by hormone treatment to 'superovulate': that is, to release a batch of eggs, rather than the usual one. All of these would then be fertilised *in vitro* with her partner's sperm. The resulting embryos would then be developed to the blastocyst stage, when they were ready for implantation. But before implantation, a few cells would be taken for analysis. Then, only an embryo that was completely free of the damaging gene would be implanted. Any other healthy blastocysts could be stored by deep-freeze – even with existing technology – and perhaps implanted later. By this means, carriers of single-gene disorders would be guaranteed a baby that was completely disease-free, and their lineage could be purged for ever of the damaging gene (except, of course, of mutants such as Duchenne which are liable to recur).

Some religious groups see ethical problems in such an approach, partly because of the general 'unnaturalness' of IVF, but mainly because the selection of healthy fetuses implies the rejection and hence the destruction of others. Some Catholics in particular feel that this is commensurate with murder, since they maintain that 'life' begins at the moment of conception. This seems to me to be sophistry. The benefits gained in the relief of suffering and of anxiety seem far to outweigh the theoretical, theological objections. The principal problem – once the technology is established – is logistic. Thalassaemia, for example, afflicts tens if not hundreds of millions of people. To offer such a service to all of them could cost more than the GNP of their countries.

Perhaps, however – nothing seems entirely unfeasible – it will one day be possible to screen suspected undesirable genes routinely, even within the gametes themselves. Even the strict Catholic objection would then be removed, for there would be no fertilisation until the health of the gametes was established.

Some affected children will slip through the screening net, however, and many have already done so. For them, an increasingly feasible option is to make good the genetic damage.

GENE THERAPY

Gene therapy carries two connotations, one somewhat sinister, the other less so. The first is to alter the gene content of the germ line (the cells that generate gametes) so as to alter the genetic content of all future generations. This is sinister partly for practical reasons: genetic engineering techniques are not likely to be so good in the future that we can say, 'By this intervention I can guarantee to correct such-and-such a defect but cause no other, unforeseen effects.' It is sinister too for reasons best discussed in the last chapter: that human beings should not assume so much power. In fact, for reasons of this kind, medical ethical committees from all around the world have already rejected the notion of engineering germ lines.

The second approach is *ad hoc*: simply to correct the defect within the affected cells. Such alterations will not be passed on to succeeding generations: according to the principle first appreciated by Weismann and discussed in Chapter 1, changes in body cells are not conveyed to sex cells. The idea of practical gene therapy is purely to help the affected person, as with any other form of therapy.

Astonishing though it may seem, gene therapy of this kind is already happening. The first, predictably, was carried out in a child with ADA deficiency. At the time of writing, early in 1993, the America's National Institutes of Health have given permission to the Cystic Fibrosis Foundation to try to carry normal genes into the noses and airways of cystic fibrosis patients using modified adenovirus as a vector. Adenovirus is one of the agents that causes the common cold – well suited to such invasion – but it has been modified to prevent causing frank disease. There is a theoretical danger that the modified virus could team up with a 'wild' type and regain its pathogenicity, and safety trials must (as always) precede therapeutic endeavours. In any case, such therapy would have to be repeated every few weeks, because the nasal lining is constantly renewed, and, at least until safety is fully established, each patient will need to spend time in isolation after each treatment, to avoid shedding the modified, transformed virus into the world at large.

Such *ad hoc* replacement therapy is liable to work best in

those tissues that are continuously renewed through the divisions of stem cells – for the stem cells can be altered, and give rise to new generations of healthy cells. As it happens, the most common single-gene disorders of all – sickle cell and the thalassaemias – are of this kind. All the blood cells, including the red blood cells, derive from stem cells – 'erythropoietic tissue' – that reside in the bone marrow. Take out some of these, equip them with genes that code for normal haemoglobin, and then put them back, and with luck, the patient should stay healthy for many months and even for life; a sharp improvement on the present treatment which in severe cases requires transfusion every few weeks (with all the discomfort, expense, and concomitant risk from hepatitis and AIDS).

There are of course many general problems, in addition to the specific kinds of difficulty of the sort identified for cystic fibrosis. The first is to identify and clone the appropriate gene. This is discussed later, though in fact this has already been achieved for thalassaemia. The second is to introduce the gene into the patient's own stem cells, which is most likely to be undertaken with the aid of a modified retrovirus vector. But the main problem, identified by Sir David Weatherall of the Radcliffe Infirmary, Oxford, who is one of the world's leading thinkers on gene therapy, would be fully to control the introduced gene, and the virus that carries it into the cell.* In particular, the engineer (in the present state of the art) cannot direct the positioning of the virus vector, so the new gene would be introduced effectively randomly, anywhere in the genome. There is a remote chance that a gene introduced in the wrong place – notably in the middle of an oncogene – could trigger the series of events that cause a cell to become cancerous.

Then there is the genetic engineer's perennial problem of gene expression. This is related not only to the gene's position within the genome – and introduced genes, as has just been noted, cannot be positioned at will, even if the engineer knew where best to put them – but also to the introns, those long and tedious sections within the genes that were once thought merely to be 'junk'. The favoured method for introducing genes at present is *via* retrovirus vectors, and they are not generally big enough

*Sir David's comments in this section derive from an interview I carried out with him for BBC Radio 3 in 1988. That is a long time ago, but the general caveats still apply.

to carry entire human genes, introns and all, into the cells. So contracted genes, exons only, are normally inserted. These may not be expressed at all, or if they are, are not subject to the same controls as a normal gene.

The logistic problem – that such therapy would be far too expensive to make an impact on the vast number of thalassaemia patients – Sir David Weatherall considers to be potentially less significant. As he points out, very high technology, once developed, is not necessarily difficult to apply. Many amazing biotechnologies are already reduced to kit form. It is not technically difficult to extract blood cells from the marrow; it is not difficult or expensive to incubate those cells with a suitable vector, and not difficult to reintroduce the cells. Such treatment would be far cheaper than the present transfusions and, combined with genetic counselling, would make a significant numerical difference to the extent of suffering.

In principle, a similar approach might be applied to a wide range of single-gene disorders. To be sure, some tissues are far harder to deal with than blood cells are, because they are not renewed throughout life from stem cells. Thus we are born with as many muscle cells as we will ever have, so damaged muscle cannot be taken out and simply incubated with a gene vector. Instead, the vector would have to be introduced into the patient's muscle *in situ*, which is clearly a much more hazardous and difficult undertaking. An alternative is to introduce fetal muscle cells, which – in early experiments – show at least some tendency to develop and function.

The main problem, however, is the one that besets all genetic engineering. It is to identify the relevant genes in the first place. This essential process should be abetted by the Human Genome Project.

THE HUMAN GENOME PROJECT

At least since the 1970s molecular biologists worldwide have been analysing stretches of DNA from a range of 'model' species, including yeast, mice and human beings, and this information has been stored in various computerised data bases, of which the largest is GenBank. As far as human DNA is concerned, most people until the mid-1980s were content simply to locate and identify genes of outstanding interest (such as the cystic fibrosis gene), and to sequence those. Furthermore, most people were content simply to sequence the exons of those genes; that is, the regions which actually coded for proteins. The introns, which are snipped out of the messenger RNA that is translated from those genes before the mRNA leaves the nucleus, were thought to be most uninteresting, and so too were the long stretches of DNA that lie between the genes.

The idea of mapping and then sequencing the entire human genome, introns and intercalatory stretches and all, was first formally mooted in 1986 by Robert Dulbecco in an editorial in *Science* (vol. 231, pp. 1055–6). Dulbecco has a Nobel prize for his work in virology, and his particular aim was to get a handle on the oncogenes, which underlie cancer.

Many people since have derided Dulbecco's suggestion, for several reasons. The sheer volume of work is stupendous: there are three billion base-pairs to be put in order, base by base. Most of the genome seems extremely uninteresting, for the genes are interrupted by introns which do not code for proteins, and are often separated by even larger tedious stretches. To sequence introns does not seem to be a priority (many suggest) while there are still so many important genes still to be sequenced, and indeed identified in the first place.

Others argue, however, that the work-load will get hundreds of times smaller as new mapping and sequencing techniques are developed, and that one obvious way to develop new techniques is by applying present methods to the task in hand. They point out, too, that although the function of introns is largely unknown, they are far from uninteresting, since they are clearly involved in the crucial issue of gene expression. Then again, by looking only at what we know to be interesting, we are assuming too much, possibly missing out on many subtleties of genome organisation and *modus operandi* that so far have escaped us. Computer analysis and comparison of the whole structure will reveal interactions and relationships at present undreamed of. Just to take a comparable example, computer analyses of DNA from various creatures have already shown a marked similarity between the cystic fibrosis gene and a number of other genes in other organisms that have a quite different function: for example with one in bacteria that repairs damage to DNA caused by ultra-violet light; and with one in fruit flies that makes eye pigment. Revelation of such relationships will tell us an enormous amount about evolution, and a great deal more besides. Finally, advocates of the whole-genome approach argue that once all the knowledge is in place, the *ad hoc* search for individual genes of interest will be virtually over. The information will be there to be read.

In practice, Dulbecco's ambition has come about, and 'The Human Genome Project' to map and then sequence the entire human genome has become a target for laboratories worldwide. Some groups have chosen to work their way doggedly through particular chromosomes, notably the supremely interesting X chromosome – though the endeavour becomes less and less dogged as the techniques improve – and others prefer the 'bottom-up' approach, beginning with the exons of genes of known interest. Already the two kinds of approach are beginning to meet in the middle.

In general, every year for the past fifteen years the amount of total DNA from all the world's organisms that has been analysed has grown by 50 per cent, which means a ten-fold increase in knowledge every five years. Walter Gilbert, a Nobel prize-winning biologist from Harvard, suggested recently in *Nature* (1991, vol. 349, p. 99) that he 'expect[s] that sequence data for all the model organisms and half of the total knowledge of the human organism will be available in five to seven years, and all of it by the end of the decade'.

In other words, the third millennium after Christ should begin with a complete printout of the human genome. What we will do with that knowledge is another matter.

WHEN WE HAVE SEQUENCED THE HUMAN GENOME – WHAT THEN?

A complete printout of the human genome will to some extent promise more than it delivers: it will be like a score of a Beethoven symphony, without an orchestra to play it; with little clue, indeed, as to the sound of an oboe, or the difference between *piano* and *mezzo forte*. But in time the meaning will become clearer. We will be able to say what does what, then to say what effect each piece has upon each other piece, and so on until we can truly read what the printout says, just as an orchestral conductor can read a score and hear the sounds in his head.

Long before we reach that point, too, the printout will be powerful, and like all power, it will lend itself both to good and not-so-good.

Good, most people will think, will be the added vigour with which we can approach single-gene disorders, both for therapy and – more importantly in the long term – for counselling. The potential for distinguishing between embryos or gametes that are free of deleterious genes, and those that are affected or are carriers, I believe is the most promising of all.

More equivocal will be the extension of such diagnoses to the genes that do not underpin simple, single-gene disorders, but which predispose to cancers, heart disease, diabetes, or whatever. Such information could in principle be used to ensure that people who do have particular predispositions – for example to particular industrial cancers – are warned against working in particular industries, which might seem good and sensible. On the other hand, the countries that are rich enough to carry out such tests are not likely to perpetuate the kind of primitive industries that lead to such cancers. Third World asbestos mine owners seem particularly unlikely to introduce them. Employers might on the other hand insist that candidate employees present printouts of their genomes, from which may be judged their likelihood of succumbing to particular disorders, or any other aptitude or weakness. More realistically, insurance companies could in principle use such printouts to load policies one way or another, with possessors of particular genes paying more. But on the other hand again, all this information will be difficult and expensive to extract, and all such information will be a matter of odds, rather than of certainty. Very probably, most people will place such information in the class of data that is more trouble than it is worth.

However, all this clearly raises ethical questions, as well as practical, and perhaps is best left to the next chapter.

CHAPTER 10

A New Created Life

When I was born – and I'm not that old! – even the nature of the gene was still uncertain: was it made of DNA or was it protein? Only forty years have passed since the structure of DNA was established, and only thirty since the genetic code was 'cracked'. The term 'genetic engineering' was coined only twenty years ago. But already 'genetic engineering' is big industry. Already doctors have embarked upon the genetic repair of human tissue, already we can speak realistically of a new 'paradigm' of medicine. One advantage of growing older (the only one that I can see) is that it begins to put time into perspective. Ageing helps you to realise that a decade is really an astonishingly short space of time. Even a century is not so long. Indeed, 200 years is a twinkling; even I knew somebody (my grandmother) who knew somebody (an old man in her village when she was a little girl) who was alive at the time of the French Revolution. We are conditioned by our western politics to think in four-year or five-year bursts; the lives of governments. But it is sensible and increasingly necessary to think in decades and centuries. So let us contemplate how far we have advanced in the past twenty years – how primordial seems the science of the early 1970s; but then think where we might be in fifty years' time, or 200, or 500. If we do not astonish ourselves, then we have not played the mind-game properly.

WHAT IS POSSIBLE?

To address the question 'What is possible?', we must make a trade-off between our own unfettered imaginations and what Sir Peter Medawar called 'the bedrock laws of physics', which cannot be transgressed. Yet the bedrock laws, although severe, need not trouble us. They include the law of conservation of energy, which says that within any system that is cut off from the rest of the Universe, the total amount of energy cannot change; and Einstein's observation that matter cannot travel faster than light. Such bounds are hardly restrictive. So in our trade-off, imagination can prevail. We may reasonably contemplate *anything*.

We might begin with comparative trivia, some of which I hinted at in Chapter 7. The bedrock laws of physics dictate that plants will grow faster in warm climates than in cold, and that their energy input is determined by the Sun (though some supplement their nitrogen supply and perhaps part of their energy by gobbling insects). None the less, there are plants that grow and flower in winter. So in trivial vein we could create and cover our houses with frost-proof vines, bearing fruit and flowers throughout the winter; fruit of any kind, from neo-bananas to post-agricultural passion fruit, engineered to grow on temperate versions of strangler fig or bindweed or what you will. Alternatively, we could develop super-versions of the London plane tree, which is already extremely tolerant of pollution: plants fitted with novel enzymes that enable them to *feed* upon the pollutants of the air. Plants to lure but then to digest insects that would otherwise be pests, and then nourish themselves from their corpses, could be with us in a few years, for many plants already produce insect-repellent chemicals which could be modified into attractant pheromones, while the sundew and pitcher plants have no monopoly on insectivory, for many 'ordinary' plants – from teasel to potato – already have a rudimentary ability to take nitrogen from insect corpses, which could be raised to specialist level by genetic engineering. These engineered plants, which lure and then feed upon insects, would either be crops – self-feeding crops – or else could be planted in hedgerows, or as copses between the crops as in agroforestry, to lure pests *away* from the crops. We could envisage such transformations on an arable scale – and the insectivorous potato is far from fanciful – or on the smaller-scale level of horticulture. Horticulture is far from trivial. In a world in which labour is super-abundant, it could again become one of the world's principal sources of food, and horticulture even for the frivolous purpose of flower production has transformed the economy of entire countries, from Holland to Ecuador.

One last, dramatic fantasy before we leave plants. In a few hundreds of millions of years' time, or perhaps less, the atmosphere of this planet could be changed in ways that make it impossible for animal, or conventional plant life to continue. I am fantasising fairly wildly, but the Earth's atmosphere is known to have changed dramatically in the past (notably by a gain in atmospheric oxygen) and there are known geological mechanisms that could bring about further reversals. We might, of course, in the future, develop systems for atmospheric regulation; indeed it will be surprising if such a system does not begin to come on line within the next few centuries. On the other hand, it may prove impossible to reverse big changes on a global scale.

What we might do, however, is create life forms that could cope with the new, post-human conditions. This would do us no good, but it might help to promote further evolution. In particular, I have in mind plants that have been endowed with the genes from chemosynthetic bacteria, of the kind which now, for example, process H_2S (hydrogen sulphide) instead of H_2O (water) for the provision of energy. A chemosynthetic plant – using the Sun's energy to process atmospheric components other than water and carbon dioxide – would effectively be a new *kingdom* of organism, as significantly different from present-day organisms as plants are, say, from animals. In contemplating the future possibilities of genetic engineering, we can afford to think radically.

We can and must take fewer liberties with animals than with plants. Animals are sentient beings, and their bodies are more integrated than plants'. If a plant is bred with a weak stem – well, it can be grown up a stick. If an animal is bred with weak legs (and many modern livestock have extremely weak legs, because they are bred and fed to 'outgrow their strength') it suffers pain, and also may die prematurely because its immobility leads to infection and starvation. Modern broiler farmers strike a precarious balance between profitable growth rate and unprofitable loss through premature death. With modern breeding and reproductive technologies abetted by genetic engineering, we could envisage sows like termite queens, nothing but a giant womb, unable to move, fed artificially, and producing perhaps hundreds of young per year. The milkiest cows now produce 4,000 gallons (20,000 litres) of milk per lactation, an amount unheard of thirty years ago. Why not more? Why not reduce the cow to a giant udder, swollen with fluid like a honey-pot ant?

The answer is obvious: because such visions are cruel and hideous. But provided we do not break the laws of physics – and although the porcine termite and the bovine ant raise technical problems of heat disposal, there is no transgression of absolute laws – such targets are achievable. Pigs are already bred to produce thirty piglets a year (while their wild counterparts give birth to only four or five); commercial turkeys are already produced exclusively by artificial insemination (since they are far too bulky to copulate); poultry farmers already accept high mortality as part of the normal debit; and poultry, pig and dairy farmers already consider high morbidity 'normal' – so we cannot anticipate that any hideousness we may conceive will be self-correcting. Visions that we may perceive to be irredeemably nasty are what many agriculturalists are working specifically towards.

Yet similar technologies, in a change of context, can be benign. A few more decades down the line and genetic engineering and tissue culture could provide colonies of microbes – possibly bacteria but more probably fungi – fitted with the genes of animal muscle, and growing *in situ* into beef-steaks or chops, chicken-breasts or Peking duck. Why not? In these preliminary years animal genes have already been put into yeasts, and fungi are already cultivated and textured as meat substitutes. In short, we could in principle produce meat – any kind of meat – *without* incommoding sentient creatures. We might rather reserve the few farm animals that remain – cosseted, long-lived, fed on meadow grasses and windfall apples – for banquets, since they would very properly cost twenty times more than day-to-day cultured flesh. Again, there are books to be written on the conceivable economic and social knock-on effects of such technologies, both good and bad.

Bolder in scope, under-financed but supremely important, are the possibilities for conservation. At present there are significant moves worldwide – in London's Institute of Zoology; in San Diego, Cincinnati, and Berlin – to store the semen and to some extent the embryos of exotic animals, and increasingly of endangered species. This should soon be extended to include eggs rescued from the ovaries of deceased females.

It is easier, however, simply to deep-freeze the flesh or blood of endangered animals. The question then is what to do with the DNA it contains (the white blood cells of mammals contain nuclei with DNA, though the red cells do not). Even with present-day technology – as explored in Chapter 8 – such DNA is useful for analytical purposes: to explore the relationships and hence the evolution of animals (including those that may not be saved from extinction), and to define the range of present-day genetic diversity for comparison with populations of the future or with the past. Already, in a comparable exercise supported by the Smithsonian Institution in Washington, Dr Rob Fleischer is comparing the genetic diversity of present-day Hawaiian geese with what is contained within the semi-preserved skins of nineteenth-century museum specimens.

Beyond monitoring, however, lies the possibility for positive action. In theory, when future wild animals lose genetic diversity, the alleles they have lost might be identified and replaced – by copying the genes that are now being put in store. This will require a great deal more technology than exists today, not least the ability to control gene expression (probably implying much more knowledge of introns, and means of inserting entire genes, introns and all), and the ability to place the gene in precisely the right locus, with the help of exquisitely targeted vectors (of which more later). A variation on this theme – in some ways simpler, in some ways harder – might be to alter the genes of future animals to restore diversity to the level revealed by the DNA that will be held in store. Of course, if we do not take the opportunity to save the tissues of present-day animals, the information will be for ever lost. But the storage of existing DNA – urgent, cheap, and potentially vital though it is – is not being adequately funded, which effectively means that it is not being carried out.

Bolder still – far bolder - is the notion of restoring entire extinct animals from samples of stored DNA. Perhaps the ultimate fantasy is to restore the dinosaurs, as envisaged by Michael Crichton in his gripping and very nearly plausible *Jurassic Park* (Arrow Books, London, 1991).

Crichton addresses many though not all of the problems. Problem one, of course, is to find dinosaur DNA. DNA is extraordinarily resilient, but when flesh has been replaced by stone – which is what is implied by complete fossilisation – it is gone for ever. Crichton, invoking ideas from the Extinct DNA Study Group at the University of California, Berkeley, suggests that dinosaur DNA could be found in the bellies of blood-sucking insects, not fossilised but embalmed in amber – that is, in the petrified resin of ancient trees. Such DNA would of course be present only in minute quantities, but can readily be multiplied to any desired quantity by the polymerase chain reaction (Chapter 8).

The next problem (which need not trouble future biologists working from material from present-day animals) would be to fill the gaps, for the DNA of ancient creatures, though highly intriguing, is very fragmented. Crichton provides a partial solution: to collect as many scraps as possible and see where they overlap, so that little scraps can be convincingly joined into big scraps, and then replace what is gone for ever by DNA borrowed from comparable, living creatures. This is plausible insofar as many genes controlling basic functions are common not only, say, to all vertebrates, but in some cases to all living things, even though each class or species may have its own variants.

The snags, however, are manifold. First, restorers will not be able to tell what extra DNA they have to supply unless they can read the DNA they already have: that is, infer its function from a knowledge of its structure. This becomes more and more plausible as knowledge increases, but there are still some obvious problems. For example, a gene of a given structure may perform different functions in different animals. Thus the gene which in mutated form causes cystic fibrosis in humans helps to determine eye colour in fruit flies. Second, some genes that are unique to the extinct creature are bound to be missing, and it is hard to see how restorers could guess the structure or function of what no longer exists. In short, it is very difficult indeed to see how future restorers could know what it is they are trying to restore unless they have a plan of the complete genome to guide them – which, of course, is precisely the thing they will be lacking. Thus future attempts to restore dinosaurs are likely at best to resemble Sir Arthur Evans's reconstruction of Knossos: interesting, but basically patchwork and probably wide of the mark. But Knossos is only architecture. An animal is an organism. The difference is very important, as will be discussed later. A patchwork city can still be grand, but a patchwork organism is a grotesque.

If, however, restorers were less ambitious, and were content to work with more recent animals, then they could be in with a chance. We need not give up hope of reconstructing the complete genome of the dodo, the amiable mega-pigeon from Mauritius which was killed off by sailors only 300 years ago, or even of the woolly mammoth of Northern Europe, exterminated by our stone-age ancestors only about 10,000 years ago. But if we had a complete genome, what then?

The immediate problems are of two main kinds, though the two are linked. The general problem is that DNA found in, say, frozen muscle flesh will already be extremely differentiated. Only a minority of its genes will have been expressing at the time of death. Most will have been firmly switched off. But if we want to restore a complete animal, then we need to restore totipotency. How this might be done is anyone's guess. But in a century or so, de-differentiation could be routine. The technology could be available in kit form.

The second difficulty is to find a suitable environment – cytoplasm – in which the newly restored DNA can express. In Crichton's novel, dinosaur DNA is placed in the cytoplasm of fertilised modern crocodile eggs. This is eminently plausible – indeed, it is difficult to envisage any other route – but none the less raises an (almost literally) chicken-and-egg problem. As you may recall from Chapter 3, the first divisions within the zygote (the fertilised egg) seem to be initiated by the cytoplasm itself, whose structure was endowed upon it by the mother. The genes of the newly formed zygote nucleus do not come on line until several cell-divisions have passed: apparently awakened by cytoplasmic signals. We may note, too, that mitochondrial genes, though few in number, play significant parts in the life of an animal, and are again conferred exclusively by the mother. In short, the egg cytoplasm is not simply a substrate for the DNA to wallow in. The cytoplasm and the nuclear genes are working partners. Perhaps crocodile egg cytoplasm would provide a suitable partner for Crichton's restored dinosaur DNA, and perhaps not. Similarly, Asian elephant egg cytoplasm may or may not be a suitable partner for mammoth DNA. We will have to wait and see.

Finally, and intriguingly, the restoration of extinct animals raises an ecological/evolutionary problem. For animals are not fixtures and fittings. The essence of an animal is that it changes as the generations pass, largely by the natural selection which moulds it to suit its environment. It is not good enough simply to restore 'a' dinosaur, or 'a' mammoth. True restoration implies the re-creation of a gene pool capable of changing with time. Furthermore, this new creature – this new, adapting gene pool –

will have to become part of the ecosystem in which it is placed, and so will influence the fate and the evolution of all around it. Restoration is not a trivial thing, in short. Whether it is a *good* thing, I will discuss in the next (and last) chapter.

An apparently quite different but none the less related fantasy (so far) is to clone favoured animals (or self-important human beings) from their somatic (body) cells. *En passant*, we may note that this would pose some interesting ethical problems for those religious groups who object to the inevitable sacrifice of young embryos during *in vitro* fertilisation on the grounds that each embryo is a potential human being. If cloning from somatic cells becomes possible, then *every* cell is a potential human being. It might be considered unethical to throw away, say, a severed finger on the grounds that it could, in principle, have populated an entire continent.

But this is going too far. Clearly, the most obvious problem posed by somatic cloning is to de-differentiate already differentiated DNA, which will be achieved sooner or later. Meantime, a form of somatic cloning is already becoming possible by culturing embryonic tissue, before the DNA is irrevocably switched off. Conceivably, rich people could have a portion of their embryonic offspring removed (at a stage when the missing portion would be restored naturally), place the tissue in culture, and, if the child turns out to be gifted, multiply the tissue and produce a clone. Some agricultural research is already aiming to multiply 'elite' livestock in just this way. Whatever can be done with cattle can almost certainly be done with human beings. The embryonic tissue route to human cloning should, in short, be technically feasible within a few decades.

More obviously benign would be a major assault on alleles that are frankly deleterious, and which lead to gene-based disorders. We discussed the most immediate possibilities in Chapter 9. Among the obvious problems – once we have established which genes are responsible for what disorders, and where they are positioned in the genome – is to find ways of introducing replacement genes reliably, in ways in which they express perfectly, and are in exactly the right part of the genome to help ensure both that they do express properly, and do not interfere with other genes. For this we need to envisage a new generation of vectors roughly analogous with the 'cocktail' vaccines of the present day, which carry out several different functions simultaneously. For the vectors of the future should be able to carry entire genes, introns and all, plus extra control regions; they should also include stretches of DNA comparable to genetic probes, which recognise the specific locus where the gene should be inserted and bring the vector safely alongside, like a space shuttle docking with a satellite; they might also ideally contain enzymes able to snip out the offending mutant DNA; and, like a virulent (though non-infectious!) disease organism, they should seek out and infect every relevant cell in the body.

More fancifully still, we could envisage such cocktails sweeping through the entire genome, recognising and correcting *all* acknowledged deleterious alleles as they went: a spring-clean of the genome. This is a very tall order indeed, but it does not transgress any bedrock laws. If we could clean up the genome in this way, we might for ever delay the onset of senility – assuming, that is, that ageing is caused by an accumulation of deleterious genes, as Sir Peter Medawar envisaged. Alternatively, we might prolong our lives indefinitely by holding up the genetic program as it unfolds, and staying as long as we choose at any particular state of maturity.

More immediately, and more down-to-earth, techniques that at present are astonishingly high-tech will be reduced to recipes, carried out with the help of standard kits. This is already the pattern: a huge variety of high tech is already available off the shelf, and is carried out by people who need have little or no idea of the underlying principles. With such kits leading both to efficient monitoring and to routine therapy (for example for the thalassaemias) we can expect to make huge inroads on the incidence of single-gene disorders, just as vaccines have vastly reduced infections. There is a knock-on effect with vaccines, as mass though not universal vaccination leads to 'herd immunity'. As noted in Chapter 9, a comparable knock-on effect in the field of gene therapy would be achieved by selecting the embryos from among the offspring of carrier parents that are completely free of the gene. Two hundred years after the discovery of vaccination, smallpox was completely eliminated. Might cystic fibrosis be wiped out in two centuries' time?

The final lure, though – the final prize or the final nightmare, depending on your point of view – is the creation of life; ultimately and it seems unsurpassingly to create, from laboratory reagents, thinking creatures that have free will; to re-enact the opening chapter of Genesis. This indeed would be to 'play God'. Is this possible?

The commonsense answer must be 'Yes'. If we arrange the right molecules in the right order and give them a suitable source of energy, self-perpetuating processes ought to begin which we should acknowledge as 'life'. But there is more to this issue. First we should ask, 'What exactly do we mean by "life"?'

WHAT IS 'LIFE'?

Some biologists favour a minimalist definition of life. For example, Graham Cairns-Smith of Glasgow University suggests that to qualify as living, systems need fulfil only two conditions. First, they should be self-reproducing. Second, as the replications continue, the succeeding generations should be capable of change, and specifically should change in response to the pressures of the environment. In other words, there should be 'descent with modification', leading to better and better adaptation. What is alive, then, says Cairns-Smith, is whatever can evolve – implying both self-replication and change – by the Darwinian mechanism of natural selection.

This definition has its attractions. It is irreducibly simple – in general a virtue in scientific definitions – and yet it seems to allow the evolutionary progression from bacterium to hippopotamus. It is also highly intriguing, in suggesting that most of us define 'life' far too narrowly. For there are certain clays that reproduce themselves – effectively by adding bits to themselves from their surroundings, as any crystal may grow, and then fragmenting. But these particular clays can take many different crystalline forms, as snow-flakes do. Which particular form prevails in any one environment depends upon the conditions; for example, one form might be swept away by a stream more easily than another, and so come to grief, while the one that better fits the available niche continues to multiply. Thus, in any one place, these versatile clays can evolve, growing, splitting and modifying to conform to the demands of the environment. If evolution by natural selection is the criterion of life, these clays should be considered to be alive.

Space probes have apparently shown that there is no 'life' on Mars. But all they have really shown, says Cairns-Smith, is that Mars has no life of the kind that most of us conventionally recognise. If the probes had sought evolving clays, they might well have found life of a sort.*

Yet I do not feel that such a loose definition truly fits the bill. Perhaps clays do evolve, yet they do not seem to be 'alive' in the sense that dogs, mushrooms and oak trees are alive. These creatures do evolve, to be sure. But they also exhibit a further hierarchy of crucial qualities.

The first of these extra qualities is thermodynamic. Clays come into being by simple chemistry. Passing atoms stick to the crystal clay lattice that already exists, effectively because it requires less energy for them to do this than it does to maintain their independence, and any collection of chemicals will interact in ways that slide towards the lowest energy state, just as water will always tend to run down hill. This does not break the bedrock law of conservation of energy, because the energy saved as the water flows or the chemicals interact is made manifest as heat.

But dandelions, rabbits, dogs, and all the other creatures that we acknowledge to be 'living' do not grow in this passive, thermodynamically simple way. The complex molecules of which they are made can be put together only by adding energy that is expropriated from the outside world. The dandelion harnesses the energy of the Sun and the rabbit seizes the energy contained in the complex molecules that the dandelion has created with the Sun's help. The performance of this thermodynamic trick – harnessing outside energy to create complex molecules – is at least as fundamental to life as the ability to evolve by natural selection. Clearly, however, systems that have mastered this trick can produce far more molecular variation for natural selection to work upon than any clay can do, for clays can assume only those molecular forms that can be assembled 'passively'.

The second, extra, crucial quality of *bona fide* living things I will call *dialogue*. This has two important connotations, the first of which is evident when we contrast the molecules in a cell with

*I discussed these ideas with Professor Cairns-Smith at the International Society for the Study of Life meeting held at the University of California, Berkeley, on 21–25 July 1986. The meeting was hosted by Berkeley and NASA.

the atoms in an ordinary crystal. In a crystal, the different atoms are simply arranged in a regular, three-dimensional pattern; the pattern being the one that requires least energy to maintain. Each atom is surrounded by several other atoms, of the same type or of various types, at a fixed distance from each other. Of course, the different atoms influence each other. Each one is surrounded by an electric field, which affects the electric fields around all the other atoms; and alterations in the electric fields around atoms imply a change of properties. But the atoms in a crystal are static, and each one is measurably influenced only by the ones with which it is in immediate contact. Two atoms a millimetre apart in a crystal seem as indifferent to each other as two bricks in a wall.

In the sharpest possible contrast, all the molecules in a living cell are affected, directly or indirectly, by all the other molecules in the cell, or at least potentially so; and each in turn affects all the other molecules. The atoms in a crystal are static, and its neighbours are unchanging. In a cell, some molecules do stay more or less in the same place, but many others are in constant movement, so that each molecule experiences a constantly changing set of contacts, like the guests at a well-hosted cocktail party. In a crystal, the effect of any one atom upon another is 'passive': each one influences the shape of its neighbours, but that's an end to it. A significant proportion of the contacts between the molecules of a living cell have far greater import than this. They result in change, to one or both of the contacting molecules. The changes are of various kinds. A molecule may simply be split into two as a result of the contact. Or it may add or shed water. Or it may add some extra chemical group, so that a derivative of carbohydrate, say, is given nitrogen in a suitable form and becomes an amino acid. Or – crucially – one molecule may simply pass energy to the other, the one that does the passing then assuming a less energetic form, while the receiver becomes energised. This is like winding a clock, as if a spring were coiled and made ready for action. Commonly, energy is imparted to molecules in living cells by adding an extra phosphate radical. It takes energy to add the phosphate and it sits awkwardly, and when it is finally released, the energy that held it uncomfortably in place is unleashed, for whatever purpose is required.

The second connotation of dialogue is that of *heterogeneity*.

The different molecules in a cell do not talk simply to others of their own kind. In fact they *rarely* talk particularly to others of their own kind. Crucially, and necessarily, a living cell contains molecules of many different kinds interacting with each other. The full significance of this will become evident later.

The final additional quality of living cells, which places them above evolving clays, is the concept of *organism*. This is really a development of the concept of dialogue. A crystal may grow, as a cell or an animal may grow, but a big crystal is just like a whole number of small crystals stuck together. Any one part of the crystal is like any other part. A fragment of a crystal, chipped off, survives just as well as the whole crystal. A dog – an organism – is not like this. Each part of the dog is unique. Each part complements all the other parts. A dog has a tail, but the whole dog is not a multiplicity of tails. A small dog does not grow into a big dog by acquiring more tails. A tail has no ability to survive by itself; a tail removed is a tail doomed. In an organism, in short, each part contributes to the whole; no one part is autonomous; the whole is greater than the sum of its parts.

Where does all this leave our initial question, 'Can we create life?' Well, if we are content with a loose definition – 'What is alive is what multiplies and evolves by natural selection' – then we could merely envisage the fabrication of some chemical system that might be hardly more than a clay. But if we want to create something more convincing than this – a dynamic, dialectic, energy-processing *organism* – then we clearly need to do a great deal more. If we set our sights on the higher target, how might we proceed?

A NEW CREATED LIFE

In principle, future creators of life could either approach the problem 'top-down', or 'bottom-up'. The top-downers will simply begin with an existing cell, and alter it and add to it, rather as custom-car buffs modify standard automobiles. Present-day genetic engineers are already doing this. Clearly, top-downers could produce organisms that are very different from those of today; as I suggested earlier, they might create a new kingdom by grafting the genes of chemosynthetic bacteria into plants.

The bottom-uppers could be even more adventurous. They will be the purists. They will contrive to begin with laboratory agents, and add them together and go on adding until the system starts to stir in ways that we acknowledge to be living. In essence (though not necessarily in detail) they will be trying to replicate or mimic the processes by which life is thought to have originated on Earth, probably around 3,500 million (3.5 billion) years ago. It is worth asking, then, what that first appearance of life involved.

THE ORIGIN OF LIFE ON EARTH

At the 1986 meeting of the International Society at the University of California, Berkeley, there were disagreements even on the most fundamental issues, notably whether the atmosphere of the Earth when life began, 3.5 billion years ago,* was chemically of the kind that would remove oxygen from systems – i.e. was *reducing* (as is commonly believed) – or was of the kind that tends to add oxygen – i.e. was *oxidising*. This would make a profound difference to the details of life's beginnings. The notion that life may, in fact, have drifted in from outer space was not considered to be of huge interest, although it may sound intriguing. We are interested, after all, in the conditions that brought life into being in the first place; and if those conditions were not in fact found on Earth, then they must have existed somewhere else, so the nature of the question does not change.

Each individual who attended that Berkeley meeting was presumably impressed by different ideas. I was particularly struck by Graham Cairns-Smith's ideas on clays as outlined above, albeit with the points of disagreement. But I was perhaps most intrigued by the ideas of Dr Lynn Margulis of Boston University. She it was who stressed the notion that the living system must contain different *kinds* of molecule, working together, the idea that I referred to above as 'heterogeneity'.

This simple observation has profound connotations, both practical and philosophical. On a purely practical level, many scientists have struggled in their search for life's origins – and continue to

*Latest studies from Australia suggest that life may have originated on Earth as long as 4 billion years ago.

struggle – because they assume, logically but I believe erroneously, that they ought to be looking for one single kind of molecule that is capable of multiplying itself, and also of assuming various forms so that it can evolve in response to environmental pressure. But it is very difficult to envisage a molecule that really can do all that is required. DNA cannot replicate itself: it needs proteins to help it. Proteins do not form spontaneously; they need DNA – and sources of energy – to come into being. As we mentioned in Chapter 2, RNA seems to some extent to combine at least some of the various necessary qualities, which has caused some scientists to speculate that RNA might indeed be the 'basic' molecule from which others evolved. One problem with that, however, is that RNA is itself a thermodynamically unlikely molecule, which cannot come into being unless aided by other molecules that can process energy. So RNA could not have arisen spontaneously.

But this problem is overcome if we stop thinking in this blinkered way: that is, stop thinking that life must have begun with a single entity that became more elaborate. Instead, we can reasonably envisage that in parts of the primitive Earth – around hot-springs? around undersea volcanoes? on the wet surfaces of clay? – conditions were ripe to bring many different, complicated molecules into being. This is not too fanciful. Amino acids, for example, are present in meteors, presumably formed spontaneously in outer space, and they can easily be made in the laboratory in conditions that could well have obtained on ancient Earth.

The important point here is that no one of the molecules in the stew would have had all the properties that could give rise to life. But between them, different *groups* of molecules did possess those properties. Some might have been effective catalysts, able to cause other molecules in the stew to behave in novel ways: for example, to split. Others – crucially – would have the ability to expropriate chemical energy from inorganic materials in the surroundings: for example by splitting hydrogen sulphide, in the way that some bacteria do even now. Some of these, we must assume, could not only release that energy, but use it: turn themselves from low-energy into high-energy molecules.

So in this complicated but primitive stew we already have two of the essential qualities that we associate with modern life. We have dialogue, for the different molecules in the stew are interacting, and we have acquisition of energy, as some of the molecules in the stew achieve high-energy status by borrowing chemical energy from materials in their surroundings. This combination of catalysis and energy input promotes and powers the elaboration of more and more complex molecules.

So far, of course, we have nothing so grand as an organism. We simply have an inchoate cocktail party of molecules. But after a time, by chance, and as complexity increases, reactions begin to occur in particular sequences. One molecule causes another to do something which, in turn, causes another to do something else, and so on. This is something more than a chain reaction, in which the same reaction occurs repeatedly. This is a chemical *cascade* – a sequence of different reactions – of the kind that still plays an enormous number of roles in today's organisms. Blood clotting, for example, requires just such a cascade of changes. Second, or perhaps as a special example of the cascade, we can envisage groups of molecules passing energy to each other, like children in a game of pass the parcel. Again, such sequences play key roles in modern organisms; for example, energy is bundled from molecule to molecule in the chloroplast during photosynthesis.

Most of the time in the primitive stew, these cascades of activity just fizz and then peter out, running through the assembly of molecules like small bush fires of energy. But after many and many a summer, another crucial step is taken. The end of one of the cascades joins up with its own beginning. The cascade thus becomes a closed, feedback loop, and such a loop, constantly gathering energy and elaborating itself, becomes a self-perpetuating system. Such a system is bound, quickly, to dominate the rest, literally consuming the other, less orderly molecules in the stew. Such a system is self-contained. Natural selection ensures that the self-contained system becomes more efficient, and this inevitably involves greater elaboration. As the system becomes more complex and more self-contained, it begins, inevitably, to qualify as an organism.

Each and all of the above events may seem very unlikely. They do not seem to happen in the present-day world, after all. But it is perfectly reasonable to envisage such events taking place on the primitive Earth, before there was free oxygen in the atmosphere to destroy complex molecules as they came into being, and before there was an ozone layer to protect the Earth from the energetic (if somewhat too energetic!) bombardment of ultra-violet radiation, and before there were other life forms about to gobble any presumptuous interloper. We know, too, that whichever way life arose, it *was* unlikely; thermodynamically and in other ways, life is a difficult trick to pull off. We should not forget, however, the sheer scale on which events as outlined here might have taken place on the primitive Earth. There might have been billions upon billions of rich organic soups. More to the point, we should not forget the sheer amount of time that life could have taken to get going. The Earth itself is about 4.5 billion years old. Life seems to have begun about 3.5 billion years ago. The earliest bacteria-like organisms seem to be around 3 billion years old. In short, to get from a primitive stew to an organism of the kind that we would still consider to be extremely primitive took hundreds of millions of years: as long as it has taken to progress from jawless fishes to us. Unlikely things are indeed unlikely, but in an immensity of time and space, unlikely things are likely to happen sooner or later. This is an important biological principle, to which I will return in another context in the next chapter.

The details of the above – the notion of the cascade that becomes the feedback loop; and the feedback loop that becomes the organism – is my elaboration. At least, I do not know who else has said it. But the notion that the first life form *did not begin as an elaboration of one kind of molecule, but as a co-operative between different kinds of molecules with different properties* I ascribe to Lynn Margulis. Practically, it is important because it stops us trying to envisage primitive molecules – such as primordial forms of RNA – that might have combined all the necessary qualities, and allows us to focus upon mixtures of molecules, which interacted, and then engendered cascades, and then formed feedback loops, and so on. Philosophically the notion is important because it shows that life, *intrinsically*, depends upon co-operation. Darwin was aware of the value in nature of symbiosis: of the co-operation between unlike creatures. But in essaying natural selection he was obliged to stress the role of competition. Subsequent commentators have seized upon the competition and tended to underplay the role of symbiotic co-operation. The notion that co-operation was in there from the start – is indeed one of the crucial mechanisms of life – is pleasing. But of course we cannot escape competition. Once a successful co-operative arose, it was immediately in competition with other successful co-operatives.

This view of life's origins has two further implications. First, it is clear – as emphasised in Chapter 2 – that modern life revolves around the dialogue between DNA and proteins; or, we might say, the discourse between DNA, RNA and proteins. This fact has led some scientists to search specifically for the origins of these three molecules. But this is surely a mistake. The dialogue of DNA and proteins is only one example of a potential infinity of dialogues. It just happens to be a particularly successful dialogue, one which, once it got going, decisively out-competed the rest. But DNA and proteins must each be seen as highly evolved molecules – which took hundreds of millions of years to evolve – or rather to co-evolve, for the evolution of each clearly depended upon the other. We must envisage, too, that organisms of a kind were already well established before this particular duo, DNA and protein, began to emerge from the pack. Both kinds of molecule are extremely thermodynamically unlikely, and could not have arisen except in the cosseted confines of a system already complex enough to qualify as an organism. We may also observe, more generally, that the most ancient organisms that still exist on Earth – such as the bacteria that are confined to marshes because they are poisoned by newfangled oxygen, which appeared in the atmosphere a mere two billion years or so ago – are still highly evolved and 'advanced' organisms, even though we tend to call them 'primitive'. They were the Rolls-Royces of their day, out-competing and obliterating the thousands of hopeful life-experiments that probably preceded them. And, primitive though we may think them to be, they took a billion years to evolve.

Second, this view of life's origins suggests that if we created life bottom-up, as life must have been created 3.5 billion years ago, then we need not pre-judge the issue. We need molecules to act as energy-processors, and others to act as catalysts, and as the system grows in complexity we need others again to act as stores of information – genes – to direct the rest. But we need not suppose that the catalysts have to be proteins, or the genes have to be made of DNA. Perhaps – in theory undoubtedly – other quite different molecules would fit the bill. In short, if we built life bottom-up, we could perhaps create life forms that were quite different not only in form but also in mechanism from any that have ever lived on Earth, or indeed in the whole Universe.

That last observation leads to a still broader question. If it is possible in theory to create life forms from molecules other than DNA and proteins, is it possible to create life forms from elements other than the carbon, oxygen, nitrogen etc of which modern organisms are compounded? This question in fact leads us to still broader issues, which I will discuss later under a new heading. But the simple practical answer to this particular question seems to me to be 'No'.* Molecules cannot fulfil all the criteria of life – versatility; the ability to adopt different forms in different circumstances; the ability to multiply and to process energy – unless they are extremely complex, or at least potentially so. Molecules cannot become extremely complex unless they contain atoms which each are capable of joining up with quite a few other atoms simultaneously. The Universe contains about 100 different stable elements, and of these, only two have atoms with the kind of structure required to form very complex molecules. These are carbon and silicon. Of the two, carbon is far more versatile because it is smaller. Carbon therefore can give rise to proteins, fats, carbohydrates, nucleic acids, each in thousands or millions of different forms, all the myriad 'organic' materials of which life is made. Indeed the word 'organic', as used by chemists, has none of the mysticism that non-scientists attach to it; it simply means 'compounded from carbon'. Silicon, though also versatile, is more cumbersome, and the best that can be compounded from it are the evolving clays. Life, then, wherever it occurs in the Universe, seems bound to be organic: carbon-based. In its way, though, as I will discuss later, this idea – that life depends upon particular chemistry – is just as mystically satisfying as the idea that life is an abstract thing that could be formed with any suitable juxtaposition of elements.

But however life is created in the future, and whatever form it takes, the newly created, self-perpetuating dialogue will almost certainly be the first such event since Earthly life first began 3½

*Not everyone agrees with my contention that life is bound to be organic, but I feel that those who disagree do not attach the same exacting criteria to the definition of life that I do. If your definition is loose then it can fit just about any complicated and interesting collection of molecules.

(or more) billion years ago, for living things, once established, would surely have consumed any presumptuous molecules that endeavoured to repeat the original successful experiment. It is a spectacular thought that the organic dialogues which chatter and converse within each of us have continued, uninterrupted, since near the birth of our planet, through millions upon millions of generations of creatures that took a million different forms from bacterium to fish to us, each individual dying but passing on a fragment of itself.

This overall view of life – involving processing of energy, dialogue, heterogeneity, feedback loops, and then the basic qualities of multiplication and descent with modification – is elaborate. But it is still only chemistry – and is that enough? Is life really only chemistry? When future scientists have assembled suitable molecules in an appropriate way, will they then have made a system capable of life, or will they merely have created a system analogous to a radio receiver, which is dead unless it is given the appropriate signals to receive? In other words, is life an innate quality of (organic) chemical materials, waiting to be realised and unleashed, or does it need some extra 'vital spark'? If the chemical systems do need some extra 'spark', is it within the gift of mortal scientists to supply that spark? Is 'life' perhaps some aethereal quality, perhaps literally God-given, that God may condescend to instil into our laboratory stews, or not? On the other hand, if we do create life, could that life evolve as we have evolved, into creatures with consciousness and 'mind'?

The question also arises, whether all that happens in a properly elaborated living system can be explained simply by the forces of everyday chemistry, or whether those systems, in their full flowering, acquire some extra set of properties that distinguishes them absolutely from non-living systems. Such cogitations have run through science, philosophy, and theology, for many centuries. I am not going to attempt a history, but there are some highly intriguing modern twists which I am sure will prove extremely important over the next few decades and which immediately affect our picture of the gene, and of how genes influence the form and function – the phenotype – of the body. Thus they also affect our view of genetic engineering; for the relationship between the gene and the final form is probably not so straightforward as the term 'engineering' implies.

GENES AND LIFE

I said in an earlier chapter that the paradigm of all modern biology is neodarwinism: the fusion of Darwin's ideas with Mendel's. Molecular biology has fleshed out that paradigm, apparently showing how it is that Mendelian genes are translated into living bodies – the phenotype – upon which Darwinian natural selection can act. Basically, genes make proteins, which shape bodies, which are then exposed to the rough and tumble of natural selection. The ideas that follow do *not*, I believe, significantly challenge the supremacy of basic neodarwinism; although some of the protagonists of those ideas believe that they do. What they do significantly challenge (I believe) is the addendum – the mechanical explanation; the simple idea that genes make proteins which make phenotypes. As seems to be transpiring, the relationship between the gene and the final form may be more indirect, and far more subtle, than this simple notion implies.

The existing paradigm – or at least, the molecular addendum to the paradigm – is simple and satisfying, and takes us a very long way, but it also leaves gaps. Various people, including many biologists, have pointed out in various ways that we are making a giant leap when we assert that the row of genes on a chromosome somehow translates into the subtly interacting three-dimensional totality of the organism. To be sure, the gap is plugged to some extent by the modern corpus of embryological theory which is now showing that as an organism develops, so groups of cells send chemical messages to other groups of cells to tell them what to do; and how gradients of hormones and inorganic ions are created along the length of limbs, so that cells closest to the body that are exposed to high concentrations behave differently from those at the far ends, which are exposed to the weakest concentration. But many feel that such explanations do not quite meet the case. Bodies seem too complicated, too subtle, to be shaped merely by such influences.

Defenders of present paradigms argue simply that if there seem to be gaps in present explanations, it is merely because we do not yet have enough facts. We are, they suggest, thinking along the right lines, and there is no reason to suppose otherwise. But many modern thinkers share the doubts of philosophers from the past, who feel that the shape and form of the final body cannot emerge simply from instructions contained in each individual cell, and who feel more generally that there must be more to life than can easily be explained by chemistry; or at least, if there is not, then there must be more to chemistry than has so far met the eye. What follows is a rapid, cavalier, and probably brutal overview of a range of ideas for which I am sure I will be severely castigated, but which I hope gives at least some of the tone of the detracting voices.

It seems to me (in this cavalier vein) that the detracting ideas fall into three main categories. First there are those which see the body not as a living thing in itself, but as a vessel of a very special kind, which is able to contain the quality known as life; or which, to take a more modern analogy, is essentially like a radio receiver, able to pick up appropriately vital vibrations from the aether, in a way that dull objects such as stone cannot do. The ancient concept known as 'vitalism' is of this kind; 'life' being something which, in the language of theology, is 'breathed into' suitable structures – organisms – and which at some arbitrary time (death) is withdrawn. Descartes's concept of 'mind' is of a similar kind; the brain does not generate the mind, but instead acts as a receiver for the quality of mindfulness that in some manner or other pervades the Universe.

For my part, I feel that such ideas are too far in the realms of mysticism to detain us at this point. They may be right; but there are more economical ideas, which are more accessible to scientific study, and it seems worthwhile to concentrate upon those first. However, at least one modern thinker has presented a body of ideas that seem to me very like those of vitalism: Rupert Sheldrake in *A New Science of Life* (Blond and Briggs, London, 1981) argues that organisms (and indeed non-animate entities such as crystals) communicate with each other *via* 'morphic resonance' – fields of influence of an unspecified nature; and that each thing that has ever existed communicates with other things of like kind *via* morphic resonance and tells them how they 'ought' to be. The idea is strange but interesting; the kind of idea that one feels in one's bones is wrong but if correct would be revolutionary, and is worth a smidgeon of the research budget just as it is worth putting two shillings each way on a 1000 to 1 outsider in the Grand National. You never know.

The second body of ideas does not suggest that organisms are radio receivers, tuned to pick up signals from the aether at large. It does suggest, however, that different parts of the organism signal to each other in ways that are not to be explained purely in terms of chemical signals – that is, by hormones or gradients of inorganic ions, of the kind studied by conventional embryologists. Three modern thinkers who doubtless will be outraged to be lumped together in the way I am about to do but who seem to me to be arguing along such lines are the American philosopher John Searle; the British physicist and philosopher Danah Zohar; and the British mathematician Roger Penrose.

Searle to be sure argues not about life *per se*, but about mind; but the analogy seems to me to work perfectly. Searle's ideas are best understood by contrasting them with those of the British pioneer of computers, Alan Turing. Turing argued that thinking – which in this context we can equate with 'mind' – is simply an ability: something that any suitably designed machine could do. By the same token, the ability to move may be expressed by steam engines, internal combustion engines, or windmills; it makes no difference what the engine is made of, or how it is powered. A box – a thing with the quality of boxiness – can be made from wood or cardboard or steel. So too a brain – a thing with the ability to think – could in principle be made from organic neurones, or from transistors, or from silica chips. To be sure, as we have observed in an early chapter, the human brain works differently from the computer (although the full extent of the difference was certainly not appreciated in Turing's day). But the difference could, in theory, simply be a matter of circuitry. Those who follow Turing literally (and there are many such) would argue that with different circuitry, a computer could think in the way that the human brain does, and that if the circuitry was sufficiently complex, the computer could take on all human qualities, including the phenomenon of consciousness; that is, not only of thinking but of knowing that it was thinking. Thus the suitably elaborate computer would not simply be a 'thinking machine'; it would truly have the essential qualities of 'mind'.

In his series of Reith lectures on BBC Radio 4, John Searle presented the quite opposite argument (*Minds, Brains, and Science*, BBC Publications, London, 1984). Consciousness, he said (which in this context can be equated with 'mind') is a quality of the organic brain. Computers may look as if they are conscious, but that is an illusion. In reality, they are simply grinding 'uncomprehendingly' through problems, just as a Chinese clerk who did not comprehend English might nonetheless (with suitable clues) arrange English words into what appears for all the world like an English sentence – even though the clerk did not understand that sentence.

Controversy followed the lectures. Many critics simply did not see why Searle's argument should be correct. For example, suppose some microsurgeon replaced just one neurone with a silicon chip that did the same job, would this alone transform the human brain from a conscious mind to an uncomprehending computer that was merely going through the motions? Presumably not. But suppose a million neurones were thus replaced, or a billion, or three quarters of the brain. At what point would the brain cease to be conscious, or 'mindful'? Why should it cease to be mindful, if each substitute chip did the same job as the neurone it replaced?

I believe that Searle is right, though I do not believe that the arguments he presented carry the day by themselves. As a philosopher and a non-scientist, Searle presented his case on grounds of logic and metaphysics; but such arguments, in this context, are not sufficient. Whether brains are conscious and computers (or other machines) are not is, in the end, a question of *fact*, and it can be judged only on empirical grounds.

Thus, if we argue that brains are not just computers, we have to address the question, 'What specifically are they doing that computers are not?' And if we are to argue that brains are *intrinsically* different from computers – that they are not merely super-advanced computers of the kind that will be built one day – then we ought to be able to show that brains actually perform *physical* feats that are beyond the abilities of computers – just as the ability to stretch, say, is beyond the physical ability of concrete. So what *physically* is a brain doing that a computer cannot?

The answer may lie in the ideas of Roger Penrose, as described in *The Emperor's New Mind* (Vintage Books, 1990). In that book, Penrose argues that human brains might be employing mechanisms that derive from quantum physics. For he suggests that different components of the human brain (and by implication of animal brains) can and do operate in coordinated fashion, and do so *simultaneously*. Such a degree of simultaneity cannot be achieved in an electric circuit, such as exists in a computer; for in a circuit there is a *sequence* of events, with one part telling the next part what to do, and so on. A small electrical circuit works very fast, so that different parts seem to act all at the same time; but in reality there is a time-lag between one part acting and the next. In the brain, says Penrose, different parts act *literally* together. It is as if they were both marching to the same drum – except, crucially, that there is no drum. In nature, such simultaneity of action, and such apparent coordinatedness, is seen when two photons escape simultaneously from the same atom; from that time on, no matter how far apart they are, the behaviour of those twin photons appears coordinated, even though no discernible signal passes between them. At least by analogy, then, Penrose suggests that what really happens in the brain can be explained only by recourse to quantum phenomena, as observed in the behaviour of photons.

If this 'quantum' view of brain function is correct, then it implies that the brain is *innately* different from the computer, for the computer does not operate in this way – and cannot do so, for it is not built for such feats. This idea would also vindicate (though of course not prove) that what a brain does *depends upon* the fact that it is made of neurones which in turn are made of organic proteins, and that it is *not* simply a machine that could be made of something quite different. For if we put Searle and Penrose together, the suggestion is that consciousness is a manifestation of particular quantum phenomena; quantum phenomena that cannot, in fact, be exhibited by anything except a brain. As Searle says, a computer may achieve some of the results that a brain achieves. But consciousness (which in this context we may equate with 'mind') is a special phenomenon which is and can be achieved only by quantum effects which in turn can be achieved only through the particular chemistry of a living brain.

This bold view of mind and the brain can be expressed more generally; to embrace the notion of life and of the organism. We know that different parts of an organism such as a human being communicate with each other by all kinds of perfectly discernible means; notably *via* nerves and hormones. Yet many people – including some modern biologists – feel that there is *more* coordinating the different components of the body than is so readily discernible. They ask, for example, how the cells of an embryonic hand 'know' that they are supposed to be forming themselves into a hand; and they do not quite believe the current explanations, which suggest that the different cells are exposed to different concentrations of growth factors which cause them to behave in different ways. To be sure, such growth factors are discernible, and do differ in concentration in different parts of the limb; but by themselves, these crude guiding signals seem (to many critics) to be insufficient. Again, some of these 'new-wave' biologists feel that the different parts of the organism are coordinated by 'forces' that are analogous to, and very possibly are the same as, those associated with quantum phenomena. Danah Zohar, in *The Quantum Self* (Bloomsbury, 1990) explores this idea more fully.

Such ideas – if correct – provide us with a sharp and pleasing model of life, and of what would be required to create it. First, we would have to acknowledge that living things cannot, in fact, be created from any old group of chemicals, even in theory.* Only those elements that were able, in suitable juxtaposition, to provide the required quantum effects (if such they are) would qualify. Life would be even more dependent upon the specific element carbon than we have already suggested. Mind would indeed be dependent upon brains made of organic neurones. Non-organic systems – such as clays – may emulate some aspects of life, and computers may simulate some aspects of mind. But in each case, the resemblance is superficial.

Note, however, that neither Penrose, Searle, nor Dohar envisage that the 'extra' forces that mind and life may require have to be imbued into the system by some outside agency, Godly or otherwise. We would not have to suppose that the future scientist, having made his potentially living stew, would also have to provide any extra 'spark', in the way that Mary Shelley's Dr Frankenstein kick-started his monster with extraneous electricity,

*Danah Zohar has told me that she does not, or at least not unreservedly, support my (tentative) suggestion that *only* complex organic systems would be capable of behaving in a 'quantum' fashion. She suggests that at least in principle, other systems might provide the necessary effects. This could-or-couldn't discussion is at present irresoluble (at least by me). I intend to pursue the hypothesis that *only* organic systems can do what is required, which is as good a working hypothesis as any.

or in the way that vitalists seem to envisage. We simply have to suppose that when the stew reaches a particular degree of complexification, and when it contains the right components in the right juxtaposition – the energy harnessers and the catalysts – then the coalition would reach a threshold point, and the quantum phenomena that are truly the characteristic of life would begin. By analogy (and only by analogy) nuclear chain reactions begin in uranium when there is a 'critical mass', but not before.

It is impossible to say at precisely what stage this threshold would be reached: perhaps at the stage when cascades begin, or perhaps at some later stage when the cascades have become loops and the loops have become an organism. But once they begin, the quantum phenomena serve to consolidate the whole; they coordinate the behaviour of all the different components, so that they truly work in harmony. But, to hammer the point, such unifying phenomena that finally consolidate the organism could not arise within any assemblage of molecules. Only those with a particular chemistry – elaborations of a theme of carbon – can achieve this. Life and mind, then, emerge as particular qualities of particular components of the Universe.

However, although some people may feel that present explanations of life's processes are inadequate – that embryology based only on hormones and gradients does not really explain how bodies form so perfectly – they may nonetheless reject notions that invoke forces, such as quantum effects, for which there is no direct evidence. Such critics might be inclined to say – as I do – 'No: I think the answer does lie in chemistry; but I think there is a great deal more to chemistry – or at least to the behaviour of materials – than has so far met the eye.' My own thinking in this field is primitive. The truly exciting thinkers in this area are those who now focus on the modern mathematical concept of complexity, the subtleties of which are beautifully described by Roger Lewin in *Complexity* (Macmillan Publishing Company, New York, 1992). Relevant in this context are the ideas of Brian Goodwin, of Britain's Open University.

The general notion of complexity can be seen as the mirror image of chaos. Chaos has been one of the mathematical revelations of the past decade. The point is that systems that are literally chaotic and which in detail are unpredictable – such as the weather – can result from the outworking of ground rules which in themselves are very simple. In particular, chaos results when two or more *different* simple systems interact; thus the vagaries of weather result from the interaction of simple physical rules that have to do with the expansion of gases when heated, and the convection of water when heated, and the conductance of heat from water to land and land to air, and the extra whorls imparted to air when it hits mountains, and of water when it hits continents. *Each* of the components taken individually is simple but taken all together, the result is bewildering.

Complexity effectively describes the opposite effect; that apparently chaotic interactions can – and do – lead to simple forms. Water swirling around rocks is chaotic; but in places, for reasons that are extremely complicated but nonetheless can theoretically be analysed, here and there it forms whirlpools – vortices – that are perfectly simple mathematical forms. It is as if the chaotic form somehow 'had it in it' to form this simple and perfect shape.

Brian Goodwin, if I understand his ideas correctly, suggests that the order of the body arises from the maelstrom of genetic information in much the same way that a geometrically perfect whirlpool arises from the literal maelstrom of water in a creek. Although the system and the information it contains may *seem* chaotic, nonetheless there is an innate orderliness within it, which needs only opportunity to be realised. Thus does the genome provide instructions, which include the passage of hormonal and ionic signals from one cell group to another in the way that the conventional embryologists describe. But the instructions by themselves are inadequate, and essentially messy. What is needed is this extra phenomenon of complexity; the fact that out of systems that are *apparently* chaotic, extra properties emerge; properties of orderliness. The mess of cells that is the primordial eye, forms itself into a finished, working organ of vision in the same kind of way, and for the same kind of reasons, that turbulent water forms itself into a vortex.

Though these kind of ideas are new, and dependent upon the new mathematics of complexity, they have an ancient heritage. As Lewin points out, the essence of the idea – that form can be 'intrinsic' in apparent chaos, and that shapes form themselves – can be traced back to the eighteenth century notion of Rational Morphology, promulgated by Kant and Goethe. In their hands the idea sounds mystical; but the new maths of complexity gives the ancient mysticism a solid mathematical basis. There is nothing unusual about this. It is commonplace for physicists to dream of phenomena which mathematicians later demonstrate to be possible; and *vice versa*, for mathematicians to dream imaginary worlds which physicists later find exist in the real Universe.

There are echoes of this general notion, too, in other branches of twentieth-century science. David Bohm, British theoretical physicist, wrote of the 'implicate order' within chaotic systems. I find echoes, too, in the writings of the early twentieth-century Jesuit biologist Teilhard de Chardin who argued, *inter alia*, that since living things are moulded from chemical elements, then those elements should be seen as 'containing' the essence, or at least the primordia – the 'seeds' – of life. So life, to Teilhard, was the realisation of a quality that in most of the Universe most of the time is waiting for the opportunity to happen. In the same way (we could argue), the vortex formed by water in a creek arises, in the end, from the 'innate' properties of the water, which are just waiting their opportunity to unfold. You could not predict that water would behave as it does in a whirlpool, just by knowing that its chemical formula was H_2O. By the same token, we might never be able to predict that the fully analysed human genome would create a human being, because the final form of the human being depends upon complex interactions that in practice can be analysed only in retrospect.

The present-day scientists and philosophers cited above have at times been lumped together as 'new-wave' thinkers; which perhaps is as good a label as any. I have somewhat shamelessly contracted some of these new-wave ideas to make a neater scenario than most of the people I have cited would acknowledge. But I also believe it is a valid exercise when formulating new hypotheses in science to extrapolate ideas (your own or other people's) to see where they might lead. Some have accused the new-wave thinkers simply of wishful thinking; of a desire to re-introduce the thrill of mysticism into a Universe already made plain by science. Purists suggest that to invoke vague new forces (or not-so-vague forces which we loosely call 'quantum phenomena') is simply to resort to magic, as a pre-scientific apothecary might do: dragging up strange influences to explain whatever we do not fully understand. One of the strengths of complexity theory is that it makes no recourse to any 'forces' that are not observable in day-to-day life – for chaotic water demonstrably does form itself into orderly whirlpools.

To be sure, say the critics, present-day explanations based on chemistry do not explain all aspects of life, but that is simply because the chemistry is not sufficiently advanced. Give us a few more years, and the i's will be dotted, and the t's crossed. But in support of the new wave, we may simply observe that present-day explanations of consciousness are clearly inadequate (as Penrose makes very clear), and that current explanations of the mechanisms that cause assemblages of cells to mould themselves into hands or feet or eyes and ears, all of the right shape and size, are not convincing at all. At the very least, the notion that different parts of the organism may be coordinated by mechanisms reminiscent of quantum phenomena, or by the kind of interactions that produce simple order out of complex disorder, is intriguing; and by committing ourselves unquestioningly to the 'life is just chemistry' model, we may be missing out on physical phenomena of (literally) vital significance.

Note parenthetically though, to reiterate my earlier point, that notions invoking quantum effects or complexity theory do not weaken the central importance of Darwinian natural selection, or of Mendelian genetics, or of the neodarwinian synthesis of the two. They certainly *do* say that a gene does not simply add a component to the phenotype in the simple way that the addition of a wing-mirror adds to a Ford Model T. They suggest that what genes really do is to change the ground rules; impose limits on the forms that the finished body may take. As we will discuss further in the next chapter, this throws doubt on the aptitude of the expression 'genetic engineering', which implies a degree of exactitude that is not achievable except in the simplest of instances. But it does not throw doubt, I suggest, on the importance of natural selection. Natural selection decides which of the forms that the phenotype is allowed to take, actually survives in the real world. In short, as Darwin said, it is natural selection that leads to adaptation.

For my part, I remain agnostic as to whether life and mind do, or do not, require extra forces beyond those of known chemistry. But I certainly now subscribe to the view that life does depend on carbon – life is not an abstract trick that any machine can pull – and that mind does depend upon the brain, and cannot be achieved by systems that are not organic. I am also convinced that the notion of complexity must soon be written into the biological orthodoxy. But how do all these speculations affect our question, 'Can life be created?' Well if anything, and perhaps paradoxically, they seem to me to reinforce the answer, 'Yes'; the same answer that the conventional notion that 'life-is-chemistry' would lead us to. Following Teilhard, for example, we should not envisage atomic elements as passive spare parts, like bits of a motor-car engine, that need to have life breathed into them even after they have been assembled. Atomic elements have plenty of fizz in them even before they are assembled. Put them together in the right way – proteins, nucleic acids, and all the other bits and pieces that you find in a truly living cell – and the complex will indeed 'take off', if given the right raw materials to draw upon and a regular input of energy, which all living things require. If it is indeed the case that living systems give rise to, and are coordinated by, physical (quantum) forces that are not observable in the standard biochemical laboratory well – so be it. If the molecules are properly assembled, then such forces will become manifest, just as chain reactions will begin in uranium once there is a critical mass – just because that is in the nature of uranium.

So perhaps in a hundred years, and perhaps less, there will surely be a 'Life-Creation Project' comparable with the Human Genome Project of today. We may be sure that the new life forms will be pressed into service, post-haste, as industrial clean-up agents, tidiers-up of genomes, churners out of nutrients or fragrances, or whatever. How those new organisms will be allowed to evolve in the centuries that follow – well: our distant descendants must decide. Whether this creation, or any of the other possibilities, are a good thing or a bad thing, I will discuss in the next and final chapter.

CHAPTER 11

Sense and Sensibility

It is surely appropriate to be thrilled by the new genetics. The knowledge of life's mechanisms, and the powers that derive from that knowledge, are stunning. The new technologies would enable us, if we deployed them properly, to alleviate some of the worst of human suffering, to feed the world's population – sustainably, and in a benign environment – and would do much to help us to save our fellow creatures from extinction. Yet we should be worried. There is much that could go wrong. In truth, most of the issues that seem to bother most people seem to me to be relatively trivial; for example, I am not particularly bothered that evil people can use powerful technologies for evil purposes, for that is no truer now than at any time in the past. What bothers me, rather, is that we simply do not have the mechanisms – economic, political, or indeed religious – to do good, even though we may wish to do so. It bothers me, too, that very few people seem aware of this, and that we trundle on with the same old mechanisms and attitudes in a world that has been irrevocably changed, and is now very close to its physical limits. The new knowledge is exciting, and the new technologies are powerful, but if we are to use them well we have first to dig deeply into all aspects of our lives and attitudes.

What, first of all, are the kinds of things that could go wrong?

WHAT CAN GO WRONG?

One relatively trivial issue that has assumed the proportions of a world debate is that of patents: whether it is or should be possible to patent a transgenic animal. It is easy to festoon such discussion with mysticism (as God made most of the mouse, shouldn't He have the patent?) but in the end we are talking legal nicety. If we work in a capitalist economy (and there is nothing intrinsically evil about a capitalist economy) then individuals who invest labour and capital must be sure that others will not simply expropriate their efforts. It is bizarre to patent a mouse, but it may be convenient legal shorthand to do so, rather than to invent some arcane (and hence extremely expensive and unwieldy) device for patenting particular processes that led to that mouse. Patenting is bad if it prevents the flow and utilisation of knowledge, but it was, after all, originally intended to do the precise opposite – to relieve inventors of the need to protect their inventions physically from prying eyes, because once a device is patented the inventor can sleep easy, knowing his ideas cannot simply be stolen. This matter is secular: a matter for lawyers versed in the dotting of 'i's and the crossing of 't's. What *does* matter is the *welfare* of the transgenic mouse, and in particular, whether it suffers as a result of its exotic gene.

In general, indeed, the biological issues are far more significant: of welfare, both of animals and people, and of ecology. These issues arise primarily out of medicine and agriculture.

MEDICINE

Molecular genetics *is* the new medical paradigm, and genetic engineering in particular can do much to alleviate human suffering, but all powerful technologies carry some risk. The main risk from gene therapy, though very slight, is probably that of cancer. Oncogenes lurk in the genome waiting to be triggered into life; a misplaced foreign gene, grafted randomly into the middle of an oncogene, could well provide one of the series of spurs that are needed to precipitate cancer. More accurately targeted vectors could in future minimise this risk; and, of course, risk has always to be balanced against the danger of withholding therapy.

The problem that has sometimes been mooted – that sufferers from hereditary diseases would be allowed to reach reproductive age and breed – has been dealt with in Chapter 10. Homozygous sufferers are rare compared to heterozygous carriers, so for them to have children would make very little statistical difference. Sufferers might well choose *not* to have children, however, since all their children would at least be carriers. However, men suffering from X-linked disorders, such as Duchenne muscular dystrophy, could still have healthy sons, for a man with an affected X chromosome could still pass on a normal Y chromosome. The technology to separate male and female embryos is already with us. In future, it should be possible to sex and separate sperm. This, then, should soon cease to be an issue.

Perhaps most worrying in the immediate future, because potentially least tractable, are the conceivable epidemiological risks: the risk of precipitating disorder among the population at large. For example, trials are already in train (spring, 1993) to deliver genes that could correct the damage to the lungs in cystic fibrosis (or, conceivably, in emphysema) by including a suitably loaded vector in an inhaler. But the dose would have to be repeated, because lung cells are constantly replaced. Chapter 9 described how patients will be isolated after each treatment to prevent spread. But as this and other conditions are treated more and more widely it will be difficult and in practice impossible to prevent at least some loaded vector finding its way into the outside world. What effect such vectors with their cargo of genes would have upon people who do *not* have cystic fibrosis (or other conditions) can at this stage only be guessed. Probably none, but this is not the kind of possibility to be lightly dismissed.

Epidemiology – the study of the spread of disease – is in effect a branch of ecology. After all, ecologists have recognised in recent years that pathogens are as much a part of the ecosystem as any creature: pathogens seek, as all creatures do, to solve their own survival problems, and as they do so they profoundly affect the lives and populations of all around them. But the main ecological problems will be raised by transgenic organisms produced for agriculture, rather than for medicine.

AGRICULTURE

Agriculture is the world's biggest industry (by far), and it is biology-based, so it is liable to be the greatest perpetrator and user of genetic engineering. It therefore offers the greatest scope for disasters, both in ecology and in welfare.

To be sure – though much has been made of them – the risks to human welfare from transgenic organisms in agriculture seem slight, though they are worth listing. Notably, animals that have been fitted with genes to produce extra growth hormone (say) could contain surplus quantities in their meat or milk, and have side-effects that could be, well, interesting. Food producers do of course argue (a) that this is unlikely, b) that such hormones (in meat) would be de-activated by cooking, and (c) that the meat would be screened to ensure that it contained no residues. To be sure, an overdose of growth hormone occasioned by a surfeit of beefsteak seems low on the list of life's hazards, lower, perhaps, than the danger of nutritional deprivation if meat, being non-engineered, is therefore too expensive. Consumers and rational lookers-on might reasonably argue on the other hand (i) that a small risk is unacceptable if no-risk is possible, that (ii) meat in most of the world is a luxury, and that people who are mal-nourished are not suffering specifically from a lack of beefsteak, and (iii) that while it is theoretically possible to screen meat to remove or pre-empt all possible hazards, we cannot take it for granted that this will be done. Consumers have, after all, been caught before.

More general and more worrying is the fact that the behaviour of a gene, once transferred from one genome to another, is innately unpredictable; an issue I will pick up on again later. Any one gene may behave differently (and even have quite different phenotypic effects) in different genetic backgrounds; all genes tend to be pleiotropic – having more than one effect – and genes introduced into a foreign genome can affect the behaviour of other genes, especially if introduced randomly, perhaps in the middle of some existing gene. Thus the genetic engineer might fit some new gene into a cow or a potato, blithely expecting it to increase milk yield or pest resistance, and find that that gene does many other things as well. Some of those 'other things' could, in theory, be bad. Thus, as was mentioned in Chapter 7, wild potatoes contain many genes that produce toxins (such as solanin), and domestic potatoes still contain some of those genes even though they are mostly turned off. Many other crops, from parsnips to spinach, are known to produce toxins in low quantities – thanks to genes inherited from wild ancestors – and we do not know how many other potentially harmful genes our garden vegetables as a whole might contain, because we cannot at present detect the ones that are switched off. But in theory, a newly introduced gene could trigger quiescent, ancestral genes that have the ability to produce toxins. Of course the breeder can screen for such dangers, but the screening is less than straightforward unless he has a clear idea of what kind of toxin he ought to be looking for, which might not be the case. Again, poisoning by transgenic parsnips is not among life's deepest pitfalls. But we should not assume too blithely that homely garden veggies are as benign as they seem.

Such points are merely caveats, not out-and-out condemnations. Risk is always a matter of pros and cons, and genetic engineering could equally well reduce existing dangers. In much of the Third World, for example, cancers of the fore-gut – oesophagus, stomach, and liver – are far more common than in rich countries. The cause lies in large part with mycotoxins: toxins produced by fungi in crops. Genetic engineering that increased fungal resistance would save many more lives than it was ever liable to endanger.

Of much more tangible significance are the welfare implications of genetic engineering for farm *animals*. Already we can see all too clearly that inappropriate breeding can lead to animal suffering. Farmers sometimes argue that if an animal can survive, grow and breed rapidly, then it cannot be unhappy, but in fact a high degree of pathology and commensurate suffering are perfectly compatible with life. We know this from human experience; many single-gene disorders are compatible with life, and yet are extremely unpleasant. The genetic disorders in domestic pets range from dislocated hips in Alsatians to in-turned eye-lids in chow-chows. As was shown in Chapter 10, present-day farm animals are also suffering, as over-blown turkeys are unable to mate and are crippled by hip degeneration, while broiler chickens simply outgrow their strength. Genetic engineering may not raise completely novel problems, but it will make it possible to produce more radical change, more rapidly; and rapid, radical change can be the hardest to legislate against. Transgenic chickens with nerveless beaks that can be painlessly amputated are all too credible: people steeped heavily in agricultural 'efficiency' and in very little else slide too easily into such enormities. As I suggested in Chapter 10, we might in the fulness of time produce sows like termite queens and cows like honey-pot ants. All the countries of the world, including animal-loving Britain, already tolerate horrendous cruelties to farm animals. The extra possibilities for radical change and ultra-cheap efficiency can easily tempt us to tolerate even worse.

But the greatest problems springing from genetic engineering in agriculture could be ecological: the influence of transgenic organisms escaped into the environment. Agriculture is not the only possible source, but it is the most likely one, because transgenic crops (and the microbes that may support them) are intended to grow outdoors, or at least in greenhouses that can never be fully secure.

ECOLOGY, LINEARITY AND GENETIC GARDENING

The word 'ecology' was derived in the nineteenth century from the Greek *oikos*, meaning household; the same root as in 'economy'. It is the study of how living things interact with their physical environment and with each other. In its roots, ecology was natural history. Now it is truly a science: highly mathematicised, and subjecting its many hypotheses to the rigour of experiment.

I say all this simply because, in many people's minds, 'ecology' is equated with a particular romantic movement: like 'Green', some feel it to have mystical and even anti-scientific overtones. So when some people object to genetic engineering on 'ecological' grounds they simply mean, when you boil it down, that transgenic organisms are 'unnatural'; and 'unnatural' organisms, some feel, are bound to pose special dangers. In fact, naturalness or unnaturalness is part of the issue, though perhaps only a small part. But you do not have to mystify ecology one whit to perceive severe ecological dangers, if engineered organisms escape or are released into the outside world. Biologists are, of course, alert to the risks, but the question remains whether those risks can, or in practice will, be contained.

The fact that transgenic organisms are 'unnatural' is of ecological interest not because this turns them into faeries or wood nymphs that are beyond the reach of normal physical laws, but for two hard-headed reasons. First, exotic genes introduced into new organisms are not likely (in the immediate future) either to contain the same introns as they do in their original form, or to be introduced into a precisely prescribed part of the genome. Thus they are unlikely to behave exactly in their new host as they did in their original context. Of course, by the time any transgenic organism escaped (or was released) into the environment we would expect to know a great deal about it. But a new gene that had simply been dumped in a new host genome might (a) mutate in unexpected ways (and all mutations are unexpected), and (b) continue to affect, and be affected by, its fellow genes as the generations pass. After all, if the population reproduces by sexual reproduction – which is obviously the case with animals and most plants – the exotic gene constantly finds itself in new company, because a shuffling of genes is the whole *point* of sex. So an exotic gene dropped into a genome, and hence into a gene pool, is an unknown quality, whose full significance can be seen only as the generations unfold.

Second, transgenic organisms that escape into the world at large would, to a greater or lesser degree, be 'exotic' within the ecosystem in which they find themselves. They obviously would not have evolved there *in situ*. Of course, it is sometimes argued in a general way that a creature that has been bred and raised artificially is *ipso facto* effete, and cannot possibly compete in the great outdoors. Many an ecologist would love that to be true. To be sure, creatures that have been grossly overbred might be unviable: puffed-up turkeys that cannot mate, presumably, and overblown beef bulls that can hardly walk – although even these would presumably be leaner and meaner if they were not effectively force-fed. Less extreme domesticates are often supremely versatile, and extraordinarily destructive. On islands worldwide pigs and goats that were domestic for many generations have wrought terrible havoc: the pig alone in Hawaii has probably wiped out literally scores of native species. And the ravages of ex-domesticates are not confined to innocent islands. Former ships' cats, probably released in the seventeenth century, before James Cook arrived, are still steadily working their way across Australia, devastating the small native animals. Ex-domestic buffalo are currently trampling much of north-west Australia to pieces. Donkeys left over from the California Gold Rush – 'burros' – are ousting bighorn sheep from the Rockies. Once-domestic plants are at least as bad. Mesembryanthemums from southern Africa are now universal in warm countries throughout the world, all at the expense of local flora and hence of local fauna. Rhododendrons from country gardens are now ousting the native woodland plants through much of Wales and elsewhere in Britain. And so on.

To be sure, creators of transgenic bacteria for laboratory purposes often incorporate some specific weakness – which usually means some special nutritional requirement – so that the organisms *cannot* survive if they escape into the wild. But here there are at least four caveats. First, bacteria are the supreme mutators. Second, as we have seen in earlier chapters, bacteria have ways of passing genetic information from one to another, *via* plasmids, and so make good each other's weaknesses: not, to be sure, in a spirit of benevolence, but simply because the genes in the plasmids are selfish too, and have found that judicious jumping of ship is a good survival tactic. Third, any organism is liable to find ways of getting around any weakness. Michael Crichton in *Jurassic Park* envisaged that the new-created dinosaurs should be made with a more than usual dependence on the amino acid lysine, so that they could not survive in the world at large. But he was right to envisage, too, that they would get around such problems; in his novel they did so by eating lysine-rich beans. Bacteria are not as clever as dinosaurs, but all creatures are resourceful.

Finally, of course, organisms designed for agriculture – including bacteria – are *intended* to live in the outside world. Thus in Chapter 7 I mentioned nitrogen-fixing bacteria, *Rhizobium*, engineered to confer pest resistance on their leguminous hosts. Ecologically, such creations could prove highly significant. Professor John Postgate of the AFRC Unit of Nitrogen Fixation in Sussex has argued that the new nitrogen-fixing plants would not achieve 'rampant colonisation' because although they would solve their own nitrogen problems they would soon run out of something else, such as phosphate (*New Scientist*, 3 February 1990, pp. 57–61). This is true but, with respect, is not the point. An 'escaped' nitrogen-fixing plant might not smother the entire world but it could (and probably would) invade many a delicate ecosystem that is now colonised by rarities (with all the genetic diversity that they comprise), just as gorse (which is nitrogen-fixing) has invaded vast tracts of New Zealand, to the detriment of the native plants. Cereals are grasses, and as such are excellent 'pioneers'. A nitrogen-fixing grass would be a formidable competitor indeed.

Indeed, some engineered plants are designed specifically to survive *better* in difficult conditions than 'ordinary' plants may do. I love the idea of producing sorghums for the Sahel that are supremely drought- and pest-resistant. But such plants could wreak havoc in natural desert communities. In this particular case I am sure the risk is worth taking. But the risk exists.

However, the overriding ecological consideration is the one that emerged from the Royal Society Discussion Meeting of 26–7 February 1986, on 'Quantitative Aspects of the Ecology of Biological Invasions'. It is quite simply that the introduction of exotic species, whether previously domestic or not, is probably second only to straightforward habitat destruction and fragmentation as the principal cause of species extinction. To the introduced ex-domesticates in Australia we may add the cane toad (introduced early twentieth century), the rabbit and the fox (both nineteenth century) as horrendous sources of destruction. All three were introduced for reasons that seemed innocent enough at the time: the rabbit for food, the fox for chasing, the cane toad to clean up pests in the sugar-cane fields of Queensland. Rats and mice are pestilences worldwide. In twenty years, Nile perch introduced into Africa's Lake Victoria have wiped out about 200 of the 300 native species of cichlid fish. The flora of New Zealand has largely been wiped by bracken and gorse. The brown trout of Britain have been largely displaced by introduced rainbows and hybrids. And so on and so on and so on. Engineered organisms that escape into the environment may not in practice be worse than any other creature that escapes, but we should never underestimate the sheer potential awfulness of *any* creature that escapes, whether it is a vector carrying a gene, a bacterium, a sorghum or a chicken. Quite simply, you just cannot tell what it will do.

Overall there are two broad principles to keep in mind when considering the ecological effects of newly created and exotically transferred genes. The first is what physicists call *non-linearity*, which means that although a given event may have certain consequences, those consequences are not related to the initial event in a simple, 'linear' fashion. Sometimes a huge event has only small consequences, but sometimes small events can multiply and multiply and have huge consequences. 'The butterfly effect' is the somewhat fanciful metaphor: the idea being that a butterfly taking wing in China may set off a tropical storm as the perturbations of its wings are multiplied.

The ecological impact of new genes is non-linear at three levels. First, for all the reasons we have outlined above – pleiotropy, uncertain expression, the effects of genes on other genes, the need for genes to operate in concert parties – we cannot precisely predict the impact of any one gene on any organism into which it is introduced. Second, because of sex and recombination, we can predict even less effectively the effect of the gene within the gene pool as the generations unfold. And third, ecology itself is a non-linear series of events. We absolutely cannot predict the impact of any one new creature on any one ecosystem. No-one in their worst nightmares conceived that rabbits in Australia would ever be more than an occasional addition to the pot.

The second overarching principle is that although individual disasters may well be extremely unlikely, something is bound to happen sooner or later, because nature is so profligate, and has so much time to play with. It seems, for example, that elm-tree seeds crossed the Atlantic *by air*, even after the ocean had assumed its present size. For an elm seed to make this flight requires a combination of air-currents so unlikely that the odds against it are, literally, on a scale that is astronomical, greater than the number of stars in a fair-sized galaxy. But if you multiply the number of seeds on a tree by the number of trees, and then multiply that by the millions of years in which such an event could have taken place, you soon find that nature is quite capable of dealing in astronomical numbers. As we saw in Chapter 6, plant breeding is largely a matter of dealing sensibly in numbers that approach the astronomical, as millions upon millions of individual plants are screened for some special set of qualities. So, of course, it is *extremely unlikely* that any engineered gene will cause any particular organism to behave in a particularly unsocial fashion, and thereby have any noticeably adverse effects on the human species, or on the ecosystems of the world, but in nature, events that are extremely unlikely, happen. There can be few events that are more biologically unlikely than the present epidemic of AIDS.

Deeply misleading, I feel, is the metaphorical expression 'genetic engineering'. It is altogether too macho, too self-consciously no-nonsense, and it does not reflect reality. Engineers (at least when operating in well-worked fields!) are precise. They achieve exactly what they intend, entirely predictably. Genetic engineers *at best* are like gardeners, who plant a seed and must then stand back and let nature take its course. The difference is that gardening is an ancient craft, with millions upon millions of precedents to draw upon. Genetic engineering is new; the genetic engineers are like eighteenth-century gardeners planting unknown seeds upon new-discovered islands. Visit those islands now – New

Zealand, Hawaii – and you can see the results of their efforts. 'Genetic gardening' would be a more appropriate term, with all the uncertainties that implies.

These, then, are among the things that can go wrong. But another set of problems will result – are already resulting – even when things go right. They are of many kinds: so what follows is a mixed bag of possibilities.

THE PROBLEMS WHEN THINGS GO RIGHT: A SMALL MATTER OF DEATH

Even with present-day therapies, life expectancy in western countries increases remorselessly. Many of today's seventy-year-olds are still looking after their parents. But the new medicine promises – or threatens – far more than this. I suggested in Chapter 3 that the changes of old age and ultimately of death arise as poorly selected genes, shuffled to the end of the genetic program, begin to come on line. It must become possible, sooner or later, to suppress or replace those errant genes, or to halt the program at an earlier stage. Then, life might be prolonged indefinitely, and halted, perhaps, at more or less any stage the person desires. The social implications, good and bad, are obvious: presumably (to take a relatively trivial example) the concept of 'retirement' will have to be abandoned. More far-reaching, however, are the demographic implications. Present projections suggest that the human population of the world will reach around 10 billion by the middle of the twenty-first century; roughly a doubling of present-day numbers. But this estimate is based on present-day longevity, with just a few years added. If centenarians become the norm, then the projections must be raised, and the restraint required to bring the population down again will have to be increased commensurately.

Indeed, in a future world in which population is finally perceived to be an urgent issue, we will find ourselves balancing the desire for babies and the general need to 'rejuvenate' society against the demands of septuagenarians to live for another three or four decades. As I see it (what is the alternative?) we will have to re-think our attitude to death. But this is a huge nettle that present-day westerners are extremely reluctant to grasp. Western medicine is among the most sacred of our sacred cows, and the whole ethos of that medicine is that death must be postponed, effectively at all costs, even when hope and dignity have gone. To be sure, many people in the western world (including many doctors) have long since become dissatisfied with this blanket repulsion for death. In the 1990s, euthanasia for terminally ill patients is becoming accepted.

But other cultures in other times do not and did not share our extreme antipathy to the idea of death. Ancient Romans were among many who considered that a dignified death was a bonus, even if it came earlier than might have been biologically possible. In the future, as genetic manipulation gets under way, we will have to accept that death is proper even for people who are not infirm; for even in extreme old age, there may be no infirmity. Unless life is rationed, there will be no room for babies. But will we ever again come to terms with death, as the Romans did?

The matter of death is clearly one of biology and also one of law and ethics, economics, politics and religion. An unrelated matter in this miscellany, but which again has huge economic, political and ethical implications, is how, in the future, we will be able to deliver the new technologies to where they might do most good. This is yet another issue with which the world as a whole has yet to come to grips.

APPROPRIATENESS, AND THE HUGE PROBLEM OF DELIVERY

I have two fundamental conceits: one is that technology can do good, if applied appropriately (which surely is not too contentious); the other is that in politics, two of the most important considerations are autonomy and democracy. Unfortunately, as things stand, the application of technology, even for purposes that are overtly benign, can and does interfere both with autonomy and democracy. Before the world gets too much older, these conflicts must be resolved. But first they have to be addressed.

Autonomy has all kinds of connotations. It implies personal freedom; it implies choice; but above all, it implies that individuals and the societies to which they belong are in command of their own destinies. If they are a farming society, it is because they want to be a farming society. If they abandon wayside tit-bits in favour of hamburgers and hot-dogs, it is because they favour the breach with tradition. And so on.

In practice, of course, it is impossible for all individuals and all societies to do exactly what they want to do all of the time. In practice, individuals have to fit in with other individuals, and societies with other societies, or everybody suffers. In practice, then, as Thomas Hobbes pointed out in the seventeenth century, the nearest we come to perfect freedom is in creating contracts (mostly unwritten contracts). In effect we contract to give allegiance to governments, but only insofar as the governments truly look after our interests. A contract entered into voluntarily is not perfect freedom. But it is autonomy. It does imply that we decide what happens to us.

Very clearly, in the modern world, most individuals in most societies have very little autonomy. Even in free and democratic Britain, most people's lives are mostly determined by circumstances imposed from outside. Yet we can say in truth that late twentieth-century Britons have a lot more autonomy than most human beings have had in the history of the world, or at least since hunting-gathering gave way to agriculture. Very clearly, too, most societies on Earth have far less autonomy than seems just. The countries that used to be part of other people's Empires are still, to a large extent, economically and politically dominated by the old Imperialists.

Technologies in general may either increase autonomy or reduce it. Sometimes an individual or a society might choose to sacrifice a little autonomy in exchange for greater comfort, or a greater chance of survival, or whatever; and if the individual makes that choice for him or herself, or the society for itself, then that is an autonomous decision and is fine. But in general, technology that increases autonomy may be said to be good, and technology that reduces autonomy is bad.

We cannot blandly assert that any particular technology is bound to increase or decrease autonomy, for that is largely a matter of how the technology is deployed. But some technologies are more liable to promote autonomy than others. The philosopher and theologian Ivan Illich explored this issue in *Tools for Conviviality* (Calder & Boyars, London, 1973). In general – obviously – the 'convivial' technologies are those that increase personal freedom, and also increase personal power by improving contact between people, and yet remain in the control of the user. The bicycle and the telephone are convivial; and so, we might add since Illich's day, are many of the technologies of information technology, such as desk-top publishing. In totalitarian societies, the first freedom to be curtailed is the right of assembly, and the first point of call for the despot is the radio station. The telephone enables people to assemble without attracting the armed guards. Clearly, what Illich means by 'conviviality' is very close to what I mean by 'autonomy'.

To be sure, many have pursued such thoughts over the past two decades (such discussions were particularly fashionable in the 1970s), but I fear that many false conclusions were drawn. It seems reasonable to argue, for example, that technologies that are cheap, small-scale and 'low' (not requiring a scientific input) are the most likely to contribute to autonomy: technologies such as the windmill and the bicycle. Such technologies have often been termed 'alternative' (alternative, that is, to nuclear power-stations and motor cars), and 'alternative' has most often been equated with 'appropriate' (that is, appropriate to the particular needs of poor societies, primarily though not exclusively in the Third World). This argument does make a lot of sense, and indeed, as the French philanthropist and historian Jean Gimpel is now demonstrating, 'mediaeval' technologies such as floating water-mills can make enormous impacts in poor countries.

But we cannot afford to be simplistic about this. To be sure, the bicycle in principle is low-tech. But bicycles that actually *work*, and which can be made and maintained cheaply, tend to have high-tech components. The nylon bush or even the steel ball bearing are not trivial technology. The railway train is often considered more benign and appropriate than the motor car, but trains do not operate efficiently unless they are run on a very large scale, which means by central government.

But most significant, and increasingly obvious, is that many of the most appropriate technologies – those that solve the task in hand most efficiently, and with least threat to democracy and autonomy – are liable to be very high-tech indeed. The computer is an obvious example. Many political thinkers (Mrs Shirley Williams, the British ex-cabinet minister, comes to mind) have expressed their fear of the computer: it will put all our personal records on disk, and enable governments to tighten their grip on us. But computers also provide the brains behind many of the most efficient and obviously convivial information technologies, including desk-top publishing. In general, despots throughout history have managed very well without high-tech. It is democracies that are difficult to run, because they demand that so many different opinions and pieces of information are taken into account. Computers are moral idiots, and poor or useless moral arbiters, but they are very good at storing information. Democracies are innately complex, but dictatorships are simple; and in principle, computers can do more to plug the gaps in democracies than they can to abet despots. You do not need a computer to run a police state. You just need informers, and an army of disaffected strong men with heads that go straight up at the back. The Romans at their worst ran a perfectly efficient police state that lasted several hundred years.

The new high tech of biology may be equally appropriate, in the sense that it may promote conviviality and autonomy. Thus, for example, traditional village life in India, which serves its people well, depends very much upon cattle. But the cattle of India are severely compromised by chronic and endemic foot-and-mouth disease, which means that many of them are permanently weakened, and do only half the work they should even though they eat as much as a fully fit animal. A foot-and-mouth vaccine that was effective, safe, cheap, and could survive the journey across country with less than reliable refrigeration, would *support* the traditional village life, because it would make the cattle so much more effective. But such a vaccine represents very high tech: probably one that cannot be provided without genetic engineering, to produce a cocktail of synthetic antigens.

Similarly, 'appropriate technology' enthusiasts often argue that traditional crops were superior to modern crops because they did not need so much protection from pesticide, pesticide that is polluting, and which poor farmers cannot afford. Furthermore, traditional societies often derive 70 per cent or more of their calories from staples such as sorghum or rice, which means that the grain forms *most* of the diet. So the flavour and texture of the grain matters a great deal, and modern varieties often have different flavours and textures.

All these caveats are important – and yet there are counterpoints. For in practice, traditional crops often fare reasonably well simply because they tend to be landraces, and so to be genetically variable (see Chapter 6). Because of this, they will not be devastated totally by any one pest (see Chapter 5). Modern crops tend to be far more uniform, and in some cases almost totally uniform, and can be wiped out 100 per cent by a disease that just happens to hit their weak spot. None the less, Third World farmers commonly lose a huge proportion of their crops to pests and diseases; a two-thirds loss is far from unusual. The ideal cannot be simply to survive the severest body blows, but to avoid the blows altogether. Pesticide may well be inappropriate. But better than mere genetic variability are genes that confer *specific* resistance. In fact, the ideal Third World crop should be exactly like the traditional crop in flavour and texture and all the other qualities that the consumers and farmers hold dear, but would also contain not just one but a battery of different resistance genes. To provide such crops requires very high technology indeed.

Indeed, what seems to me to be one of the most stunning paradoxes of the 1990s is that *the problems of Third World communities often demand much higher technologies than those of the affluent West.* In Britain, strictly speaking, we do not need multiple-gene-resistant high-yield crops. We have high-yield crops already, mostly produced by conventional breeding, and can perfectly well afford the most modern pesticides which are not particularly polluting. We have little need for foot-and-mouth vaccines, because strict laws have, in fact, banished the disease from our islands altogether, except for the occasional imported epidemic which we deal with *ad hoc.* It is the poor societies, surrounded by diseases, ploughing poor soils with oxen in extremely hostile conditions, who need the super-vaccines and the super-crops.

But herein lies the problem. High tech is expensive, and poor people by definition have no money. Therefore, as things are, it is impossible to apply our new technologies where they might do most good. To be sure, the rich countries do 'give' to the Third World. But they give astonishingly little, relative to their wealth, and they conveniently forget (as many a Third World leader has bitterly pointed out) that their own wealth depends in part, and often in large part, on their previous imperial status, when they took what they wanted from the tropical world and gave very little in return. Modern countries give Aid, which has become a key component of 'Third World' economy. But the Aid given by governments always comes with strings attached. It was originally conceived in the 1960s in the United States as a means for disposing of grain surpluses, and immediately became an agent of political control. Technology that is achieved by Aid, however potentially benign, does not contribute to autonomy.

All this leads me to the crux of my argument, which is that *although the new high technologies have such power to do good, it is at present impossible to deliver them to the people who need them most without compromising those people's autonomy*. We have the means to do good, but we are prevented from doing good, except in ways that are themselves bad. To solve this problem is, I suggest, one of the key issues of the modern world; in fact we could argue that it is, politically and economically, the most important issue.

Of course it may be argued that high tech, once developed, *can* be cheap, provided it is produced on a big enough scale. The pocket calculator is high-tech, and you can buy one for £5.00. There can hardly be a South-east Asian village that is not alive with the sound of transistors. But there is a world of difference between high tech that 'trickles down' arbitrarily from the tables of the rich, and high tech designed expressly to solve the identified problems of the poor. Yes, the rich world is equipped to provide the Third World with transistor radios – those that spin off from its own markets – but it is not equipped to provide transgenic sorghum, for which the rich world itself has no use.

The second truly serious issue, which applies equally to all countries in the world, is that of democracy.

DEMOCRACY AND THE CONTROL OF SCIENCE

'Democracy' means very little unless the people in a society control the forces that influence their lives. Politicians have such influence, and in the West we go to great lengths to elect them, and this, despite the many imperfections, is obviously democracy of a sort. But politicians, taken all in all, have relatively little effect in the modern world compared to the huge ground-swell of change that is brought about by advancing technology. The factory, the railways, vaccines, the motor car, the tractor, artificial fertilisers and synthetic pesticides, antibiotics, television, the jumbo-jet, and now the computer, are the agencies that have changed the face and the economic structures of the world over the past 150 years, and the technologies of genetics are rapidly joining the ranks of those transforming innovations. The lasting legacy of Marx does not lie in the economic systems that were created in his name, but in his observation that, *in reality*, people's lives are dominated by economic structure, and not the other way around; and that economic structure is dominated by what he called 'means of production' and we may call technology. In short, it is *technology* that influences our lives, much more than politicians. The politicians are the puppets of the technology, like the rest of us. When the technology is 'high-tech' – tech that is the scion of science – then we can legitimately say that it is science that really drives our lives.

So the question a democrat ought to ask is, 'Are we as a society truly in charge of science?' Only the most Panglossian or naive could possibly answer 'Yes'. Again, of course, we face a paradox: unless we give science its head, and allow scientists to follow where the ideas take them, we will never know what fruits – what potentially convivial high technologies – they might provide. I also feel (of which more later) that to put a brake upon science would be a cultural insult comparable with the burning of books. I also despise the science policies of the 1980s in which science that was not expressly designed to provide immediate wealth became unfundable. As a Bangladeshi scientist once said to me, any science that already has a target in mind is already technology. So the imposition of *direction*, it seems to me, has to come *after* the pure ideas have been developed, but *before* the applications have progressed too far to call a halt.

You might of course say – as many a modern politician does – 'Ah, but the market can decide what technologies we develop from the science, and what we do not!' But markets are slow to react. Markets can be manipulated. Markets may operate well in societies that are already just, but they do not create justice: left to themselves they are bound to cater primarily to those who have the means to pay. Markets are not moral: most of us dislike the idea of the battery chicken, but buy cheap eggs in any case. Most of us would probably buy cheap milk from cows built like honey-pot ants.

Surely it is true, too, that if we were truly in command of the technologies that emerge from science, we would not now be anticipating the greenhouse effect; there would not be a hole in the ozone of the sky, growing bigger; we would not be wondering if the world can truly contain the projected 10 billion population of the mid-twenty-first century; we would not endure traffic jams; Third World cities would not be knee-deep in rubbish, and Third World forests would not be wiped by pesticide, and so on. If we were truly in command, then surely we would not have created the world we have, and left undone so many things that obviously need doing; not unless we were overwhelmingly perverse.

There is one further economic issue, more general in nature, and pertinent to new technologies in general and to the idea of progress. It is that most governments equate 'progress' with 'increase in material well-being', and equate that with 'increase in disposable wealth'. In other words, the supposition is that, as time passes, we will all grow richer; and that it is the function of technology to achieve this end. It also seems to be accepted that wealth and effort must go hand in hand, in a linear relationship: that more effort ought to mean more wealth, and that extreme effort ought to mean extreme wealth. In the 1980s, the rubric of Mrs Thatcher's Britain and Ronald Reagan's America was that without the prospect of extreme personal wealth, there would be no 'incentive', and without incentive, all economies would grind to a halt.

But personal wealth depends upon resource, and it is now obvious that the resources of the world are limited. There are many ways to express this. If Indians used as much fossil fuel *per capita* as Americans now do, we would go through the world's oil in a few decades. Clearly, only a minority of people can own the vast estates that many people, worldwide, regard as their birthright. If people continue to exploit the 'Gold Coast' of Queensland for fishing and casinos, the extinction of species must accelerate. And so on. We have long since reached the stage, in fact, where extreme personal wealth can be seen to be desperately anti-social, and highly destructive of the planet, in just the same way (but probably to a greater extent) that over-population is threatening. Yet the desire for personal wealth continues to be – or is acknowledged to be – the principal driving force of the world's dominant economies.

This, then, is the huge dilemma of the late twentieth century, one which must be solved in the twenty-first century if humanity is to survive satisfactorily until the twenty-second. The new technologies offer us wonderful powers to do good. But for *economic* reasons, we cannot deliver those technologies where they will do most good. Furthermore, although most of the world these days aspires to be democratic, we lack the mechanisms to direct the scientific and technological forces that in practice influence our lives most profoundly. And finally, our economies are driven by a motivation – the desire for personal enrichment – that can be seen, all too obviously, to be destructive; but as things are, the new technologies are themselves driven by, and in their turn feed, that motivation. We are driving a powerful train, but it is a runaway train, and the tracks are not of our making. So what's to be done?

TAKING CONTROL

In practice, societies are controlled, day to day, by their economies and by their laws. In the West at least we seem to have plenty of bankers, answerable to treasuries, able to finance this piece of work but not that; and even as I write hundreds of lawyers and scores of politicians and committees worldwide are discussing laws, regulations and codes of practice to direct the new science and technology, and especially the new genetic technologies. So that's that taken care of. Isn't it?

Well no, actually, I don't think it is. I have already identified deep flaws in the economic framework within which the bankers are operating. We need a quite new economic structure – a new economic 'paradigm' – if we are ever to be able to direct the new technologies where they are most needed, without bringing the recipients under a new imperial yoke. In short, the bankers are operating against the wrong *background*.

So too, I suggest, are the learned committees. They contain politicians of course – but politicians are answerable to their parties and their political ideologies, all of which were designed in another age, in response to problems that were qualitatively different from those of today. The committees include lawyers, but lawyers are not prophets. It is not their job to decide what is right or wrong, but only to express, as clearly as they can manage, the will of the society they serve, so that that will may be carried out as consistently as possible. It is obvious that laws work properly only when they do express the will of the people. Most people do not commit murder, not because there is a law which says that murder is illegal, but because most of us feel, in a general way, that killing is wrong; the law merely underlines that belief. Most people, however, drive their cars unlawfully fast on motorways; because if the car is big and the road is clear, why not? Try as they might, the police cannot stamp out the tendency to break the speeding laws because, deep down, people do not think there is anything wrong in driving fast unless there is obvious danger to life and limb.

But in trying to frame rules for the control of technologies, lawyers operate in a vacuum. People want the good things that technology offers, and obviously hate the side-effects, but they have little or no understanding of the science that provides the new technologies, and they notice that the technologies exist only when they already encroach on day-to-day life. In this crucial issue, the wishes of the society that the lawyers seek to express are extremely vague.

In short, just as the bankers operate against an inappropriate economic background, *so the politicians and lawyers are operating against an inappropriate cultural and ethical background.* This cultural and ethical background we may call the *ethos* of the society. If ever we hope to deploy science and technology well, to do good and to avoid evil, then it is the *ethos* that we must first repair. Again, I would not presume to prescribe how all of society for all time should organise itself. But again, I would like to offer some observations. The first is that we cannot hope as a society to control science and technology well – we cannot, in fact, have an appropriate ethos – until and unless all of us are a great deal more aware than most people are (including almost all politicians) of what science is, and what it can do, and what it cannot. In other words, we cannot as a society be truly in charge of science, until we become scientifically literate. So what does this imply?

THE NEED FOR SCIENTIFIC LITERACY

To become scientifically literate it is necessary first to love science – but to love it in the way that good parents love their children: wanting them to do well, admiring their achievements, but aware of their shortcomings. If we do not love science, we cannot hope to understand it: it is very hard to take a concerted interest in something you hate. If we do not understand, we cannot hope to control it; and if we do not control it, then the world as a whole will be in deep trouble and (on a large point of detail) democracy will continue to be a nonsense.

I find science very easy to love; it is what I always wanted to study. Many others are deeply antipathetic, mainly I believe for two quite separate reasons. The first lies with the scientists themselves. They have often appeared arrogant. In this century they have been responsible for some truly repellent but in the end very limited ideas. I think particularly of behaviourism – the form of animal psychology that as a matter of philosophy regarded animals purely as machines, and insisted that it was muddle-headed to do otherwise. To be sure, such purism (a form of positivism) was necessary if intellectual progress was to be made: the pleasant but essentially poetic romanticism of the nineteenth-century animal psychologists had to be tidied away. But the mistake was to confuse the positivist model of reality with reality itself. Just because it was experimentally convenient to regard animals as machines, that did not mean that they *were* machines. By the same token, the computer can in some ways serve as a useful 'model' of the brain, helping to throw light on aspects of function, yet we know that brains taken all in all are very different from computers. But many behaviourists acted for a time – several decades – as if animals were nothing but machines, and this was a supremely limited and philosophically unjustified view which in the end delayed intellectual progress, and also (more seriously) has resulted in some hideous cruelties, as the idea that animals are sentient beings has often been dismissed as mere 'sentimentality'. It is the case, for example, that the worst practices of modern factory farming were put in place (in the 1950s and '60s) at a time when behaviourism was at its height, and now they have a momentum that is difficult to reverse in this age which we may hope is more enlightened. This kind of scientific arrogance is diminishing, I believe, but it still exists in many contexts, and it is no less abhorrent.

Scientists also, at least in the discourse between themselves, often seem to go out of their way to be tedious. To be sure, the technical terms that so many non-scientists find off-putting are usually necessary. There was nothing in common parlance to express the idea of 'gene' before the word was coined; and certainly not for 'allele'. If you want to refer to genes and alleles, you have to give them their own names, and the same goes for ten thousand other entities and ideas that science has provided. So technical language, used properly, is *not* to be confused with 'jargon', which literally means the chatter of birds, and is properly used only to describe the kind of shorthand in which all professionals speak. To some extent this shorthand is forgivable; it is what arises rapidly in all closed communities, from primary schools to football teams to armies, and is a way both of bonding, and of conveying familiar ideas quickly. But in another way this insider-talk is less forgivable. It is used to repel boarders, and scientists have often tried by the unnecessary use of technical colloquialisms to keep the unwashed at bay.

But the vocabulary of science is not the issue. What truly repels outsiders is the contorted syntax of formal science; the use of the 'past passive' and the third person to give the impression that science is carried out not by human beings but by some unseen hand. The style has filtered down to the lowest levels, as if there was some innate virtue in it: the first sentence I ever wrote after my first formal science lesson at the age of eleven was, 'A test tube was taken and a piece of wood inserted . . . ' Many a child, quite properly, has given up science after minimum exposure to such nonsense. The basic idea, of course, is to give the impression that the work of science and therefore its conclusions are 'objective', which doubtless is a noble and necessary intent, but is itself a lie.

Yet the notion that science *is* entirely objective has been perpetrated by successive generations of philosophers, as if there was simply some 'method' which, when applied, provided inexorable truth. To be sure, there is a pattern to science which is not seen so clearly in other disciplines, and has been described most convincingly by Sir Karl Popper. Scientists progress (says Popper) by making guesses (hypotheses) about the way the world works, and then testing those hypotheses by experiment. In effect, the hypothesis must contain predictions about how the world would work in particular circumstances if the hypothesis is correct; and the experiment creates those circumstances, to see if the prediction holds up. Anyone in any field can produce an hypothesis, but this ability to test the hypotheses by experiment, says Popper, is unique to science. This basic *modus operandi* of science is extremely powerful. The findings and ideas of science are *not* (as Popper emphasises, and history has shown) unequivocal and perennial truths. They are only the best ideas that scientists are able to put forward on the evidence and with the means available at any one time. But they have a robustness, none the less – a testability – which is not so easy to discern in the ideas of history, say, or literary criticism.

Yet some people find this robustness repellent, possibly because they confuse 'robust' with 'unequivocal' and 'objective', and therefore conclude that the ideas of science are innately 'arrogant'; and possibly because they mistrust ideas which, they believe, are arrived at in a disembodied way by people who, they imagine, simply grind out ideas according to some algorithm (which is a pleasant scientific term applied to formal methods for solving problems, methods which the user does not have to understand, but produce the answer anyway). To those who still believe that the theories of science are framed without emotion by automata working to a formula, I commend the latest book by the Nobel prize-winning particle physicist Steven Weinberg, *The Dream of a Final Theory* (Hutchinson Radius, London, 1993). In it he describes how scientists in his own field – the most rarefied of all – finally come to believe in ideas not because of some key experiment (rarely can there be such a thing in fields that are at all complicated) but because, in the end, they are overwhelmed by a sense of 'beauty': a thrilling sense that the idea *must* be right.

Biologists also work in just this way. There was no one, key experiment which showed that genes were made of DNA (although the evidence inexorably grew through the middle decades of this century), and there can be no key experiment to 'prove' the central ideas of Darwin. But there is and has been a growing sense that the ideas must be right, and what in the end conveys the sense of 'rightness' is an emotional response. Without that response, none of us can begin to make a judgement on what is true and what is not. A computer trying by itself to understand this messy Universe would be all at sea, precisely because it has no emotion to guide it, and a soul-less tally of evidence cannot of itself make a case. Science, in short, must strive for 'objectivity' because, in the end, it is seeking explanations of what actually is happening in this Universe; and ideas that belong *simply* in the imagination, and do not relate to the real Universe, are not the business of science (although they may be the business of maths!). But science none the less is the most human of activities. The hypotheses that are its raw material spring from goodness knows where – from flashes of inspiration, whatever they are, and sometimes (literally) from dreams. And the final confirmation that such-and-such an idea is, in fact, 'true' – or as near to the truth as we can approach at any one time – is provided by a surge of emotion no different, as far as one can feel, from the rush that fires an artist.

Yet some people object (perhaps taking their lead from Plato, did they but know it) that to contemplate the mechanisms of the Universe is an inferior pursuit, and that those who concern themselves with such matters are *ipso facto* rude mechanicals. With this goes the notion that to explain the workings of the Universe is somehow to diminish it, to 'rob it of its mystery'. The first of these notions is merely snobbish, and like all snobbery is extremely silly, and is rooted in intellectual and emotional laziness. The second idea – that to explain the Universe is to diminish it – is, I suggest, blasphemous; an insult to God. God is not a conjuror whose tricks seem cheap when once they are revealed. The more you know about the workings of the Universe, the more extraordinary and wonderful it becomes. I like the seventeenth-century notion espoused by such luminaries as Isaac Newton, Robert Boyle and John Ray: to explore the workings of the God-given Universe is a tribute to God, a proper use of God-given intellect. To them, science was itself an act of reverence. Quite right.

Yet here we seem to face another conflict: science, being exploratory, tends also to be exploitative. Science is used to change the Universe, and to change the Universe seems most un-reverential. Many are repelled by the role of science as the restless meddler that will not leave well alone.

So am I, yet I do not think that is the fault of science. The trouble, rather, lies in our own materialism. To a large extent society has tended to support science primarily as a means of providing new technologies, or of solving material problems. This was true also of the seventeenth century, despite the theological predilections of many scientists: astronomy in seventeenth-century Britain was supported largely as an aid to navigation, to bring order to the trade routes. In the hyper-materialistic 1980s this trend was exaggerated. Science in Thatcher's Britain at least, and to a greater and greater extent elsewhere (I was disappointed to find exactly the same trend in Australia, for example, which has so many brilliant scientists), was supported only insofar as it supplied high tech, which in turn was supposed to prop up and reinforce the existing industries. Hence science became at once an agent of material change, and also of economic conservation, which in two ways is a betrayal and misuse of science. Yet the fault lies in the simple equation: science = technology = industrial wealth = *status quo*. We must begin again to perceive (as did Newton, Boyle, and Ray) that the true role of science is not to change the Universe, but more fully to appreciate it.

Once we grasp this, we have the essence of scientific literacy. Once this is built into the ethos of our society, we will begin to be able to control science, as a parent can control a well-loved child. Until we have this affection for science, and the 'feel' for what it is and what it can do, it will for ever be wayward. Delinquency in children is generally the fault of the society.

Scientific literacy also has another aspect. The point is not merely to appreciate the ideas of science, it is also necessary to place those ideas in perspective, to integrate the ideas of science with those of philosophy as a whole. It grieves me mightily that science is still taught only as a vocation, as if to apprentice scientists.

The flow should be two ways. Clearly, for example, philosophers interested in epistemology – the way in which ideas are arrived at, and the way in which we decide what is true – have long taken a keen interest in science, and there is obviously much fruitful discourse to be had between epistemologists and psychologists, and between physicists who talk of 'beauty' and poets. I also believe, however, that in a general way scientific literacy colours – and I mean enhances, not 'discolours' – all ideas and perceptions. Particularly pertinent to this book is that the ideas of biology have contributions to make to moral philosophy, and outstanding among the ideas of biology are those of sociobiology, outlined in Chapter 4. This is an extremely touchy point, which must be expanded upon.

SOCIOBIOLOGY AND ETHICS

All pioneer thinkers are liable to be ill-served by their disciples.

Many who came after Darwin said things in his name that he himself would never have approved of. That most gentle and enlightened of men could not have approved of 'social darwinism', the notion that because in nature the fight seems to go to the strong, this is what *ought* to happen in human societies. Many who followed Ed Wilson into sociobiology committed the same kind of solecism. Because it seems to be in the interests of men to scatter their seed widely, some enthusiasts took this as proof that men *ought* to philander. Because it seems to be in the interests of women to persuade a male to help them to look after their offspring, those same instant moralists suggested that women, by contrast, *ought* to be monogamous; and, for good measure, that as the gathering half of the hunter-gathering team, they ought to stay at home and attend the kitchen sink. Such naivety was rightly criticised, and was an aspect of what the critics called 'genetic determinism': the idea that we, as human beings, are somehow obliged to do what we are genetically programmed to do, and so we might as well lie back and enjoy it.

In fact, those who criticised naive sociobiologists for indulging in 'genetic determinism' were resurrecting an ethical issue which has troubled moral philosophers at least since the time of Saint Paul, but which most moralists felt was settled. If patterns of behaviour are built into our genes, then by some definitions at least, such behaviour could be called 'natural'. The ancient moral question is, 'What is the relationship between "natural" and "good"?'

Saint Paul himself was actually somewhat equivocal about this (although he is consistent in his puritanism). He hated the unnaturalness of homosexuals – 'the men, leaving the natural use of the woman, burned in their lust one toward another; men with men working that which is unseemly . . . even their women did change the natural use into that which is against nature'. (Romans, 1:26–27.) But in general he seems to prefer no sex at all – 'It is good for a man not to touch a woman' (Corinthians, 7:1), though if it all gets too much for you, 'it is better to marry than to burn [with lust]' (Corinthians, 7:9). To Paul, in short, 'unnatural' is definitely bad, though you also have to restrain what is natural. It was very difficult to win with Saint Paul.

In the eighteenth century David Hume defused this debate.

What is 'natural', he effectively said, has nothing to do with the case. 'Is, is not ought': what *is* has nothing to do with what *ought* to be. More specifically, at the turn of this century, G.E. Moore examined various criteria upon which ethical systems had been based, one of which was biology; for example, the natural selection-based morality of Herbert Spencer. But, said Moore, to base morality on biology was to commit 'the naturalistic fallacy'. Immanuel Kant at about the same time as Hume arrived at a comparable position by a different route. Anything you *want* to do cannot really be said to be 'ethical' at all, said Kant. Doing what comes easily can hardly be said to be 'good', even if the end result is helpful.

All that seems to put genetics firmly in its place. If we have genes that make us aggressive, that does not justify aggression. Our biological 'urge' to spread our genes does not justify promiscuity. In fact, Paul and Kant might agree, the more you suppress such urges the better.

Yet two points remain. The first is that although we should avoid Moore's 'naturalistic fallacy', we cannot ignore our own biology totally. As the modern philosopher Michael Ruse points out, an ethics suitable for us would not suit termites, nor *vice versa*; the fact that we are *human* animals is relevant. Second, if it is really true that we should strive to override our own 'nature' – as Saint Paul advocated, and as Hume and Moore certainly would not deny – then we should understand exactly what we are up against.

On the first of these points, I like to offer the example of human abortion. Quite rightly, it is the subject of a debate that probably will never end: the 'pros' and the 'antis' are agreed at least that the issue has to be treated very seriously indeed.

What would our attitude be, though, if human beings were as fecund as pigs or dogs? Of course you might argue – as biologists have often argued – that this could not be. An animal as bright as we are is bound to be born with a big brain, and a suitable skull to hold it in; and animals with big skulls are bound to be born one at a time, or two at most, and only very occasionally three. But in practice, of course, women given fertility drugs have often given birth to quins or even sextuplets, and although the babies need special care at first it is clear that with a little biological modification, women *could* be fecund. If they were, biologists would argue with equal conviction that intelligent creatures were bound to be born in batches, because play is essential to early mental development. Besides, pigs and dogs are themselves among the brightest of animals. In short, there is no insurmountable conflict between intelligence and fecundity.

But if humans were as fecund as pigs we would not have the *luxury* of taking abortion seriously – or even infanticide. We could not afford to be circumspect. Human beings as fecund as pigs could triple their population in a couple of years, multiply a thousand fold in a few centuries. In the absence of efficient contraception – which is a recent invention! – we would be obliged not simply to accept, but to contrive, a very high infant or fetal mortality rate. The notion that this was wrong would never have occurred to us. To the edict 'thou shalt not kill' we would have added the chilling codicil, 'anybody else's children'. It is a distasteful thought, but it makes the point. Our biology is relevant, and sometimes, as it happens, it saves us from certain kinds of awfulness.

What, then, of the second notion, that we should learn from our own biology, if only to find out what it is we should be careful of? What of the crude 'sociobiological' notion that as men should scatter their seed, it is proper for them to rape, and as women are destined for child-care, they must stay at home?

We can destroy such arguments by biology alone, without appeal to moral philosophy. If it is a good sociobiological tactic to rape, why isn't rape the norm among animals – given that animals have no moral philosophers to guide them? To be sure, some male animals do seek to spread their genes very widely, and gather harems, like stags – though rape is not a common tactic; some females are feckless, like hen hedge-sparrows. But many creatures are monogamous and faithful, from dung-beetles to gannets. Every creature wants to spread its genes, of course. But rape, mayhem and cuckoldry are not necessarily favoured by natural selection. Rape is not merely evil, it is also, usually, biologically inept, which is why only very few animals practise it, and usually only then in special circumstances. Indeed, we can make a strong argument from the game theory outlined in Chapter 4 that the best way to propagate your genes is to help to create a peaceful and stable society, because this gives your offspring their greatest chance of reaching reproductive age. But all this simply provides yet another argument *against* biology-based ethics – for the fact is that you can with ingenuity extrapolate almost any moral lesson you choose from biology, and half of the possible lessons contradict the other half.

But when you spread the net very broadly, and speak not of cases but of universal principles – when, indeed, you apply John Maynard Smith's notion of the evolutionarily stable strategy, or ESS – the notions that emerge are very interesting indeed. The 'utilitarian' philosophers of the early nineteenth century argued that a 'good' society was one that brought the greatest benefit to the greatest number. Game theory shows that if you quantify benefit, this is achieved by an all-dove society; the one which, incidentally, was also advocated by Christ. In the short term, however, as Saint Peter pointed out, aggressive individuals can flourish in an otherwise all-dove society. In fact, as Maynard Smith has shown, aggressive individuals are bound to increase among doves until there are so many of them that they get in each other's way; and the point at which that happens – the evolutionarily stable point – is clearly different from the all-dove point that is most beneficial to the society as a whole.

If the ESS society offends the utilitarians and Christ's Sermon on the Mount, it also offends Immanuel Kant in a different way. For Kant argued as a 'categorical imperative' that no system was truly ethical unless it subjected everyone to the same behavioural rules. But at the ESS, there is a balance between two different behavioural strategies, aggressive and peaceable. According to Kant's philosophy, then, the ESS – the state of affairs that nature reaches if left to itself – is categorically bad.

Christ wanted us to move towards the all-dove society, which is also the most biologically 'efficient' society, and which therefore, at one level, is good biology. This, however, defies the natural slide towards the ESS, and can also put those who strive to be dovish at a disadvantage against putative hawks. Kant demanded that we should all answer to the same principles, and also proclaimed that ethics begins at the point when we start to behave against our own inclinations. The all-dove society of Christ is the only one that is biologically possible but also fulfils Kant's criteria.

The notion of ESS also offers a sharp political insight. Hawks in an all-dove society reduce the sum-total of happiness. They are also, at any one time, in the minority. Yet, because doves are doves, they can do nothing to stop the rise of hawks. Is it foolish to draw an analogy with human society? Most of us in any one society at any one time are doves, or at least, we behave dovishly. The hawks are in a minority. Yet we cannot stop the hawks from flourishing.

The hawks include the politicians; those who aspire to lead. Hawks create policies that enable hawks to flourish. Thus, although doves predominate numerically, and although doves in a hawk-dove society do worse than in an all-dove society, we allow the hawks to create policies that are largely hawkish. This seems to me in large part to explain why we allow the minority of people who control technologies to do things in the world that adversely affect the majority, while failing to do things that are good. There is no mechanism to stop them. We need not, to be sure, invoke the notion of genes in making this point; this is just game theory, objectively applied.

So – absolutely not do I wish to suggest that Moore's notion of 'naturalistic fallacy' is wrong. But I do contend that the ideas now emerging from biology have much to teach us. They help us better to understand our own natures, and if we choose on moral grounds to override those natures – as well we might – then it is as well to know that in some cases we must take special care, because we may well have a predilection to transgress. 'Know your enemy', as Napoleon said, even if – or especially if – the enemy is within. Game theory applied to sociobiology helps us to see how easily we can allow our societies to be overtaken by selfish people who do bad things, even though most of us are unselfish, and intent on doing good. We need not be fooled by the assertion of the emergent hawks that their hawkish ideas are necessary, or are for our own good; they are necessary merely to reinforce the hawkish domination, and their net effect may well be to reduce total happiness. Game theory exposes the underlying con trick of power politics, and reinforces the value of democracy because (it seems to me) the majority opinion of the majority of doves ought to result in dovishness. It is surely no coincidence that most democracies are more benign than most dictatorships.

Archaeology and palaeanthropology, which describe and seek to explain human evolution, offer more insights to the moralist. The chief lesson, yet again, is that we are much more the victims of our own inheritance than we realise, and that it can be even more difficult than we suppose to do good, because we must first overcome that inheritance. The key consideration is the evolution of attitude.

THE EVOLUTION OF ATTITUDE

In Chapter 9 I discussed briefly the early evolution of human beings on the plains of Africa, and in Chapter 4 we saw that evolution does not affect only the physique. Behaviour evolves too. Behaviour that favours survival, survives. Behaviour that does not – or is out-competed by something more appropriate – goes to the wall.

I suggest that all behaviour is underpinned by *attitude*: whether in general the animal behaves aggressively, or boldly, or timidly, or modestly, or arrogantly; hawkishly or dovishly. Attitude is also subject to natural selection. Boldness in the face of the enemy may succeed – may indeed be the only survival tactic. But if you pick the wrong enemy, it may end in disaster. The fight, or at least survival, in many circumstances may go to the coward.

Like all animals, human beings have in general survived either by pursuing a strategy of exploration and exploitation, or by pursuing a generally conservative strategy: doing only what was known to have succeeded in the past and attempting to leave the environment as they found it. These, then, are key attitudes: explorativeness-exploitativeness versus more cautious conservativeness. In general, hunter-gatherers have been obliged to be conservatives. If they hunt for only a few hours a week (as the hunter-gathering Kung bushmen of the Kalahari were shown to do in the 1960s), and catch only the occasional antelope, then the prey can survive indefinitely. If they step up their efficiency by better technology, or work harder – the kind of things of which a modern industrialist would approve – then they over-exploit, and drive their prey to extinction. Many of those hunter-gathering peoples of whom we have direct knowledge – ¡Kung, Australian aborigines, New Guinea Highlanders, North American Indians – seek to ensure that they do take only what they need by *revering* the creatures they exploit, and apologising to the Gods that protect those creatures for each one they kill.

Because hunter-gatherers must operate only within the limits of their 'prey-base' – over which they have little or no control – they have very little leeway. They do not attempt to produce food surpluses (in fact they *produce* absolutely nothing) and do little in the way of food storage. Because they have no leeway, they cannot afford to make too many mistakes. This means they are not in a position to experiment. At any one time, they should do only those things that they know, from experience, will succeed. Accordingly, the archaeological record shows that for a long period of human evolution – the time occupied by the hunting-gathering *Homo erectus* – the culture and way of life altered not one whit. Perhaps life was tedious; indeed, this period is known to archaeologists as 'a million years of boredom'. But the population stayed steady, the prey-base survived (despite the vicissitudes of Ice Ages) and the people survived.

Farming changed everything. Farmers do not rely on the food that the environment happens to be able to produce. Farmers work specifically to increase that output. In general, furthermore, the amount of food that the environment will produce is directly related to the amount of work that the farmers put in. So farmers, unlike hunter-gatherers, do not spend nine-tenths of the week telling stories. They spend nine-tenths of the week working. The result is probably most unpleasant: the archaeological evidence is beginning to show that the farmers of the Middle East around 10,000 years ago – who may well have been among the world's first *full-time* farmers – had a very hard life indeed. But natural selection is not concerned with pleasantness. Hard work and farming produce more food, and therefore enable more people to survive. Big populations are bound, sooner or later, to oust smaller populations, so the productive farmers were bound, sooner or later, to squeeze out the non-productive hunter-gatherers. The farmers might be miserable and the hunter-gatherers might be jolly (or not; it really makes no difference to the argument). But the farmers must win, provided they farm properly – and energetically.

Thus, I maintain, the effect of farming on human beings and on the world at large was even more profound than is generally appreciated. That it had huge ecological consequences is obvious. That it enabled the human population to expand after at least a million years of stability is also widely acknowledged. Less well recognised, but hugely important, is that it brought new natural selective pressures to bear on *attitude*. Natural selection no longer favoured the conservatism and essential passivity of the hunter-gatherer; or at least, conservatism and passivity began inexorably to be outgunned by the exploitative and innovative attitude of the farmers. Industrialisation, which developed 10,000 years or so after farming, was an extension of farming; it built upon, and quickened, the farmers' exploitativeness and innovativeness. So much so, indeed, that the industrialists' energy made even the farmers seem rustic and conservative by comparison, which perhaps has tended to blind us urban moderns to the extraordinary adventurousness and energy of agriculture.

For 10,000 years the simple equation applied: the harder the farmers worked, the deeper they dug, the more forest they cleared – the more food they produced, and the more the human population could grow. But the equation cannot be applied for ever. The world is finite. The resources of agriculture – soil, water, nitrogen, phosphorus, even sunlight (per unit area per unit time) – are limited. After 10,000 years of relentless energy, from millions upon millions of people, we can begin to see that the limits are in sight. To be sure, there are respites; new crops and techniques, of the kind mentioned in this book, can solve some of the problems of particular societies in particular places for a few years. But the limits become ever more apparent none the less.

In fact, I suggest, the situation that the human population now faces as a whole is exactly comparable with what our ancestors faced in the millions of years before agriculture became the norm. We have to learn to accept, for the first time in 10,000 years, that our resources are limited, and that output cannot therefore be increased infinitely, in direct proportion to the effort expended.

Unfortunately, we have inherited an attitude, honed and reinforced over 10,000 years, which says that the race is bound to be won by the energetic; that the exploitative-innovatory mode is the one that succeeds; that any other approach is second-rate and doomed. What is doubly unfortunate (although absolutely inevitable) is that all our political and economic systems spring directly from that attitude. Exploration, exploitativeness, sheer hard work, are rewarded. Those who sit back or who strive to leave things as they are are swept aside, just as the old hunter-gatherers have been swept aside.

Indeed it has become the case that *everything* in modern societies is now honed by the agricultural attitude of exploration and exploitativeness. These have become the principal virtues; and if not always perceived as virtues, they are certainly seen as the *sine qua non*. People in most industrial societies take it as read that the specific role of industry is to create wealth; that the specific role of technology is to serve industry; and that the principal role of science (far overshadowing all other considerations) is to create new and more 'competitive' technologies. Several points follow from this. One is that this underlying attitude is no longer appropriate even in simple survival terms, because it can succeed only so long as the resources can continue to expand. A second is the one we explored earlier: that the political and economic structures that derive from the exploratory-exploitative attitude prevent us from doing even those things that we can perceive are worth doing in the interests of morality, and – now – in the interests of survival. A third is that science is tarred with a brush it does not deserve. The notion is promulgated that science is simply a materialistic pursuit whose ultimate purpose is simply to create wealth. Newton and Boyle's notion that science is innately reverential has been buried deep. Sensitive souls sheer even more emphatically from scientific literacy.

So where does all this leave us? I have argued above that we need a new economic 'paradigm', if we are to use our new-found abilities to do good, and to avoid doing harm. I have argued that economies must in turn be led by the ethos of the society; they must not be allowed, as has become the fashion, to determine the ethos. I have argued that this ethos must include two key components: scientific literacy, and an attitude towards each other that is dovish, and towards nature that is reverential. But I have also argued that it will be even more difficult to achieve the necessary components of scientific literacy and of dovishness than we might suppose. For one thing, science is locked into a system which robs it of much of its attractiveness; it is not allowed to be romantic and reverential because it is locked into the imperatives of exploitativeness. And dovishness is difficult for the reason identified both by Saint Peter and by John Maynard Smith: hawks, in an all-dove society, are bound to succeed up to a point, so that all societies seem doomed to contain a proportion of hawks.

Taken all in all, then, I find it very difficult to be optimistic. Science, including the extraordinary science of genetics with all its ramifications, is brilliant. The technologies that derive from science, and in particular the technologies that derive from modern genetics, have the power – or much of what is needed – even at this late hour, to pull the human species and our fellow creatures from the brink that we are surely approaching, and that many other species have already passed. But we are locked into systems that prevent us from doing what is required; and, I believe, the last 10,000 years of our cultural evolution (doubtless underpinned by some genetic shift) have moved us towards a mind-set, an attitude, that is no longer appropriate. The present-day ethical committees that are supposed to keep us all on course are deeply flawed. The lesser point is that they put people into false roles, and in particular give lawyers, who are dotters of 'i's and crossers of 't's, the status of moral philosophers and even of prophets, which is not their metier. The greater point is that the present ethical committees operate at the behest and against a background of social and economic systems that themselves are inappropriate. We need to change radically – literally to our roots, down to the subliminal level of attitude. But we have no mechanism to make such a change.

I feel, as a final conceit, that modern societies, like all societies at all times, need a religion: which has two key roles. First, I suggest that Christ and game theory are correct: the meek shall inherit the Earth; the all-dove society does bring the greatest benefit to the greatest number. But at a purely practical level, Saint Peter (and game theory) are right as well: hawks are bound to arise, and are bound to push their way to the front, until there are so many hawks that they get in each other's way.

So to create an all-dove society we have to behave in ways that on the face of things seem foolish. We have to give up hawkishness, which, for anyone surrounded by doves and therefore by easy pickings, is a self-sacrificial and apparently irrational thing to do. The first role of religion, in short, is to provide a set of incontrovertible reasons – taboos, commandments, edicts – that will persuade us to behave irrationally: against our own interests, for long-term goals that are difficult to envisage.

Then we may observe that it is literally impossible to think straight without sensibility. At one level, as Steven Weinberg said, aesthetics is a necessary component of science: without an aesthetic response, we cannot judge what is true. At a quite different and more obvious level, we rely in the end – as David Hume said – upon *feelings* to tell us what is right and wrong. Sensibility is as much a part of understanding as is intellect. The second key function of religion – its *raison d'être* – is to cultivate sensibility by all the means at its disposal: meditation, prayer, ritual. I am quite sure that the insights arrived at by people who employ these techniques – and I do not claim that they are more than techniques – are qualitatively different from, and in significant ways superior to, those arrived at by people who merely *think*, in the manner of today, in committees. Lawyers are not moral philosophers, and moral philosophers on their own need the extra insights of prophets, who practise methods of contemplation that are not those of simple cerebration. Moses, Isaiah, Christ, Mohammed, Buddha, Gandhi, the present Dalai Lama, have provided the world with profound and resonant moral insights, but they would not have arrived at those insights unless they had employed the special methods of their religions, including their long periods of deprivation and solitude. The methods, I believe, engender particular states of tranquillity in which it becomes possible to conceive and appreciate ideas that would not otherwise be apparent, and yet are invaluable. If the prehension of such ideas in such states of mind seems like revelation, then so be it. I don't care whether it is, or is not. What matters is that it works. So I would welcome the re-creation of a priestly caste, focused upon the problems of survival, but employing – in addition to their intellect – the methods of Isaiah.

At a purely practical level, religions have generated two key concepts that are highly pertinent to our present state, and are quite different in kind from any concept I am aware of in any secular field. The first of these is the Greek notion of *hubris*, which I alluded to in Chapter 3: the notion that it is not given to us, as human beings, to usurp the power of the Gods; and that if we attempt to do so, retribution will be swift and sure. I would guess that the Greeks originally arrived at this idea – or, more probably, inherited it from some pre-literate people of whom we now have no knowledge – for the same kind of reasons that I feel are pertinent now. That is, that if ancient people ever attempted to do more than they could truly control, then disaster would follow.

Excessive agriculture could be punished on a local scale even then – even though people then were far from the global exhaustion that is beginning to face us now. Too much ploughing could lead to erosion; erosion led to famine, and famine was followed by pestilence. The sequence must have been plain. The lesson was easily applied to generals who tried to conquer too much, and to leaders who thought they were above the law.

The second crucial concept is the Jewish one of blasphemy: an offence against God. Like hubris, blasphemy is beyond mere sin, and way beyond mere crime. I doubt if many readers of this book believe literally in hell-fire (I hope not!) but I suggest that all of us at times have felt the opening of awesome pits, in the face of acts so heinous as to seem transcendental. 'Ethnic cleansing' has been a blasphemy.

In our secular western society – or at least in sophisticated Britain – we have lost the concepts of hubris and of blasphemy. Yet, I suggest, the kinds of prospects opened up by the new technologies of genetics can be of such enormity that our response to them cannot be properly expressed by any other means. Why do we feel such disquiet at the resurrection of the dinosaurs? To be sure, the animals might escape and do damage, as in Michael Crichton's *Jurassic Park*. But there is more to our misgivings than that. Why are we repelled by the notion of genetic engineering applied to human intelligence? Again, we can envisage untoward after-effects; but again, the sense of foreboding that we feel seems far to outstrip mere practicalities. In each case, the act simply seems too enormous for human beings to contemplate. It is not simply a challenge, like putting a man on the Moon. It is hubristic, and it is blasphemous.

In short, the science is wonderful and the new technologies are astonishing, but if we are to deploy them for good, we have also to dig deep, and go on digging, through to the roots of our own psyche. Only when we are straight in our own heads, and have structured societies that are able to override their own innate tendency to be overtaken by hawks and hawkishness, can we hope to create the kind of world that can be sustained, for only the meek *can* inherit the Earth.

That is a difficult idea, but one generated independently both by a prophet, and by science. That, I suggest, is a powerful combination.

Index